Marquis Berrey
Hellenistic Science at Court

Science, Technology, and Medicine in Ancient Cultures

Edited by
Markus Asper
Philip van der Eijk
Markham J. Geller
Heinrich von Staden
Liba Taub

Volume 5

Marquis Berrey
Hellenistic Science at Court

DE GRUYTER

ISBN 978-3-11-065888-0
e-ISBN (PDF) 978-3-11-054193-9
e-ISBN (EPUB) 978-3-11-054015-4
ISSN 2194-976X

Library of Congress Cataloging-in-Publication data
A CIP catalog record for this book has been applied for at the Library of Congress.

Bibliographic information published by the Deutsche Nationalbibliothek
The Deutsche Nationalbibliothek lists this publication in the Deutsche Nationalbibliografie;
detailed bibliographic data are available on the Internet at http://dnb.dnb.de.

© 2019 Walter de Gruyter GmbH, Berlin/Boston
This volume is text- and page-identical with the hardback published in 2017.
Typesetting: Meta Systems Publishing & Printservices GmbH, Wustermark
Printing and binding: CPI books GmbH, Leck
♾ Printed on acid-free paper
Printed in Germany

www.degruyter.com

Acknowledgments

In true Hellenistic fashion Fortune gave birth to this book. When I was bringing a dissertation to completion I was asked to lead a discussion on Alexandrian science before a seminar of Ptolemaic archaeologists. My audience raised a series of questions I could not answer to their satisfaction. How and where do we locate scientific practitioners in ancient daily life? What was the particular interest of the political order in science? How did ancient scientists take part in the intrigues of court life? In what way did court life impact their science? My initial answers, I saw, depended upon the assumption that a model of science within an egalitarian social order can be applied to hierarchical ones. But that assumption seemed deeply flawed the longer I studied the problem. So out of my chance encounter, this book was born. I can only hope that my previous audience will pardon a seven-year wait and find my answers here more to their liking.

The book was written over a period of time when *Brill's New Jacoby* had not yet entirely supplanted *Die Fragmente der Griechischen Historiker*; my references indicate the state of scholarship in June 2016, when the draft was finished. A section of chapter 4 was previously published as Berrey 2017; I thank the editors and publisher of the *Journal of Juristic Papyrology Supplement* for permission to republish a lightly edited version of that article.

I am fortunate too to have had time and institutional support to finish this book, so different now from the dissertation I completed at the University of Texas at Austin. My research was also supported for a year by the Stanford Humanities Center, where the stimulation from commensality and collegiality inspired much of my analysis. The Classics Department at the University of Iowa arranged for a reduced teaching load. Library staff at all three institutions was unfailingly attentive to requests for books and journals. Audiences in Chicago, Iowa City, Palo Alto, San Antonio, San Francisco, Warsaw, and Durham listened to and offered suggestions on sections of the manuscript.

Friends and colleagues saved me from many mistakes but I bear the responsibility for those that remain. Marco Michele Acquafredda has been an encouraging and patient editor. Markus Asper, Stephanie Craven, Lesley Dean-Jones, David Depew, Jennifer Gates-Foster, †Carin Green, Stephanie Malia Hom, Rob Ketterer, James Lock, Reviel Netz, James Patterson, Thomas Rose, and John Ryan have offered advice and improvements at one time or another. By forcing me to question my own assumptions and improving the argumentation I owe more to David Riesbeck and Luis Salas than I can recall.

Finally, for a work concerned in part with the financial conditions of scientific research, I must make clear my own debts and record my thanks to the people of Iowa and Texas. Their support for historical scholarship at large gives me hope that the book may find a welcoming general audience.

The encouragement of my partner Zoe Petersen allowed the manuscript to be brought to completion. It is right that I dedicate the book to her.

Contents

Acknowledgments —— v

Introduction —— 1
    A Theory of Court Science —— 8
    Genealogies of Presentism in Science —— 10
    Writing Progress, Embeddedness, and Emergence —— 16
    Social Dynamics of the Court and its Consequences —— 22

1 Simmias the Elephant-Hunter and Other People at the Court of Ptolemy —— 29
    The King and His Family —— 33
    The *Philoi* or Friends of the King —— 41
    Bureaucrats, Military Officials, Judiciary, Minor Functionaries —— 48
    Priests —— 52
    Flatterers, Entertainers, Lovers —— 53
    Poets —— 55
    Prose Authors: Antiquarians, Geographers, Historians —— 64
    Prose Authors: Philosophers, Grammarians —— 69
    Scientific Prose Authors —— 72
    People missing from the historical record —— 85
    Appendix: Doubtful and Excluded persons —— 86

2 Kingship, Symposia, Gift-Exchange: Parameters of Friendship —— 89
    Scientific Selves, Scientific *Personae* —— 91
    Kingship —— 95
    Courtiership —— 103
    Symposia —— 109
    Scientists in Gift-Exchange —— 117
    Kings in Gift-Exchange —— 122

3 An Entertaining Genre —— 127
    Origins —— 128
    An Entertaining Genre —— 130
    Belatedness and Praise —— 133
    Text and Image —— 139
    Authorial *Personae*: Expertise —— 150
    Authorial *Personae*: Courtiership —— 154
    Conclusions —— 161

4 Technology and Performance in Eratosthenes and Andreas — 163
    Eratosthenes — 163
    Andreas — 179
    Technology and Performance — 190

5 Herophilus' Pulse and Archimedes' Mechanized Mathematics — 191
    Herophilus' Pulse — 191
    Aristoxenus and Rhythm — 196
    Herophilus' Water-Clock — 202
    Emergence — 206
    Archimedes' Mechanized Mathematics — 209
    Appropriating Mechanical Discourse — 215
    A Close Reading — 219
    A New Approach — 223
    The Limits of Court Science — 225

Epilogue — 227
    Towards a History of Scientific Interdisciplinarity — 227
    What is Interdisciplinarity? — 228
    Ancient Scientific Interdisciplinarity — 229
    Just-so Stories — 233
    A Future History — 237

Editions of Primary Sources — 241

Bibliography — 249

Index Locorum — 259

Index Rerum — 271

Introduction

One way to tell the story of Hellenistic science at court is to open with theft and close with two assassinations. In 245/4 BCE the Greek-Egyptian queen Berenice II, wife of the pharaoh Ptolemy III Euergetes, dedicated a lock of her hair to the temple of her deified mother-in-law, Arsinoe-Aphrodite-Zephyritis, as a vow fulfilled for her husband's safe return from his invasion of Syria. Soon afterwards the lock was stolen from the temple, a disrespect to the gods and a potentially ominous sign, but the court astronomer Conon of Samos diffused the tension by claiming to see it rise apotheosized as a new constellation in the night sky.[1] Twenty-seven years later in 217 BCE, war again took a Ptolemy northward to Syria, this time Ptolemy IV Philopator, son of Euergetes and Berenice. Shortly before the battle of Raphia in Syria one of Ptolemy's generals who had earlier turned traitor snuck into the king's tent to murder his former master.[2] Ptolemy Philopator was not inside but the royal physician Andreas of Carystus was: Andreas died in the failed assassination attempt. Five years later in 212 BCE on the western Mediterranean island of Sicily, the Greek city of Syracuse (until recently under the king Hiero II) had been fighting the Roman siege for two years with help of strong walls and the engineering machinations of the mathematician Archimedes, kinsman of the king. When at last the Romans entered the city, so the apocryphal story goes, a Roman soldier found Archimedes drawing mathematical diagrams and against his commander's orders assassinated him, Archimedes all the while asking for time to finish his proof.[3]

The tragic close to this violent version of our story might furnish historical exempla for a political narrative about the expansion of Roman power over the petty squabbling of weak leaders who exhausted themselves and their kingdoms in war. Yet *Hellenistic Science at Court* is not a book about violence between warring kingdoms in the Hellenistic period, but a book about the growth of knowledge about the natural world. Conon, Andreas, and Archimedes were scientists embedded within a social order of monarchs and courtiers. They set their interpretations of the natural world to their own use but also to the use of their patrons. What was the relationship between science and monarchy in the third century BCE in the Hellenistic world? And why does it matter?

Consider the case of Conon as a brief example. Conon was a native of the island of Samos who lived in Alexandria, the famous capitol of Ptolemaic Egypt. A mathematician and astronomer by trade, he wrote on mathematical spirals and conic sections, observed star-risings in conjunction with weather signs, and corre-

---

[1] Callimachus *Aetia* fr.110.
[2] Polybius 5.81.
[3] Plutarch *Marcellus* 19.9; Jaeger 2008: 77–100.

sponded with Archimedes.[4] The episode of the lock occurred near the end of his career. When Berenice dedicated a lock of her hair to the temple for the fulfillment of a vow of thanks for her husband's safe return from his invasion of Syria, the lock was soon stolen from the temple but Conon claimed to see it rise as a new constellation, the Coma Berenices, technically called a catasterism. The court poet Callimachus of Cyrene turned this conceit into a famous poem, which the later Roman poet Catullus adapted and translated.[5] In the poem it is Conon's deep knowledge of the celestial world which permits him to identify the new constellation: Conon appears as a dedicated, objective observer of natural phenomena. The surviving line of the opening of Callimachus' poem refers to Conon's observations: "Looking at every boundary in the lines and how [the constellations] move".[6] Commentators have rightly understood the phrase "in the lines" to refer to both the imaginary visual lines which connect the fixed stars in the heavens as well as a star-chart drawn on papyrus.[7] Conon did write a star-chart, since the later astronomer Claudius Ptolemy preserves observations from it.[8] Conon was clearly a serious student of the stars in the night sky.

Yet Callimachus' desire to credit Conon with a motivation for knowledge is contradicted by other sources. In historical accounts from later ancient authors Conon was caught up in the contemporary social situation in Alexandria.

> When the king took it [sc. the theft of the lock] badly, just as we said before, Conon the astronomer wishing to enter into the king's favor (*gratiam regis*) said that the lock seems to have been fixed among the constellations and pointed to seven definite stars, unenclosed by a figure, which he claimed was the lock.[9]

---

[4] For his life and work, see *EANS* 486, *RE* 11.2: 1338–1340 s. v. 'Konon (11)', and Knorr 1978. He is listed as **no. 137** in chapter 1.
[5] Callimachus' version is poorly preserved; Catullus' adaption is extant. Among many pieces see Harder 2012: 2.793 ff. I take her warning that Catullus 66 should not be considered a simple translation, yet she has also given good reason to suppose that Callimachus *Aetia* fr.110.6–7 may be a condensed couplet. Callimachus is listed as **no. 110** in chapter 1.
[6] Callimachus *Aetia* fr.110.1: πάντα τὸν ἐν γραμμαῖσιν ἰδὼν ὅρον ἧ τε φέρονται. All translations are my own, unless otherwise noted. My translation practice distinguishes between nouns and adjectives implied by the grammatical gender but unexpressed by the Greek text, which I mark by the angular brackets ⟨ ⟩, and other words unexpressed by the Greek text but necessary in English translation, which I mark by the square brackets [ ]. I quote the cited Greek or Latin in parenthesis ( ).
[7] Harder 2012: 2.801–802. Catullus' translation of Conon's knowledge seems to have elided the more technical terminology in Callimachus and even Catullus portrays Conon as a serious, objective observer of celestial phenomena.
[8] Ptolemy *Phaseis* 67.7–8 credits Conon with these observations not from the vantage point of Alexandria in Egypt but from Italy and Sicily.
[9] Hyginus *De Astronomia* 2.24, 1003–1007 Viré: *quod factum cum rex aegre ferret, ut ante diximus, Conon mathematicus cupiens inire gratiam regis dixit crinem inter sidera videri collocatum et quasdam vacuas a figura septem stellas ostendit quas esse fingeret crinem.*

The motivation for the discovery comes from Conon's desire for royal favor (*gratiam regis*). The uncategorized natural world is made to bow to royal power and stamped with the mark of the Ptolemaic queen. Another source tells a similar story.

> Conon the astronomer, offering a favor to Ptolemy (Πτολεμαίῳ χαριζόμενος), catasterized Berenice's lock from these unnamed stars.[10]

Conon, the commentator states, was motivated less by his knowledge of the natural world and more by his relationship to Ptolemy. The two technical phrases in Latin and Greek – *gratiam regis* and Πτολεμαίῳ χαριζόμενος – suggest that Conon hoped to enter into a patronage relationship with Ptolemy III Euergetes, transacted by the exchange of gifts. Perhaps, in potential return for financial support and courtly honors, Conon literally offered Ptolemy III Euergetes the stars.

Furthermore, the position of Conon's new constellation is no chance flattery. As Daniel Selden has argued, the position of Berenice's catasterized lock is located in an area of the sky resonant with Egyptian and Ptolemaic ideology.[11] The catasterized lock appears in conjunction with other constellations identified by Egyptians as defenders of the cosmic order: these other constellations surrounded and guarded the leg of Seth, the god of disorder and chaos. The lock joins the celestial forces of good to surround and guard against a breakout of evil in the world. Just as the pharaoh, the living god, was seen as the defender of order and justice on earth, so the constellations as gods (among them now the lock of Berenice, wife of the pharaoh) preserve order and justice on a cosmic scale. Berenice's catasterized lock thus forms part of the celestial deities keeping cosmic order in the sky and over Ptolemy III Euergetes' reign in Egypt. While Selden credits Callimachus with location of Berenice's lock in the attendant Egyptian images, this would be seem to deny credit also to Conon who, when he identified the catasterized lock, surely located it in a space unenclosed by a figure among the Greek constellations Callisto, Virgo, Leo, and Bootes. Conon was aware of Egyptian mythology among the constellations no less than Greek mythology among the constellations. Conon saw both an Egyptian and Greek sky in his star charts, a hybridized mixture of cultures known also from other court poetry.[12]

To the pharaoh Ptolemy III Euergetes, then, the new knowledge about a constellation was a form of propaganda of his power as pharaoh as well as entertainment for himself and his court. To see his wife's hair rising nightly to guard the leg of Seth was a show of power and no less a source of spectacle. Conon's motivation may be thus explained in the social and aesthetic codes of the Ptolemaic court

---

10 Σ.Aratum 146: Κόνων δὲ ὁ μαθηματικὸς Πτολεμαίῳ χαριζόμενος Βερενίκης πλόκαμον ἐξ αὐτῶν [sc. τῶν ἀκατονομάστων] κατηστέρισε.
11 Selden 1998: 343–344.
12 Stephens 2003. See the discussion in **nos. 117–120** in chapter 1.

from his act of recognizing a new constellation in the night sky. Sources other than Callimachus show Conon to be something more than an objective observer of the night sky; but the entertainment value of Conon's knowledge is clear in retrospect from Callimachus' poem too.

Patronage, Egyptian mythology, entertainment – we have come a long way from our initial description of Conon's science pertaining to spirals, conic sections, and star-charts. Our sources describe Conon variously as scientifically objective or motivated by patronage; perhaps the different portrayal of Conon's motivation makes Conon in our eyes less a scientist and more a royal sycophant. Yet the social background, different as it is from our own, makes Conon's science no less real. After all, Conon undertook a procedure quite common in the sciences: the recategorization and taxonomy of hitherto-disparate natural elements into a known and recognizable named scientific object. New scientific knowledge emerged in the recognition, naming, and classification of a new constellation. Did Conon lose any prestige among his fellow astronomers for his court flattery? Hardly. The astronomer Claudius Ptolemy still thought his star-chart with the Coma Berenices worth citing nearly 400 years later.[13] (For that matter, one may still find the Coma Berenices on star-charts today.) The value of Conon's science, it seems, depended on the audience. Perhaps Callimachus, Ptolemy III Euergetes, and Berenice II herself valued the entertainment and propagandistic possibilities of Conon's science; perhaps Conon, Archimedes, and Claudius Ptolemy valued most the identification and classification of the new constellation.

So the case of Conon briefly illustrates the relationship between science and monarchy in the third century BCE Mediterranean world. Well, so what? To employ a modern star-chart to locate the Coma Berenices in the night sky needs no knowledge of the constellation's historical origin. We should not be surprised (even if we are sometimes forgetful of the fact) that science produced in history has a history. Every scientific discovery, including the Coma Berenices, passed through a historical moment when people recognized it as a fact. To be sure, it is the historian's task to recover and preserve the memory of Conon's motivation. But does the historian's task tell us anything significant about science as such?

History matters in this case because it provides an alternative account from our own of how people come to believe scientific knowledge.[14] In our society cre-

---

13 The Coma Berenices appears named in defining the ungrouped stars in the area of Leo in Ptolemy *Syntaxis Mathematike* 7 = 1.2.100–101.9–19 Heiberg.
14 Alternative, that is, to the current dominant model of the development of scientific innovation: funded by government grants underwritten by democratic nation-states, tested by experiment in university laboratories by professional knowledge-workers, and published in peer-reviewed journals. Even if the dominant model of the nineteenth, twentieth, and twenty-first centuries has lately been showing some cracks through the enlarged participation of amateurs, crowd-sourced solutions, and crowd-funded projects, it is still by far the dominant model.

dentialed scientists are the only persons socially authorized to pass judgment on other scientists' work in order to legitimize it. Yet Conon's discovery of the natural phenomenon was socially legitimized and spread by the non-scientist members of the court. If discovery of a fact about the natural world is science's originary moment, then how does one person convince others of the value of her discovery? There is intellectual space in the historiography of science not only for histories of factual discovery but histories of social belief. Conon's new constellation is a paradigmatic case of what I call "court science". The term court science describes knowledge about the natural world produced for the entertainment of the court. My identification of this kind of science is grounded in two claims: first, that the scientist is motivated by the promise of gift-exchange in patronage to dedicate, address, or perform his work for a social superior; second, that science thereby developed attempts to interest the socially superior lay audience by its aesthetic similarity to poetry or other culture performed for entertainment of the court. Conon's Coma Berenices fulfills both criteria: it was motivated both by the potential for patronage and is aesthetically similar to Greco-Egyptian hybridized court poetry. At the heart of court science is a study of the development of social belief in scientific knowledge, that is, the persuasive means by which a scientific discovery or fact comes to be realized in the scientific community and society at large.

Court science has been studied before by historians of science but not in antiquity.[15] In the coda to his path-breaking *Galileo, Courtier* Mario Biagioli entertained a hypothesis about the social shift of the institutions of early modern European science from Renaissance and Baroque courts to royal academies and corporations.

> The shift from a social system of science rooted in patronage networks to one centered in scientific institutions was also accompanied by the emergence of new scientific practices. As discussed by Shapin and Shaffer [in *Leviathan and the Air Pump*], experiments and collective certifications of "matters of fact" became central to the new scientific discourse. With the emer-

---

[15] The question of patronage in Hellenistic science has interested only a few scholars who, strikingly, have been general social and political historians rather than historians of science: Fraser 1972: 1.305–479, Green 1993: 453–496, Massar 2005. That the dominant historiographical account of ancient Greek science remains the work of G.E.R. Lloyd may explain ancient science historians' disinterest in patronage: Lloyd 1990 explicitly argues that the political *explanans* of the fifth and fourth centuries BCE, the ideology of democracy, in which a propaganda of openness makes debate and judgment available for all, led to a social context of accountability, evidence, and proof throughout Greek science. With respect for Lloyd's insight that social agonism starts a scientific trend toward apodeictic demonstration, it is not always true for court science under consideration here. Social agonism played out in Greek science in ways sometimes different than certitude of facts, as I will show. Patronage studies in the history of early modern and modern science have been better integrated into the historiography of science; see the overview of Anderson, Bek-Thomsen, and Kjærgaard 2012.

gence of experimental philosophy we see a move from a discourse rooted in entertaining disputes to much less contentious forms of knowledge.[16]

Biagioli's book is a sustained argument that the career and science of Galileo, often seen as the first modern scientist, took place within and was shaped by the court dynamics of patronage, particularly the court of Cosimo de Medici. Biagioli's coda is thus an account of the shift from the old social world of science, in which Galileo worked, to the new world of early modern and modern science, with scientific institutions, experiments, technology, and epistemological closure. But why look only forward to modernity? Biagioli's book shows in detail how Galileo functioned and succeeded in the older world of monarchs, patronage, and pre-institutional science. I believe that this older social setting for scientific thought is very old indeed, as the case of Conon shows.

Hellenistic Science at Court argues that a certain sort of science involving cross-disciplinary scientific approaches (described below) functioned and succeeded in the world of monarchy and patronage during the early Hellenistic period in the ancient Mediterranean, roughly 300–200 BCE. Picking up many of the themes of Biagioli's analysis of Galileo, I show that Hellenistic court science was shaped by the social setting of the Ptolemaic court: scientists clothed in the virtues and *personae* of courtiers engaged in the reciprocity of gift-giving; scientists sought court patrons for patronage such as kings, who in fulfillment of their noble *personae* gave support for scientific research through networks of personal trust; the pre-institutional setting of the court where mathematicians, musicians, physicians, and mechanicians met allowed for cross-disciplinary approaches to scientific problems; the new science sponsored by the court participated in broader culturally Hellenistic aesthetics of performance, belatedness, and hybridization (all defined below) across intellectual disciplines and Greco-Egyptian culture. In short, science succeeded at court when scientists acted like the poets who dominated cultural life at court.

The theme of this book is scientific emergence, the story of how scientific knowledge comes to be. The stories I tell in this book concern new scientific ways of seeing and the subsequent theoretical studies of nature which grew out of them. I focus on four: Eratosthenes' invention and deployment of a machine for determining two mean proportionals; Andreas' invention of a new orthopedic machine for returning a dislocated joint to its socket, known as a reduction; Herophilus' discovery of a new biological concept, the pulse; and Archimedes' use of the lever for determining the volume of geometric shapes. These scientific objects and concepts have biographies and *Hellenistic Science at Court* is the story of their births.

Scientific emergence is a genealogical account of the discourses that bear upon the biography of science and scientific objects. While many historians have written

---

16 Biagioli 1993: 354.

the story of science in ancient Greece as it pertains to the origin and emergence of scientific thought in the genealogical discourse of rationalism, this book is not concerned with the origins of ancient scientific thought.[17] Rather, this book seeks to account for the origin of new scientific thought in the Hellenistic period. I use 'new scientific thought' in a qualified sense: my criterion of periodization for Hellenistic scientific novelty is as much characterological as functional, as in the case of Conon's catasterized lock. How did Eratosthenes produce a machine for calculating geometric mean proportionals? How did Andreas invent a machine for reducing hip dislocations that improved on Hippocrates' bench? How did Herophilus develop his rhythmical understanding of the pulse, a novel biological concept? How did Archimedes argue that mathematical objects could be sliced, balanced, and weighed on a scale-beam? After all, these physicians and mathematicians who worked at the Hellenistic courts were concerned with improving upon the theories and practices of their scientific predecessors; Eratosthenes, Andreas, Herophilus, and Archimedes were already working within a particular scientific tradition. All these examples involve one domain of scientific activity, such as the physiology of vascular tissue and the calculation of area, placed and analyzed within the compass of another scientific field, music and mechanics respectively.[18] These court scientists' new objects and ways of seeing nature emerged from cross-disciplinary approaches.

What motivated successful Greek scientists in the Hellenistic period to adopt cross-disciplinary strategies for interpreting the natural world? It is the central thesis of the present book that new cross-disciplinary science in the Hellenistic period gained social currency and subsequent scientific success first through its entertainment value as court science. Scientists working at court had to work to justify their new scientific ideas. New objects and ideas are like children in a big city: without guardians, they are isolated and lost, always at risk of disappearing.[19] New ideas need support from their progenitors and the world around them; without social acceptance, scientific objects do not impose their existence upon the world. New ideas gain their conceptual credibility from their integration into existing facts, from the mode of their presentation, and from the social reliability of their inventor.

Since scientific emergence is an account of how new scientific ideas developed, *Hellenistic Science at Court* is concerned with the persuasive power of scientific ideas. Persuasion, it must be acknowledged, depends on a particular audience at a particular time. Chapter 1 presents a microsociology of the social networks and belletristic interests of the one hundred forty-one known individuals at the courts

---

17 The classic account is Lloyd 1979.
18 The crossing of scientific disciplines in the Hellenistic period is not the mixed sciences known from Aristotle's philosophy of science, in which physical sciences are subordinate to mathematical principles, such as optics or meteorology. For the mixed sciences see Lennox 1985: 39–49.
19 I borrow the image of Latour 1987: 33.

of the Alexandrian pharaohs Ptolemy III Euergetes (ruled 246–222 BCE) and Ptolemy IV Philopator (ruled 222–205/4 BCE). The court, a spatial and sociopolitical configuration of household, bureaucracy, political elites, and entertainers, consisted of the monarch and his family, generals, ministers of state, priests, jesters, courtesans, secretaries and bureaucrats, poets, philosophers, and slaves of all sorts – a prospective audience far from the editorial board of *Nature*. To persuade this audience that his new scientific ideas were correct, the court scientist had to hold this audience with entertainment, be patronized by its hierarchy, and advance the status of their patrons: the court scientist must act like a courtier. Chapter 2 considers the values of the behavior of Hellenistic intellectual courtiers within an economy of gift-exchange: the virtues of courtiership, a description of the ethical practices of the king in accepting gifts, the social setting of gift-exchange, and how intellectuals used gift-exchange to achieve political aims. The court scientist adapted the discourse of court friendship and its paradigms of ethical types – true friend or flatterer – when approaching superiors for patronage. Chapter 3 reads closely four treatises of court science, dedicated by Hellenistic scientists to their patron monarchs. The court scientist offered his expertise as a friend or flatterer, after the discourse of court friendship. He offered his science to be patronized by rhetorical techniques appealing to the aesthetics of belatedness, praise, and the didactic relationship between text and image. Chapter 4 analyzes the emergence of the new objects of Andreas and Eratosthenes within the context of Euergetes' and Philopator's court. Chapter 5 extends the argument to analyze the emergence of the new concepts of Herophilus and Archimedes from their social setting as court scientists at different third-century courts. A reading of cross-disciplinary science beyond the court of Ptolemy 246–205 BCE shows both extensions and limitations to the reading of court science. An epilogue envisions the *longue durée* of premodern scientific interdisciplinarity, from Greco-Roman antiquity to the rise of German research universities in the nineteenth century, as an extrapolation of *Hellenistic Science at Court*'s microhistory of ancient cross-disciplinary science.

In the remainder of the Introduction I aim to defend and justify theoretical claims articulated in my reading of Conon's Coma Berenices paradigmatic for court science.

A Theory of Court Science

The present enters the past in many ways. Historians must be on their guard against writing a history of the present using materials of the past.[20] This is cer-

---

[20] Attributing the concern of presentist historiography only to Foucault 1977: 31 might deny it to professional scholars from the start of the discipline of *Altertumswissenschaft* in the eighteenth and nineteenth centuries; see Porter 2000: 167–224. Hence my programmatic concern with presentism is doubtless a reflex of my doubled disciplinary circumstances.

tainly true for historians of science who talk about scientific progress and scientific discoveries, some of which anticipate the present world. Above in my reading of Conon's Coma Berenices I referred unproblematically to "science" and "the scientific" as if it were a well-understood phenomenon. So it is, or rather, twentieth and twenty-first century science is. Am I unreflexively imagining and analogizing historical science to our own contemporary science? *Hellenistic Science at Court* is a book about science in the Hellenistic Greek world; yet there may have been no such thing. "Science" – it has been claimed – did not exist in the ancient world, "natural philosophy" did.[21] Why use a modern term when the terms and categories of the past are better descriptors of the past? The historian who writes of "science" in antiquity commits presentisms, modern categories and concepts smuggled across the historical divide into the readings of ancient texts.[22]

Throughout this work I will discriminate "science" strictly from "philosophy". This has been a deliberate methodological choice. Philosophy played an integral part in Greco-Roman conceptions of nature and it is impossible to draw a complete picture of science in the Hellenistic period without reference to philosophy.[23] But the point of investigating Hellenistic "science" without reference to philosophy is to learn what "science" contributes to our picture of their investigations concerning the natural world. That is, the present study aims to gain a deliberately one-sided view of what Hellenistic practitioners of "science" thought that they were doing in that activity. The purpose of the presentist discrimination of philosophy from science is to specify more precisely the historically contingent category that we are studying.

After all, what is "science"? Any essentialist characterization of modern science would include objectivity, experimentalism, technology, and universalism.[24] It is easy to show that the description of the essential characteristics of contemporary science and the corresponding claim that "science" in Greco-Roman antiquity lacks these features are simplistic and tendentious at best; it is enough to enumerate some ancient texts which take an interest in essentialist criteria of science. This book discusses some of these texts, such as Eratosthenes' *Letter to Ptolemy*, a par-

---

[21] Roger French's 1994 general introduction to the Routledge series "Sciences in Antiquity" (printed only in the volumes *Ancient Astrology, Ancient Natural History,* and *Cosmology in Antiquity*) strongly denies that one can speak of "science" at all in Greco-Roman antiquity, only natural philosophy.
[22] See Flemming 2000: 3–25.
[23] Compare Nussbaum 1994. Although the contribution of the named philosophers at court was negligible, Hellenistic scientific inquiry ran principally on Peripatetic lines; cf. **nos. 123–125** in chapter 1.
[24] French 1994, who also includes secularism. Since we know nothing at the present about the religious attitudes of Hellenistic Greek scientists discussed in this book, I leave off this topic. Scholarship's discussion of these essential categories goes back to the sociologist Robert Merton.

ent of Descartes' presentation of linear curves of higher-order polynomials.[25] Others include Aristotle's *History of Animals*, Galen's *On the Usefulness of the Parts*, and Claudius Ptolemy's *Harmonics*. When seen from the past rather than the present, "the works of the scientific revolution [sc. in the sixteenth and seventeenth centuries] often look the way they do *because* they were consciously modeled on these Greek counterparts in harmonics, mechanics, astronomy, geometrical optics, and the life sciences."[26] Ancient science does take an interest in these essentialist characterizations, even if not everyone would agree that these features are essential to "science".

Essentialist qualifications aside, the point that ancient "science" does not wholly map onto the contemporary category remains. In spite of the non-congruence of categories I will continue to employ the word "science" (hereafter without quotation) throughout this book, since there remains a historiographical space for accounts of investigations and practices about the natural world before the nineteenth century.[27] Against essentialist arguments about science I argue for a modified social constructionist historiography of ancient science. In "Genealogies of Presentism in Science" I examine three potential sources of presentism in the genealogy of "science" as a culture practice, in the genealogy of nature as a given, and in the genealogy of a society that allows epistemic closure. I argue in "Writing Progress, Embeddedness, and Emergence" that all potential presentisms can be historicized within their own intellectual genealogies except one: there are important social differences between the socially legitimating audiences of Hellenistic and modern science. The different socially legitimating audience of Hellenistic science justifies the thesis of court science I will advance in the final section, "Social Dynamics of the Court and Its Consequences."

Genealogies of Presentism in Science

At stake in the question of the label science is the characterization of a cultural practice and, correspondingly, the way to write a history of that cultural practice.

---

25 See chapter 4 footnote 46.
26 Lennox 1998: 471. It is problematic to talk uncritically of the "Scientific Revolution", as in the famous rhetoric of Shapin 1996: 1: "There was no such thing as the Scientific Revolution, and this is a book about it."
27 I place myself in opposition to the trend noted by Daston 2009: 806–807: "Historians of premodern science grew increasingly skittish about calling what they studied science at all, and the word *scientist* [her italics] when applied to Archimedes or Galileo set their teeth on edge. This was not so much finickiness (although it was also that) as a desire to capture lost disciplines (*scientia*, natural philosophy, mixed mathematics) and personae (courtier, sage, philosopher) that were crucial for accurate historical reconstructions." The most accurate way to describe the investigations of Eratosthenes, Andreas, Herophilus, and Archimedes would be to use their own language; but books writ-

So we had best make ourselves aware of where we are and how we write. As mentioned, any essentialist characterization of modern science would include objectivity, experimentalism, technology, and universalism. Is science as a cultural practice merely the assemblage of these epistemic virtues?

Some modern historians have written a history of the development of these epistemic virtues. Lorraine Daston and Peter Galison's *Objectivity* traces the historically contingent development of scientific objectivity.[28] Their opening pages describe a nineteenth century British physicist peering into a microscope to study the appearance of individual milk drops impacting a glass plate.

> He lit his laboratory with a powerful millisecond flash – poring over every stage of the impact of a liquid drop, using the latent image pressed into his retina to create a freeze-frame "historical" sequence of images a few thousandths of a second apart. Bit by bit, beginning in 1875, the British physicist Arthur Worthington succeeded in juxtaposing key moments, untangling the complex process of fluid flow into a systematic, visual classification ... For Worthington himself, the subject had always been, as he endlessly repeated, a physical system marked by the beauty of its perfect symmetry ... For years Worthington had relied on the images left on his retina by the flash. Then, in spring 1894, he finally succeeded in stopping the droplet's splash with a photograph. Symmetry shattered. Worthington said, "The first comment that any one would make is that the photographs, while they bear out the drawings in many details, show greater irregularity than the drawings would have led one to expect." But if the symmetrical drawings and the irregular shadow photographs clashed, one had to go ... For two decades, Worthington had seen the symmetrical, perfected forms of nature as an essential feature of his morphology of drops. All those asymmetrical images [he had drawn] had stayed in the laboratory – not one appeared in his many scientific publications. In this choice he was anything but alone – over the long course of making systematic study of myriad scientific domains, the choice of the perfect over the imperfect had become profoundly entrenched. From anatomical structures to zoophysiological crystals, idealization had long been the governing order. Why would anyone choose as the bottom-line image of the human thorax one including a broken left rib? Who could want the image of record of a rhomboid crystal to contain a chip? What long future of science would ever need a "malformed" snowflake that violated its six-fold symmetry, a microscopic image with an optical artifact of the lens, or a clover with an insect-torn leaf? But after his 1894 shock, Worthington instead began to ask himself – and again he was not alone – how he and others for so long could have only had eyes for a perfection that wasn't there.[29]

All the essentialist features of modern science listed above characterize the story of Worthington's milk drops. The milk drops in the laboratory represent not only milk drops but also all liquids outside of the laboratory: science is universal. Worthington uncovers the nature of milk drops by changing their morphology on a glass plate: science is experimental. The microscope, the glass plate, the light

---

ten in ancient Greek do not find many readers nowadays. Historical nominalism can be carried to seemingly absurd ends.
28 I return to Daston and Galison's 2007: 191–251 notion of the ethical self in chapter 2.
29 Daston and Galison 2007: 11–15.

flash, the photograph, even writing and drawing with his pen all help Worthington see the milk drops: science is technological. Worthington changes his mind when the experiment shows things different to what occured before: science is objective.

While universality, experimentalism, technology, and objectivity are explicit in Daston and Galison's story, there is also a narrative virtue implicit in the story of Worthington's milk drops. After nineteen years of study by eye alone Worthington moved from seeing a uniform appearance of the impacted milk drops to in 1894 seeing at once by photography an irregular and changing impacted milk drop. Worthington valued systematic symmetry over other depictions until confronted with photographic evidence; he published only drawings of symmetrical splashes and kept the asymmetrical drawings in his laboratory. So when was Worthington doing science in 1893: in the morning when he looked into his microscope and drew an asymmetrical splash? In the afternoon when he drew a symmetrical splash? Or was he doing science when he published the symmetrical drawings and propagated an image of natural regularity? Was he doing science in 1895 when he trusted machines instead of his eyes? In short, does doing science mean only that progress in some form is occurring?

Daston and Galison are able to problematize immediately the historical development of the judgment of one modern scientist because they have dated evidence from notebooks and from publication papers. Imagine now the historiographical conundrum if they lacked dated evidence for which came first, the regular impacts or the irregular impacts. Would Worthington have published the asymmetrical drawings, keeping the symmetrical drawings in the laboratory, and thereby have propagated an image of natural irregularity and complexity? Would he have then claimed in 1894 that technology vindicated his eyes? In my alternate history the story of Worthington's impacted milk drops is a story of increasing improvement in scientific seeing and a lack of order and regularity in nature: in this alternate history, science is progress in technology. Worthington himself thought that the milk drops showed regularity in nature and he expected technology to vindicate the symmetrical patterns he saw by eye; the story of the milk drops was both the increasing revelation of order in nature *and* the improvement of seeing technology: for Worthington before 1894, science was progress in the revelation of regular natural order. Yet Daston and Galison's story is complete only when the known historical development in seeing technology interrupts expected patterns of order and regularity in nature: for Daston and Galison, science is objective progress in seeing.

My reading of Daston and Galison's story of Worthington's milk drops raises questions about the way we recognize science and write its story. It is difficult to tell a story of Worthington's milk drops without imposing on it some version of scientific progress, whether experimental, technological, or objective seeing. The narrative of progress, whatever its form, is a teleological account of essential epistemic virtues. If contemporary science is a cultural practice of progress with the epistemic virtues of objectivity, universality, experimentalism, and technology, then potentially any treatment of historical science that tells a story of progress in some of these epistemic virtues commits presentism.

The essentialist characterization of science presented above (so called because it begins from definitional criteria of the cultural practice of science) has left largely unexamined the object of the cultural practice of science: where was nature in Worthington's milk drops? It was in the universal: the milk drops in Worthington's images are universal milk drops. These photographs represent the behavior and morphology of the billions more milk drops in the world and the untold number of liquid drops beyond milk. Worthington's drops act as many liquids do. They take a consistent form, they respond to the same causal laws. Yet Worthington's impacted splashes are local milk drops, produced perhaps in the udder of a single local cow near his laboratory in England. How can the milk from the udder of one single English cow contain the whole world's liquid?

Nature is what is given, what is objectively the case.[30] Nature does not lie. The fault in the asymmetry and symmetry of Worthington's milk drops lay with Worthington and his cultural practices, not the milk drops. His behavior was irregular and variable; their behavior was regular and consistent. That is, Daston and Galison's book raises concerns not only about implicit teleology in the writing of science but category differences in different historical epochs: it raises a number of questions about nature and ways of seeing nature. Daston and Galison's book can be viewed as an extended argument that the mental categories with which modern scientists operate are historically contingent, imprinted from a certain cultural moment: for example, at some point objectivity became a virtue for scientific authority.[31] To speak dialectically, we are accustomed to think of "nature" as something immutable, a given, and to think of "culture" as a changeable object. Yet this categorical divide itself can be shown to be the product of a specific cultural moment, at least in modern European thought around the turn of the eighteenth and nineteenth centuries.[32] The concept of "nature", what is given, has a history, just as much as "ways of seeing nature" (of which objectivity is a part) have a history.

Can we therefore speak about science in antiquity? If "nature" and "ways of seeing nature" have histories, it is not clear that scholarship can claim to speak

---

30 Daston 1998; Flemming 2000: 17.
31 My best attempt at neutral language in this context is "historically contingent" instead of "historically" or "socially" constructed. The phrase "historically contingent" does have social constructionist overtones, which is the interpretative tradition to which Daston and Galison 2007 weakly belong. Nonetheless, at the present I intend no more than the uncontroversial genealogical claim that scientific ideas come into the mind of scientists at certain historical points.
32 See Daston and Galison 2007, Daston 1998. As a classicist I think more readily of the dialectic between *nomos* and *physis* of Greek intellectuals in the fifth century BCE; Kerferd 1981: 111–130. The historiographical division between an established, given nature and mutable culture exists in both ancient and modern thought.

about science and its object, "nature", in Greco-Roman antiquity: doing so imputes modern categories and values to ancient conceptions of nature and natural investigations. Any historian who writes of nature – as if it were universal, causally consistent, and separate from culture – commits presentism.

Finally we come to that realist bugbear passed over in the essentialist characterization of science, namely how society impacts science.[33] Science as a cultural practice has a history, nature as a naturalized given has a history, and surely society has a history. Worthington was not working or talking to himself in isolation: where was society in the story of Worthington's milk drops? Though hidden in the background, society still permeated the story. Worthington interacted with the scientific community from his laboratory by publication. From his postbox went his paper publications, the written record of his observations, to journal editors and expert reviewers. The expert reviewers, known or unknown to Worthington, fulfilled a social role of ensuring that Worthington's observations and analysis conformed to the observations and practices of the contemporary scientific community. Perhaps they too in 1893 had observed asymmetry in nature, perhaps they too kept them private in the laboratory, for "in this choice [to publish only symmetrical drawings] he was anything but alone." And yet when in 1894 Worthington published his asymmetric drawings, the reviewers acquiesced: "Worthington began to ask himself – and again he was not alone – how he and others for so long could have only had eyes for a perfection that wasn't there." Social expectations – that is, the scientific community's reviewers' expectations – of natural symmetry shifted over 1893–1895: idealization gave way to imperfection. Was Worthington a part of this shift or a cause of this shift? Differently put, does science impact society or does society impact science? That final question shows the impossibility of constructing a tidy dichotomy between rational, transhistorical scientific thought and contingent, historicized social values.

Since the claim "society impacts science" is too vague to admit analysis, it is useful to introduce a more detailed specification of the claims made for social influence on science. Hacking has provided a useful typology to possible perspectives on the natural sciences from the vantage point of the humanities and social sciences in his reading of the American Science Wars of the 1990s, an intellectual debate between scientific realists and their social constructionist critics.[34] Constructionists are distinguished by their commitment to the contingency of science, to philosophical nominalism (that is, denying that the world has a predeterminate character prior to human interpretation), and to explanations of intellectual stability external to science. While the debates between scientific realists and social constructionists concerned science in contemporary Western society, Hacking's typol-

---

[33] I understand "the social" to be an explanatory category, against Latour 2005.
[34] Especially Hacking 1999: 63–99.

ogy offers useful categories to compare and contrast ancient and modern sociologies of science. A brief genealogical preview of the primary evidence shows continuities between ancient and modern sociologies of science – Hellenistic science was contingent, nature as a naturalized given existed in Hellenistic science prior to human interpretation – but also an important discontinuity: Hellenistic science was not intellectually stable in the same way as modern science. I therefore postpone discussion of the contingency of ancient science and the assumption of nature as a naturalized given to focus on the intellectual stability of ancient science.

Hacking's final classification of social constructionist arguments is the explanation of stability external to science. Social constructionists argue that the causal means by which scientific debates come to an end, or have closure, are social factors such as group coherence or financial hurdles. The social constructionist tradition, led sociology-inspired historians like Shapin, has argued that modern science has social mechanisms to achieve closure. Shapin's famous thesis is that the values of honesty and mutual trust from social codes of seventeenth century gentlemen allowed the rise of scientific bodies, like the Royal Society, to arbitrate and judge disputes between scientific opponents.[35] By contrast, premodern science – or, to use sociological terminology, culturally specific knowledge production before the sixteenth century – had no such social or political mechanisms to allow for closure to scientific debates. The classicist and historian of ancient Greek and ancient Chinese science G. E. R. Lloyd has emphasized how knowledge production in both ancient Greece and China was practiced by individuals, with no institutional support, under the demands of social competition, as in Greece, or for the stability of the emperor's reign, as in China.[36] Lloyd's work has shown that, apart from mathematics, ancient science had no single agreed-upon result or methodology. It is hard to see how closure to scientific debate could exist in such societies.

Even the work of Galileo, often seen as the first modern scientist, took place in a court context without social or political mechanisms for closure to scientific debate. Biagioli's investigation of the Medici's patronage of Galileo suggests in fact that scientists attached to courts and political institutions in premodern societies functioned more as entertainers and artists.[37] Biagioli's coda quoted above considers precisely the transition from social mechanisms that prevented epistemic closure to social mechanisms which allowed it. The ancient scientific debate does not have closure in the same way that social constructionists claim that modern science has closure. There is a genealogical account of a society that has epistemic closure but its roots do not lie in antiquity. A historian who writes that ancient science was

---

35 See especially Shapin's 1994: 7–40 theoretical overview.
36 Lloyd 1987, Lloyd 1990, Lloyd 2002.
37 Biagioli 1993.

stable due to external social factors and concludes that stability allowed epistemic closure to debate commits presentism.

Writing Progress, Embeddedness, and Emergence

So there are three ways in which a historian faces the charges of presentism. A presentist treatment of ancient science tells a teleological story of progress in the epistemic virtues of objectivity, universality, experimentalism, and technology; a presentist treatment of ancient science writes of nature as if it were universal, causally consistent, and separate from culture; a presentist treatment of ancient science concludes that intellectual stability from external social factors brought epistemic closure to scientific debate. I respond to these challenges with a discussion of how to write progress, embeddedness, and emergence in ancient science.

Hellenistic Science at Court tells a story of improvement and progress; how could the genealogical account of biographies of scientific objects and modes of seeing not be progressive? After all, even Daston and Galison's story of Worthington's discovery of irregular impacted milk drops is a story about objective progress in seeing, a type of implicit narrative progress. Yet the teleological progress of a scientific understanding is a thesis about the unconditional development of scientific ideas. So implicit ideas about the development of ancient science also impact the telling of that story, the narrative of presentism. Consequently I will show first that Greco-Roman science was contingent and second that the historiographical problem of teleological narrative can be avoided.

Hacking's first division between social constructionists and scientific realists concerned the contingency of science, the thesis that the present state of science was not inevitable. Contingency takes many forms, whether historical, metaphysical, or narrative. Historical contingency is the category which concerns historical treatment. Hacking ties contingency to the notion of progress and development in science: a constructionist argument claims that the present scientific order, whatever it is, could have been different and still be considered a scientific account of the natural world. Now it is often the aim of genealogical scholarship to describe the historical (and thereby social) contingency of cultural artifacts, codes, and discourses with the implication that, by being historically and socially contingent, they are less real, less logically compulsive, less objectively so.[38] Nevertheless Daston has articulated a different view:

---

[38] Hacking 1999: 19 charts six increasing degrees of social constructionist attitudes: historical, ironic, reformist, unmasking, rebellious, revolutionary. My reading of Daston and Galison 2007 is that they are "historical" social constructionists.

> Probably most historians of science these days, if asked about an episode like the refinement of precision measurement techniques or the formulation of statistical correlation, would answer that such scientific practices are both socially constructed *and* real. That is, they depend crucially on the cultural resources at hand in a given context (mid-nineteenth-century industrializing Prussia, early twentieth-century eugenics-obsessed Britain) *and* they capture some aspect of the world; they work. But they are neither historically inevitable nor metaphysically true. Rather, they are contingent to a certain time and place yet valid for certain purposes.[39]

Many aspects of investigations of nature in Greco-Roman antiquity are not objectively real, that is, valid within our scientific system; but Daston does not say that the results of modern science are objectively true, only that "they work" and "are valid for certain purposes". Her claims are an explicit appeal to her readers' experience of the force of science in their lives. Such an appeal could be rewritten anthropologically to cover contemporaries' experience of a particular knowledge production within any given culture. For instance, in Daston and Gallison's account Worthington's pre-1894 interpretation of his milk drops' impacts as regular worked for and satisfied in some deeply cultural way Worthington and his Victorian contemporaries.

The science of Greco-Roman antiquity is historically contingent; its progress could have been different. Herophilus' use of the water-clock to measure pulse rates is the path not taken by ancient medical theories of the pulse. As I argue in chapter 5, ancient evidence shows that Herophilus employed a normative concept of time to measure pulses, expressed both through the application of the water-clock and by rhythms based on Aristoxenus' theory of the primary time-unit. But none of Herophilus' sectarian followers seem to have employed his water-clock; and Galen eventually did away with Herophilus' appropriation of Aristoxenus, opting instead for a more descriptive analysis of pulse rhythm. Despite Galen's radical reformulation of Herophilus' theory, Greek medicine could have been purely Herophilean in its analysis of the pulse; there was no metaphysical or historical necessity for the triumph of Galenism.

Contingency denies not only the historical development of a scientific understanding but also the narrative progress of that development. The story of Worthington's milk drops shows that assumed progress as a historiographic principle can take multiple forms and implicitly incorporates a definition of science as progress into history. Historical contingency is related to narrative contingency and, since ancient science was historically contingent, we cannot incorporate a teleological narrative opposed to narrative contingency in our writing of ancient scientific thought. Herophilus' attempts to time the pulse attributed more empirical phenomena to the concept of pulsation: this is a story of an increasing revelation of order in nature. Archimedes' attempts to measure the parabola's area contrib-

---

[39] Daston 2009: 813.

uted to the particular kind of the *Method's* mechanical "way of seeing": this is a story of improved seeing technology. "Scientific progress" results from the emergence of new objects and approaches in science; and the scientific progress told in chapter 5 comes from Herophilus and Archimedes, not the modern historian.

The society examined in *Hellenistic Science at Court* is the court society of the Hellenistic period, principally the courts of the Ptolemaic monarchs Ptolemy III Euergetes and Ptolemy IV Philopator. The scientists Eratosthenes of Cyrene and Andreas of Carystus lived and worked at the courts of these monarchs.[40] The Ptolemaic monarchs held court in Alexandria, the Greek colony in Egypt founded by Alexander the Great at the western edge of the Nile delta. A character in a mid third-century BCE mime praises the city in explaining why a lover has not returned from there: "All that ever is and comes to be is in Egypt: wealth, a wrestling school, power, prosperity, fame, sights, philosophers, gold, young men, the shrine of the *theoi Adelphoi*, the king is good, the Mouseion, wine, all good things you want, women."[41] The city was one of the great metropoleis of its age, growing to around 300,000 inhabitants by the year 200 BCE in the best modern estimate.[42] It was perhaps *the* intellectual center of the Greek cultural world: scientists, philosophers, poets, and literati of all sorts immigrated or visited the city and its famous repositories of Greek culture, such as the Mouseion and library. While we know very little about how either the Mouseion or library functioned or the books stored there, it is clear that the Ptolemaic monarchs' ambition was to collect and acquire all knowledge about the known world.[43] Knowledge of the natural world potentially available to Alexandrian scientists was the cumulative efforts of previous Greek thinkers. Since I take the ethnographic view that scientific objects and practices are locally, not universally, embedded,[44] any response to nominalism (Hacking's second division between social constructionists and scientific realists) must be formulated in terms of the discourses, practices, and beliefs of actors of Hellenistic court society. Such a formulation is called a thick-description of the discourses, practices, and beliefs of the social group under study. Although there is no universally Greek view of nature, it seems likely that Hellenistic period scientists believed that nature had a predeterminate character.

While philosophical nominalism has a long history back to Classical Greece where Plato denied the position when he argued that our terms cut nature at the

---

40 In the prosopographic list of chapter 1 Ptolemy III Euergetes is **no. 1**, Ptolemy IV Philopator is **no. 8**, Eratosthenes of Cyrene **no. 136**, Andreas of Carystus **no. 141**.
41 Herodas *Mimiambi* 1.26–32: τὰ γὰρ πάντα, / ὅσσ ἔστι κου καὶ γίνετ', ἔστ' ἐν Αἰγύπτωι· / πλοῦτος, παλαίστρη, δύναμι[ς], εὐδίη, δόξα, / θέαι, φιλόσοφοι, χρυσίον, νεηνίσκοι, / θεῶν ἀδελφῶν τέμενος, ὁ βασιλεὺς χρηστός, / Μουσῆιον, οἶνος, ἀγαθὰ πάντ' ὅσ' ἂν χρήιζηι, / γυναῖκες. Cf. Hunter 2003: 37's reading of the passage.
42 Scheidel 2004.
43 See the discussion of the Mouseion and library to **no. 1** in chapter 1.
44 The work of Clifford Geertz 1973, 1983 has been especially influential here.

joints, it seems more likely that ancient scientists believed that nature had a predeterminate character.⁴⁵ The Greek word φύσις, transliterated as *physis* "nature", had itself a specific nature. In an important essay, Lloyd argues that *physis* was invented as a regularized category during the Classical period in the polemic between natural philosophers (a category including scientists and philosophers) and their opponents, traditionalist healers and diviners.⁴⁶ The category of the natural became naturalized, so to speak. The historically-contingent *physis* of philosophers and scientists may have been a regular and normativized category but was still not a singular one: thinkers on *physis* disagreed with each other about the appropriate methodologies to study *physis*. The predeterminate character of nature existed in ancient thought, although we should not infer that the specific nature of ancient *physis* was equivalent to modern category "nature".

The Ptolemaic court scientist Eratosthenes, for instance, believed that nature had a predeterminate character. In a revealing passage, Eratosthenes explains how the earth's surface reveals changes from its past condition.

> [Eratosthenes] particularly says that it demands an investigation how in two or three thousand stades from the sea in the middle of the land one sees in many places a mass of shells, oysters, scallop-shells, and lagoons, just as he says are around the temple of Ammon and the road to it of 3000 stades. For he says that there is a great profusion of oysters shells, that even now many sea-salts are found, and that there are blow-ups of the sea to heights. ... In addition, he says that one can point out the wrecks of sea-vessels, which they say were cast ashore by the earth's movement, and on mast flags there are the ornamental dolphins bearing the inscription of Cyrenean sacred ambassadors. Saying these things Eratosthenes praises the opinion of Strato the *physicus* and also Xanthus of Lydia ... [A digression follows on the opinions of Strato and Xanthus that the inundation of rivers has changed the sea floor of the Black and Mediterranean seas] ... Perhaps, Eratosthenes says, the temple of Ammon, now in the middle of the land, once was at an outflow of the sea. He conjectures that the oracle was so clearly and recognizably at the sea to a great extent and that the great removal from the sea did not logically cause its present placement and appearance. Egypt, he thinks, was anciently washed by sea up to the marshes in Pelusium, Mt. Casium, and Lake Sirbonis [modern Tell el-Farama, Ras Qashrun, and Sabkhat el-Bardawil]. And even now in Egypt, when salty sand is dug up, shell-like trenches are found, as if the land was once sea and the entire region around Casium and the so-called Gerra were shallow water so that it conjoined the gulf of the Red Sea.⁴⁷

---

45 Plato *Phaedrus* 265d–e.
46 Lloyd 1991.
47 Eratosthenes *Geographica* fr. 15 Roller = Strabo 1.3.4 = 49C.10–50C.20 (122.1–124.20) Radt (my practice throughout is to cite references by fragment collection, if available, followed by a cross-reference to a source text): μάλιστα δέ φησι ζήτησιν παρασχεῖν πῶς ἐν δισχιλίοις καὶ τρισχιλίοις ἀπὸ θαλάττης σταδίοις κατὰ τὴν μεσόγαιαν ὁρᾶται πολλαχοῦ κόγχων καὶ ὀστρέων καὶ χηραμύδων πλῆθος καὶ λιμνοθάλατται, καθάπερ φησὶ περὶ τὸ ἱερὸν τοῦ Ἄμμωνος καὶ τὴν ἐπ' αὐτῷ ὁδὸν τρισχιλίων σταδίων οὖσαν· πολλὴν γὰρ εἶναι χύσιν ὀστρέων, ἅλας τε καὶ νῦν ἔτι εὑρίσκεσθαι πολλούς, ἀναφυσήματά τε θαλάττης εἰς ὕψος ἀναβάλλειν ⟨lacuna⟩· πρὸς ᾧ καὶ ναυάγια θαλαττίων πλοίων δείκνυσθαι, ἃ ἔφασαν διὰ τοῦ χάσματος ἐκβεβράσθαι, καὶ ἐπὶ στυλίδων ἀνακεῖσθαι δελφῖνας ἐπιγραφὴν ἔχοντας Κυρηναίων θεωρῶν. ταῦτα δ' εἰπὼν τὴν Στράτωνος ἐπαινεῖ δόξαν τοῦ φυσικοῦ, καὶ ἔτι Ξάνθου τοῦ Λυδοῦ ... τάχα δὴ καὶ τὸ τοῦ Ἄμμωνος ἱερὸν πρότερον ἐπὶ τῆς θαλάττης ὂν

The famous oracle of Zeus-Ammon at the Siwa oasis in the Egyptian desert was a frequent destination for Greeks consulting the gods. Whether Eratosthenes visited the oracle as a tourist or relied on a traveler's report, the oracle was noticeably surrounded by the remnants of marine life. Eratosthenes conjectured that the oracle had once been on the edge of the sea: marine life flourished then, their organic remains stayed in place after their death, and the sea level eventually shifted so that the oracle stood in desert. Here Eratosthenes agreed with the important third-century BCE Peripatetic philosopher Strato of Lampsacus that sea levels could shift due to natural processes.[48] To Eratosthenes and Strato then, the application of ongoing natural processes could create different appearances in the natural world in time. The predeterminate character of nature is not the present appearance of the natural world but the natural processes of tides, inundation of rivers, the life cycle of biological species, the sea's salt content, and so. These processes are evidently timeless and in accord with *physis*; they are objectively given. For Eratosthenes and other third-century BCE scientists, *physis* was universal and causally consistent across time and space.

We will see in chapters 4 and 5 Eratosthenes, Andreas, Herophilus, and Archimedes appropriate concepts from other scientific domains in strategies of naturalization. The integration of elements from the domain of one science into another forges a new link between an area of *physis* and a scientific domain not traditionally associated with it. The enlarged networked structure of scientific domains increases the sense of natural regularity and normativity Lloyd pointed to as "the invention of nature." Since what the givenness of nature contains is historically contingent, court scientists' strategies of naturalization mark a deeper understanding of the Greek cultural sense of *physis* as a regularized category. The emergence of new science in the Hellenistic period via naturalization marks connections between previously separate areas of *physis*.

Presentism in the historiography of Hellenistic science, differently said, is another way of envisioning continuities between ancient and modern scientific material. We have seen that ancient science expresses the epistemic virtues of objectivity, universality, experimentalism, and technology and that Hellenistic scientific writers in Alexandria write of nature as if it were universal and causally consistent. Here there are two continuities or two possible presentisms between Hellenistic

---

ἐκρύσεως γενομένης νῦν ἐν τῇ μεσογαίᾳ κεῖσθαι. εἰκάζει τε τὸ μαντεῖον εὐλόγως ἐπὶ τοσοῦτον γενέσθαι ἐπιφανές τε καὶ γνώριμον ἐπὶ θαλάττῃ ὄν (τόν τε ἐπὶ πολὺ οὕτως ἐκτοπισμὸν ἀπὸ τῆς θαλάττης οὐκ εὔλογον ποιεῖν τὴν νῦν οὖσαν ἐπιφάνειαν καὶ δόξαν) τήν τε Αἴγυπτον τὸ παλαιὸν θαλάττῃ κλύζεσθαι μέχρι τῶν ἑλῶν τῶν περὶ τὸ Πηλούσιον καὶ τὸ Κάσιον ὄρος καὶ τὴν Σιρβωνίδα λίμνην· ἔτι γοῦν καὶ νῦν κατὰ τὴν Αἴγυπτον τῆς ἁλμυρίδος ὀρυττομένης ὑφάμμους καὶ κογχυλιώδεις εὑρίσκεσθαι τοὺς βόθρους, ὡς ἂν τεθαλαττωμένης τῆς χώρας καὶ τοῦ τόπου παντὸς τοῦ περὶ τὸ Κάσιον καὶ τὰ Γέρρα καλούμενα τεναγίζοντος, ὥστε συνάπτειν τῷ τῆς Ἐρυθρᾶς κόλπῳ. I follow Roller's 2010: 224–248 identification of ancient place names.

48 Both Eratosthenes and Strato are indebted to Aristotle *Meteorologica* 352b–353a.

and modern science – presentism in scientific practice and presentism in nature – that are already securely attested in the ancient evidence, historicized within their own intellectual genealogy.

But society remains an important discontinuity. Is it true that epistemic closure is a result of external stability to science? I accept the conclusion from a generation of research in the history of science that factors external to the technical questions of science necessarily but perhaps not sufficiently allow the epistemic closure of technical questions.[49] Daston's enjoinment to her contemporary audience of the force of science in their lives recognizes the history of social belief implicit in the historiography of science. All the same, I suggest that a preoccupation with questions of closure is the province of the historian of early modern and modern science. Hacking's characterological report on the claims of social constructionists represents only the interests and preoccupations of historians of early modern and modern science: closure is their answer to the question of why science succeeds. Furthermore, Forman has argued that the essentialist disciplinary virtues of modern science derive not from their essentialist character but from the historical epoch of modernity, in which disciplinary virtues became institutionalized and canonized.[50] In his thesis about the disciplinary structures (real and imagined) of contemporary academic disciplines in the historical epochs of modernity and postmodernity Forman thus proposes to identify contemporary history of science scholarship with the sociological criterion of closure. I am unable to judge the merits of Forman's thesis because I write inside the disciplinary structures of the postmodern epoch. Yet I am sympathetic to the explicit contrast drawn between modern and premodern science. There are no social mechanisms for closure to a scientific debate in antiquity; epistemic closure does not exist in Hellenistic science. Instead, a responsible historiography of social belief of Hellenistic science will study the external factors that allowed scientific ideas to open and add to debate. The present book therefore pursues the ethnographic category of newness in science, emergence, the opposite of closure.

I understand emergence to be an epistemological category, not an ontological one. We sometimes confuse ontology for epistemology by passing from discussion of beliefs to objects without marking the difference. To be clear, this book describes emergent beliefs about objects, not objects themselves. The emergent beliefs about objects I discuss are, perhaps at first glance, of two kinds. Andreas' machine and Eratosthenes' instrument are physical machines, made of wood and lead respec-

---

[49] Social studies of science have been the vanguard form of histories of science over the last thirty years. See Daston 2009: 803.n11 for bibliography, again all from the early modern or modern history of science. From a large literature on the internalist/externalist debate in the history of science see Shapin 1992, Kinzel 2012.
[50] Forman 2007, Forman 2012.

tively; Herophilus' pulse and Archimedes' weighing method are conceptual objects, which only take physical shape in arterial motion and in the diagrammatic presentation on a papyrus sheet. "Emergence" might be applied differently to these objects: machines must be built at some point in time and may be said to emerge from their assembled parts; arterial motion is a natural phenomenon of every living being, and it is not quite clear how to apply emergence to vibrating arteries that exist from birth until death. The difficulty faced here is the application of emergence to physical objects only. We would not talk about a house existing or emerging when only the timber frames and nails lay separately and unassembled in the lumber yard.

Emergence is a notoriously difficult term. I distinguish an ethnographic sense from a philosophical one.[51] Emergence in an ethnographic sense I understand to be a historic conceptual holism which for historical actors marks the temporal moment a new belief about already-existing objects came into being; my observer's claim is to identify the manner in which the historical actors created the concept. For instance, the wormlike regular fall and rise of a vibrating artery is a transhistorical physical object; genealogical scholarship begins its work of historicization when historical actors identify emergent concepts with transhistorical objects in a holistic account. The case of Herophilus' discovery of pulse involves the following emergent concepts: the distinction between arteries and veins; the natural, regular, and non-pathological motion of the arterial rise and fall; and the various descriptive categories that Herophilus thought motion capable of assuming described and analyzed in chapter 5. Similarly, Andreas' machine involved the following emergent concepts: the lever; the threaded screw; the joint, its dislocation, its restoration. The contribution of genealogical analysis is to identify how the historical actor assembled these emergent concepts into a conceptual holism. Andreas' machine, for example, emerged when the concepts of the screw and lever were assembled into a device capable of exerting traction on dislocated limbs.

Social Dynamics of the Court and its Consequences

In "Genealogies of Presentism in Science" I argued that a history of science faces a charge of presentism in three ways: a presentist treatment of ancient science tells a story of progress in terms of epistemic virtues; a presentist treatment of ancient

---

[51] I distinguish my ethnographic sense from a philosophical one, since in contemporary philosophical literature strong emergence is often contrasted with reductionism: O'Conner and Wong 2012; Kim 2006. For an ethnographic sense of emergence I am especially indebted to Bruno Latour's genealogical description of networks of scientific concepts and objects: Latour 1987, Latour 2000, Latour 2005.

science writes of a social belief in nature as if it were universal, causally consistent, and separate from culture; a presentist treatment of ancient science concludes that intellectual stability from external social factors brought epistemic closure to scientific debate. In "Writing Progress, Embeddedness, and Emergence" I argued that a well-defended historiographical response to presentism shows an awareness of teleological progress in the narrative of scientific discovery, provides local embeddedness with a thick-description of the society and its attitudes that produced emergent science, and articulates how successful science in the Hellenistic period allows a controversy to open, thus offering a version of emergentism focused on the historical actors' assembly of conceptual holism.

The core of my response to concerns of presentism was to argue that, in contrast to essentialism, a social constructionist presentation of Hellenistic scientific history provides a historically defensible genealogical account of new scientific ideas. "Writing Progress, Embeddedness, and Emergence" has shown that Hellenistic science was contingent and nature as a naturalized given existed in Hellenistic science; it has also shown that Hellenistic science did not have closure. I therefore suggest that the successful emergence of cross-disciplinary Hellenistic science ought to be contextualized within the historicized forms and practices of the individual scientist's social position at court. In particular I argue that cross-disciplinary court science, devised first as entertainment for lay court actors, gained initial social currency among elite circles, followed by acceptance by the wider scientific community. Since we may consider the specific court situation of Hellenistic science, I now move to identify and articulate the relevant external social factors which led to the emergence of new Hellenistic science at court. These social factors are gift-exchange among social actors at court and science as entertainment. I sketch here the characterological analysis of Hellenistic court science detailed in chapters 1–3.

Court scientists worked in a community both with other scientists and many lay individuals, where the lay actors possessed more social power than the scientists. The court society of Ptolemy III Euergetes and Ptolemy IV Philopator was a society of elites.[52] The king ruled the kingdom with the help of a professional administrative bureaucracy and a professionalized military class, as well as by an extensive network of courtiers and elites. This last group acted as close staff of advisors, aides, and agents who represented the king's power through diplomacy, military commands, executive commands, and the like. The courtiers held a complex hierarchy of court titles, specifying social function and rank relative to other grandees.[53] Titles of social function existed throughout the third century BCE, but

---

[52] On court societies in antiquity see Spawforth 2007, Moyer 2011, Strootman 2011, Strootman 2014. None specifically addresses the courts of Ptolemy III Euergetes or Ptolemy IV Philopator.

[53] Mooren 1975 presents the evidence of court titles in Ptolemaic Egypt. Ultimately eight gradations of rank were fully known under Ptolemy VIII Euergetes II (reigned 169–116 BCE with several interruptions and coregents): ὁ συγγενής, τῶν ὁμοτίμων τοῖς συγγενέσιν, τῶν πρώτων φίλων, τῶν

titles of gradations of rank came into being sometime during the second century BCE. The court society under study here therefore lacks, with one exception (the bland and elastic title φίλος τοῦ βασιλέως "friend of the king"), verbal markers specifying the relative social esteem enjoyed by courtiers. While philological ambiguity reigns over the precise social position of courtiers, extensive contacts with court society and association with the ruling king chart the relative social position of courtiers with similar social functions. The deliberately elastic sense of what courtiers did in Hellenistic courts permited a wide sweep of aulic social achievements: diplomatic, martial, legal, bureaucratic, poetic, artistic, even scientific. Chapter 1 summarizes the rank, social connections, and belletristic interests of the one hundred forty-one individuals who can be located at the courts of Ptolemy III Euergetes and Ptolemy IV Philopator in the forty-one years from 246–205/4 BCE.

Gift-exchange was the mechanism of patronage among these court actors, including scientists. We are principally examining the encounter between two different kinds of persons, represented by the scientist and the king: one a person knowledgeable about a particular domain of the natural world and the other a lay person with no special knowledge of a given subject but having superior social or political authority. The patronage dynamic between the scientist and king supposed an asymmetric exchange of gifts given in the unequal and different categories of knowledge and social esteem. The result of this encounter is that the king, who possesses social prominence, gains knowledge of the natural world and the scientist, who possesses knowledge, gains social esteem associated with the king.

Yet because the gifts come from asymmetric categories, neither scientist nor king meet on equal ground. The king did not possess the technical competence to judge the scientist's ideas. Court patronage was not a meritocracy in which better ideas won out because they were technically superior. If we modern observers, casually conversing about two rival scientific ideas, were to say that one was better than another, we would mean to indicate that one idea was correct, that is, the better fit to natural phenomena. But we live in a society in which fellow scientists are the only persons socially authorized to pronounce judgment on another scientist's work. We lay people trust the institutional system of peer-reviewed publications to assure that the scientific ideas propagated in our society are, in fact, the better fit to natural phenomena. The peer-review system, in which credentialed peers in the field review and suggest revisions to papers before publication in reputable journals, is an institutional system of trust. A scientific author who intends to use data published in a journal article trusts that the journal editors have accurately printed the transmitted materials, that the editors have used reputable and

---

ἰσοτίμων τοῖς πρώτοις φίλοις, ὁ ἀρχισωματοφύλαξ, τῶν φίλων, τῶν διαδόχων, τῶν σωματοφυλάκων "the kinsman, of those equal in honor to the kinsmen, of the first friends, of those equal in honor to the first friends, the chief bodyguard, of the friends, of the successors, of the bodyguards".

knowledge experts to vet the submitted paper, and that the author has conscientiously reported the results of her instruments and experiments. The scientific author may know none of these individual people – editors, experts, printers, other scientific authors – and yet, when accepting the final product in the printed journal article, implicitly believes and trusts that every person has performed their functional role competently. In such cases, the author is said to have institutional trust, that is, a trust in the broader social system of peer-reviewed submissions in which these specified social functions are possible. In the analysis of role of the society in the story of Worthington's milk drops, Worthington relied on institutional trust. The journals in which he published disseminated his ideas to broader lay society; and we lay people, like the author, trust the social roles that the system provides. A narrow analysis of the impact of society on science in a system with institutional trust emphasizes first the professional values of the scientific community.

Peer-reviewed submission as institutional trust is a part of modern scientific life, but its roots lie in older social systems of trust in which personal relationships between individual social actors are more important than a systemic trust in a certain social practice. The older system is called personal trust. The kingdoms in the Hellenistic period too functioned with personal trust. The society of elite actors relied upon by the king to assist in governance, diplomacy, and war was bound to the king by personal trust. The king and courtiers exchanged gifts with asymmetrical expectations of moral reciprocity. The court title φίλος τοῦ βασιλέως "friend of the king" indicated a relationship that, while historically originating in true friendship between equals, became what Herman has called "ritualized friendship".[54] Ritualized friendship as a sociological term covers a spectrum of amiable relations, from familial bonds, to relationships between equals from the same social circle, to relationships between individuals originating from separate social units.[55] The trusting bond of ritualized friendship is maintained by a gift-exchange: the gifts are a means for creating a moral obligation of reciprocity. The gift-exchange economy underlies not only obligations of reciprocity between friends but also social bonds between patrons and clients. The analysis of science in a society functioning with

---

[54] Herman 1987: 10–11 calls 'ritualized friendship' the range of social phenomena denoted in ancient Greek by ξένος "host, guest, guest-friend, foreigner, stranger"; φίλος "friend"; and συγγενής "kinsman, parent": "[R]itualised friendship is here defined as a bond of solidarity manifesting in an exchange of good and services between individuals originating from separate social units."

[55] The court titles συγγενής "kinsman" and φίλος "friend" adopt the language of blood relations and social bonds from separate social units; cf. Herman 1987: 32–33. Herman 1987: 11: "[T]he words listed above imply special relations of friendship, trust, loyalty, reciprocity, and mutual aid between the people they characterise." Yet the only true equals Hellenistic kings recognized were other Hellenistic kings. Further adopting the language of blood kinship, they referred to each other as brothers; see RC no. 71. Strootman 2014: 149.n19 mistakes Polybius 4.48.5 as language of blood-relations between friends (philoi), rather than kings. Mooren's 1975 work suggests that, at least during the third century BCE, the language of blood-relations was not extended to the philoi.

personal trust emphasizes first the social network and values of the trusted audience to whom the scientist appeals.

Chapter 2 examines the economy of gift-exchange at court between intellectuals and social superiors: the ethical literature advising monarchs to choose which social inferiors to trust when accepting gifts; the gift-exchange contributions of scientists when acting as courtiers; and the forms and places where culturally sophisticated gift-exchange happened at court. The aim of the court scientist was to gain τιμή or δόξα, fame or renown, the esteem prized by courtiers and the ideology of kingship. Courtiers when aspiring to the equality denoted by the term 'friend' presented their ethical selves as true friends or flatterers.

Finally, the asymmetry in expertise between scientist and king forced the scientist to communicate his knowledge in a form appreciable by the king. Scientists intent on patronage from the king sought to present their work in an entertaining genre. Court science as entertainment encompasses both the object of science, nature, as well as the stylistic communication of knowledge about that object. The viewers of nature and technology were routinely said to feel a particular emotion, θαῦμα or wonder.[56] Eratosthenes' and Andreas' new science, like Herophilus' and Archimedes', is cross-disciplinary science. Cross-disciplinary science is the deliberate appropriation of one scientific domain into another. While Hellenistic scientists often write self-consciously within the tradition of their science to an audience of the same domain, the phenomena of appropriation from one science into another does not commit an individual science to be constrained by its tradition. An ancient audience that understands appropriation as science is a narrow audience, a scientific elite educated in multiple sciences and capable of appreciating direct and perhaps indirect allusions to other scientific traditions. Therefore, it is not apparent that a general audience would understand the appropriation as *science*; rather, perhaps, a general audience would receive the appropriation as entertainment. The lay audience of court society understood cross-disciplinary science as spectacle. A court scientist seeking to entertain a lay audience of court elites with cross-disciplinary science sought to gain the personal trust of his audience by providing entertaining learning.

The cross-disciplinary science of the court scientists is a type of court science. Court science, to reiterate, is knowledge about the natural world produced for the entertainment of the court grounded in the two claims that the scientist is motivated by the promise of gift-exchange in patronage to dedicate his work to a social superior and that science thereby developed attempts to interest the socially superior lay audience by its aesthetic similarity to poetry performed for entertainment of the court. Insofar as it is entertainment, court science participates in the aesthet-

---

[56] Berryman 2009: 50–55, Tybjerg 2003. Indeed, Aristotle *Metaphysica* 982b picked out wonder as the motivation for intellectual study.

ics of the court. While unfortunately we possess limited graphic and plastic arts from the courts of Ptolemy III Euergetes and Ptolemy IV Philopator, we do possess some poetry and literary prose.

Literature at the court of Ptolemy 246–205 BCE was interested in paradoxography (a genre in which unusual natural features are viewed with pleasure), praises of monarchs and courtiers, hybrid literary forms, texts with embedded performance of motion and sound, and referencing prior literary models and traditions, called belatedness.[57] These features are typical of broader Hellenistic literature. While briefly referring to some of the most important readings of scholarship may seem insufficient, I intend such a survey to be complementary to Netz's sustained reading of the aesthetics of third-century Greek mathematics.[58] Netz has shown that much of third-century BCE Greek mathematics is marked by an aesthetics of playfulness with narrative surprise, a flair for the unusual, and an awareness of generic boundaries in both science and literature as well as subsequent breaking of those generic boundaries. Exemplary is Netz's summary of his stylistic reading of Archimedes' *Spiral Lines*, a mathematical treatise on the circumference and area traced out by a particular spiral:

> A feature we saw repeatedly is the surprising role of calculation ... Generalizing the role of calculation, another – central – feature of the treatise is its breaking of genre-boundaries ... I have described above an aesthetic whose key components are variety and surprise. And indeed the spiral itself is marked by its multi-dimensional structure – combing linear and circulation motion, straddling the geometrical and the physical; while the key results obtained by Archimedes – a certain line being equal to the circumference of the circle, a certain area being one third that of the circle – are all important because they are so *surprising*, given the expectation of the impossibility of measuring the circle.[59]

In addition to Netz's remarks that the spiral itself encourages literary surprise and variety, I would add a performative dimension: the spiral's motion, forwards and to the side in the effort of wheeling around a point, mirrors the literary progression of Archimedes's treatise. Netz's final remarks position the object, the spiral, and the rhetoric effects of Archimedes' study, *Spiral Lines*, on the same aesthetic plane; both offer the reader a source of wonder.

Other Hellenistic literary phenomena seen in literature from the court of Ptolemy 246–205 BCE include hybridity and belatedness.[60] Hybridity is a well-documented literary phenomenon of the Hellenistic period. Kroll's seminal article em-

---

57 See the discussion of the court authors **nos. 109–131** in chapter 1.
58 Netz 2009a.
59 Netz 2009a: 12–15.
60 Netz 2009a: 5.n6 recognizes the belated aspect of Archimedes' treatise: Eudoxus' *On Speeds* is the predecessor master-text to which Archimedes' *De Lineis Spiralibus* responds.

ployed a biological metaphor to describe the authorial strategy of mixing genres.[61] Following Netz I understand hybridization to be broader than Kroll's genre-mixing: hybridization is the breaking down of traditional generic and ontological distinctions, such as that between Greek and Egyptian customs or between different sciences. We have already seen that Conon's Coma Berenices mixes Greek and Egyptian star-charts. All cross-disciplinary science in this book is an example of Hellenistic hybridity.

Scholarship has come to see literary belatedness as another major characteristic of Hellenistic poetry.[62] Belatedness is usually taken in a specifically textual sense: that Hellenistic authors are in textual debt to the writers before them. Hellenistic poets, for instance, seek to authorize their works by invoking a prior literary figure as a model. Greek scientists too often refer to previous models for their work but in a polemical way; canonically these references are read as evoking the famous agonistic social context of most Greek science.[63] Hellenistic science is imbued with agonism. The broader contemporary cultural appreciation of situating and authorizing written texts within a tradition dovetails with a scientific contextualization of new work within an intellectual tradition of a particular discourse on nature. Moreover, scientific agonism offered an opportunity for the scientist to represent himself as a *parrhēsiast*, the archetypal truth-teller. With the criticism of previous scientists making a canon of truth-telling, the court scientist imbued himself with authority about the natural world and represented himself as a courtier with specialized expertise in *paideia*, or culture, to advise the king. Chapter 3 considers the court scientist's rhetoric and strategies of seeking patronage by reading closely four court science treatises.

The second half of this book undertakes to contextualize scientific emergence as entertainment with specific case studies. Chapters 4–5 read the cross-disciplinary emergent science of Eratosthenes, Andreas, Herophilus, and Archimedes by identifying how assembly of conceptual holism participates in the aesthetics of the court by appropriating one scientific domain into another. An epilogue concludes the book by looking ahead toward a history over the *longue durée* of scientific interdisciplinarity.

---

61 Kroll 1924 speaks of "Kreuzung der Gattungen", literally the "crossing" or "hybridization" of genres, as if they were plant varietals; Fantuzzi and Hunter 2004: 18.
62 Fantuzzi and Hunter 2004: 1–41.
63 See discussion of Lloyd in footnote 15.

1 Simmias the Elephant-Hunter and Other People at the Court of Ptolemy

The court is best conceived as the broad retinue upon the king and his family. Yet not everyone in antiquity was keen on courtly service or ambitious to serve a monarch. The fourth century BCE comedic poet Diphilus, for instance, wrote:

> To serve at court, I think, is the business of an exile, a hungry man, or a whipping boy.[1]

While it is rarely advisable to take a comedian's words as straight reporting, it is unsurprising that courts were not everyone's ideal setting. Diphilus portrays the court as a place for second chances away from one's birth place or as a place for people of servility and obsequiousness without independent means of support. The couplet thereby juxtaposes the court with the traditional Greek vision of a man of independent wealth, living in his native city. Diphilus witnesses that not every Greek was drawn to the possibilities of life at court. The dichotomy of a life of independence at home and a life of servility abroad is of course a literary construct supported by traditional city-state ideology.[2] Despite Diphilus' rejection of courtly service, this chapter recounts those who participated in life at the courts of Ptolemy III Euergetes (reigned 246–222 BCE) and Ptolemy IV Philopator (reigned 222–205/4 BCE).[3] They were not all poor subservients without independent means of support nor adventurers and careerists. Indeed, among the people documented at court during those forty-one year were soldiers, ministers of state, bureaucrats, intellectuals, and, yes, exiles: the absence in the historical record of hungry men and whipping boys at court is a failure of evidence, not Diphilus' imagination.[4]

Scholars speak in the now-traditional metaphor of shipwreck when discussing the Hellenistic period. We lack primary source historical narratives for most of the period; few complete texts of literature, philosophy, science, history, or rhetoric survive; the surviving wreckage is preserved scattered as fragments or summaries in later authors. While evaluations of fragmentary testimonia are by definition limited by evidence, it is possible to draw reasonable and sober conclusions from a

---

[1] Diphilus fr. 97 Kassel-Austin = Athenaeus 5.189e: αὐλὰς θεραπεύειν δ' ἐστίν, ὡς ἐμοὶ δοκεῖ, / ἢ φυγάδος ἢ πεινῶντος ἢ μαστιγίου.
[2] Herman 1987 has especially good remarks on the social and political conflict between communal *polis* ideology and elite gift-exchange.
[3] The date for the end of Ptolemy IV Philopator's reign is in some doubt. Documents marked under his reign continue into the year 204 BCE but notice of the king's death had been kept secret for some unspecified period of time; see Hölbl 2001: 149.n38.
[4] Whipping was a common punishment in Ptolemaic Egypt: Diogenes Laertius 7.186, Polybius 15.28.2.

wider study. Here the wider study must embrace the evidence in the context in which the fragmentary authors and historical actors lived, namely court society.

The collective efforts of historians over especially the last 175 years in collecting and organizing primary documentary materials preserved on stone and papyrus have sought to remedy one aspect of the historiographical problem. An outstanding result of the effort to understand the history of the Hellenistic period is the nine-volume prosopographical series recording all the names and professions of the individuals living in Ptolemaic Egypt known from surviving primary evidence. This series, called *Prosopographica Ptolemaica*, assigns each individual a unique number (although several numbers are sometimes used) and provides a reference list of primary documentation of that individual. The print volumes (published 1950–1981) have been superseded in some aspects and incorporated into the Trismegistos database at <http://www.trismegistos.org/>. The online Trismegistos also does not include literary texts and so several elite individuals known from literary texts are missing in their prosopographic listings. Given the absence of a primary historical narrative, a descriptive prosopographical survey offers the most empirically historicized account possible of the actors of the Ptolemaic court and their interests.

Previous scholarship has not yet given a complete prosopography of the courts of Ptolemy III Euergetes and Ptolemy IV Philopator. The reasons are not far to seek. Scholarship has given much attention to the court culture and personages under Ptolemy II Philadelphus (reigned 285/4–246 BCE). The court of Ptolemy II Philadelphus, with its extravagent celebration of the royal family and the king's power, set the pattern for later Ptolemaic rulers.[5] The poetic and historical sources for Philadelphus' reign are especially rich. Yet the lives and social connections of many Hellenistic scientists are better attested under the rulers Ptolemy III Euergetes and Ptolemy IV Philopator. A study of Hellenistic science at court centered on the extant evidence needs a prosopography of court life under these pharaohs. In presenting and describing the primary evidence for individual court actors I aim to ground in historical individuals the discourses of friendship and contextual aesthetics for court patronage described as a cultural system in succeeding chapters. Throughout the remainder of the book I set a bold number next to the individual

---

[5] Hazzard 2000 argues that later Ptolemaic royal presentation largely imitates Ptolemy II Philadelphus. A word on Ptolemaic names: while scholars are unsure whether Ptolemy I Soter and Ptolemy II Philadelphus used the surnames during their lifetimes by which they were afterwards worshipped as deified gods, Euergetes already proclaimed himself and his queen *theoi Euergetai* in 238 BCE during their own lifetimes; Ptolemy IV Philopator likewise proclaimed himself *theos Philopator* during his own lifetime. Since this work focuses on the courts of Ptolemy III Eurgetes and Ptolemy IV Philopator, I will hereafter refer to Ptolemaic monarchs by their surnames. I have preferred Latinization for all Greek names but exceptions remain, especially in the case of authors on whom scholarship has already settled a name.

described as a shorthand reference to his or her culturally representative interests. (I avoid the otiose numbering in the case of the kings Euergetes and Philopator.) These chapters jointly offer a thick description of court actors, their discourses, and their practices.

A prosopography read in terms of gift-exchange and literary tradition in court friendship offers more than biographical anecdotes. As an example of my intended reading, consider the epigram said to have been written by a king Ptolemy about astronomy.[6]

> Hegesianax, Hermippus, and many others have set all (πάνθ') the portents in the sky and these phenomena in their books †but they missed the mark†. Yet light-worded (λεπτολόγος) Aratus holds the scepter.[7]

Since both Hegesianax and Hermippus (an antiquarian biographer and poet described below as **no. 118**) flourished around the end of the third century, the king Ptolemy who authored this piece, if genuine, was probably Philopator. What are we to make of this poem? As often in preserved fragments of court authors, there is a temptation to read the literary piece within the biographical context of the author. In this surviving poem the author evinces a preference for an aesthetics of *leptotēs* (λεπτολόγος), smallness or poetic fineness, associated with **no. 110** Callimachus and Aratus, among the best poets of the previous generation, against poetic comprehensiveness (πάνθ'). Hegesianax, Hermippus, and Aratus wrote some poetry entitled *Phaenomena* about the mythological portrayal of celestial observations, although little survives apart from Aratus' *Phaenomena*.[8] If the author of the epigram is Philopator, he complains of the decline in poetic skill of his own generation while praising the poetry of the past. Philopator's imitation of a previous master-poet in this regard gains credibility from the associated reports that Philopator wrote a tragedy about Adonis in the style of Euripides and established a shrine to Homer in Alexandria.[9]

Yet biographical readings of both poetry and prose have gone out of style so that as a consequence the previous reading of the fragment from the perspective of Philopator's personal aesthetics will not find many supporters. Ancient sources and modern scholarship often clash in this regard, since fragmentary literary evi-

---

6 The book of poems was called Ἰδιοφυῆ "Unique Natures", a title which is also attributed to the poet Archelaus (**no. 113** described below) who lived at Euergetes' court.
7 *SH* 712 = *Vita Arati 1*: πάνθ' Ἡγησιάναξ τε καὶ Ἕρμιππος ⟨τὰ⟩ κατ' αἴθρην / τείρεα καὶ πολλοὶ ταῦτα τὰ φαινόμενα / βίβλοις ἐγκατέθεντο †ἀπὸ σκοποῦ δ' ἀφάμαρτον†· / ἀλλ' ὅ γε λεπτολόγος σκῆπτρον Ἄρατος ἔχει. Fraser 1972: 2.1090.n459 defends its authenticity.
8 For Hegesianax's works see *SH* 465–470 and *BNJ* 45 (Hegesianax of Alexandria Troas). I assume that Hegesianax and Hermippus wrote astronomical poems, since, if their astronomical works were prose rather than poetry, the epigram's juxtaposition of their work with Aratus' loses its force.
9 See below in the description of **no. 8** Philopator.

dence of the Hellenistic courts is often preserved in later ancient authors who aim to pick out some mark of character or a certain witticism to some moralizing end. The biographical style of reading I just described is typical of P. M. Fraser's *Ptolemaic Alexandria*, a still essential cultural history about the Hellenistic city. Fraser's survey of Ptolemaic patronage emphasizes both the personal aesthetics of the monarch and the institutional structures of the Mouseion and library; for Fraser, patronage was set within institutional structures as a top-down activity of the monarch's personal preferences.[10] By combining recycled historical gossip with a top-down view of the social dynamics of gift-exchange a descriptive survey of the leading individuals of the Hellenistic courts risks construing picturesque details of characters as the decisive factors in patronage. Consequently, this survey rejects biographical readings of literary fragments and emphasizes instead the social context of gift-exchange and the literary traditions of individual actors. In succeeding chapters I will speak more of *personae* than persons within such a social context.

A contextual reading of gift-exchange and literary tradition in the above fragment of Philopator, then, sees a certain irony to the stylistic claims of poetic fineness from the perspective of Philopator. While Aratus lived and performed his poetry at the court of the Seleucids and Antigonus Gonatas, Hermippus had lived for some time at the Ptolemaic court while Hegesianax lived and worked at the court of Antiochus III the Great, Seleucid king and chief rival to Philopator.[11] Perhaps the Seleucids' previous court poets had produced better Callimachean poetry than Callimachus' own student in Alexandria and the contemporary court poet of the Seleucids.[12] The aesthetic voice of the *persona* of the king as chief benefactor and patron praises the poetry of the previous generation in an ironic recognition of the literary belatedness of his court: poetry on *physis* competitively imitates the acknowledged master-text of that genre. The interest of Philopator's *persona* participates in a larger trend; it does not drive it. Antiquarianism had become a key aesthetic at Philopator's court.

---

10 Fraser 1972: 1.305–314 is his clearest expression of this position. Consequently Fraser 1972: 1.317–331, 357 tries to situate all known scholarly activity within the context of the Mouseion and library. Some cases seemingly give Fraser the evidence necessary to credit the monarch's personal character as an *explanans* of scholarly activity. For instance, since Philopator rejected Zoilus Homeromastix's plea for financial support as recounted by Vitruvius (Fraser 1972: 1.310), perhaps we ought to connect Philopator's personal admiration for Homer with his patronizing rejection of the Homeric critic. Yet this particular piece of evidence falls apart under scrutiny: *BNJ* 71 (Zoilus of Amphipolis) shows that Vitruvius' evidence is a fabrication and that Zoilus, a fourth-century BCE rhetorician, never went to Alexandria (and hence does not belong in PP). While I admire Fraser's comprehensiveness and do not aim to replace his work, my general criticism of Fraser's views on science and patronage echoes Duindam's 2011: 8–9 remarks on power and exchange in court culture. In chapter 2 I emphasize the social practice of gift-exchange over the institutional context of the Mouseion.
11 *BNJ* 45 (Hegesianax of Alexandria Troas) T3.
12 Fraser 1972: 2.1090.n459.

Finally, it is not an easy question to answer as to who counted as a court actor. Doubtless contemporaries could have identified who was a court actor using markers such as social status, insignia and clothing, titles and honors.[13] These identifying symbols have largely disappeared from the historical record. Remaining in extant texts are titles, anecdotes about interaction with the monarch, and geographical proximity to the king. My survey builds on work of previous historians as to who was a court actor but I have not been afraid to advance my own opinions.[14] Since the prosopography is divided by social function, each individual is referenced by their *Prosopographia Ptolemaica* number (= PP). I have cross-referenced the PP number with the equivalent Trismegistos person number when it exists (= TM). Not all of these individuals when met in the historical record can be placed at the royal palace in Alexandria, although all probably visited there at some point.[15] Their dates can only occasionally be specified more precisely than under a certain monarch. For many only their social function is known, with no description of the individual possible. I have counted one hundred and forty-one total individuals known to me, although I have not listed all names nor described each individual. A brief précis to each social group prefaces its first numbered individual.

- **1– 10** *The King and His Family*
- **11– 18** *The Philoi or Friends of the King*
- **19– 41** *Bureaucrats, Military Officials, Judiciary, Minor Functionaries*
- **42–102** *Priests*
- **103–108** *Flatterers, Entertainers, Lovers*
- **109–116** *Poets*
- **117–122** *Prose Authors: Antiquarians, Geographers, Historians*
- **123–131** *Prose Authors: Philosophers, Grammarians*
- **132–141** *Scientists*

People missing from the historical record
- **142–158** *Appendix: Doubtful and Excluded Persons*

The King and His Family

At the center of court life was a royal house that celebrated royal beneficence and patronage of elite culture.

---

13 Strootman 2014: 39.n32.
14 I have found Fraser 1972, Mooren 1975, and Weber 1993 indispensible guides to organizing the vast information in PP.
15 I take Strootman's 2014: 31–33 point that the court is the retinue of the king and thus mobile, but the Ptolemies shifted the seat of state power far less than the Seleucids.

1 The career of Ptolemy III Euergetes (PP 14543; TM 12909), the king from 246–222, is much better documented than other members of his court, although primary sources for his reign are fewer than those for the reigns of his father and grandfather, Ptolemy II Philadelphus and Ptolemy I Soter. The successive reigns of these three Ptolemies – Soter, Philadelphus, and Euergetes – are traditionally seen as a Golden Age for Ptolemaic Egypt with political continuity, military success, and the peaceful encouragement of commerce and learning.[16] Euergetes' victory in the Third Syrian War over his traditional enemy, the Seleucids, at the beginning of his reign brought the Ptolemaic kingdom to its greatest extent. The Ptolemaic kingdom controlled territory from Cyrene in the west, across the Aegean, all of Egypt, and the Palestinian coast into Syria. Euergetes brought the war to an end suddenly because of revolt at home and concentrated on the domestic benefactions that marked his deified surname.[17] This short description of Euergetes' political success serves as background to a survey primarily of his interests in literature and the circulation of texts.

Euergetes tried to present himself both as a warrior-king and pleasure-seeking patron of the arts. A contemporary inscription published under the name of Euergetes describes him as a descendent equally of the god Heracles, a powerful, world-traveling warrior-hero, as well as of the god Dionysus, a patron of pleasure and learning.[18] Euergetes' tutor was **no. 109** the poet Apollonius of Rhodes, also head of the royal library.[19] There were two libraries in Ptolemaic Alexandria: a library in the temple of Serapis, evidently built by Euergetes, and the so-called royal library, which was the major and now-famous 'library of Alexandria'.[20] Whatever the aesthetic values he took from Apollonius, Euergetes certainly had a first-rate education.

Euergetes was probably an author of a historical work, if surviving material is correctly attributed.[21] He was also remembered as a collector of books and art, part

---

16 A typical survey of the periodiziation of Ptolemaic history is Hölbl 2001. Euergetes falls under "Part 1: The beginning and golden age of Ptolemaic rule." Clayman 2014: 178 concurrs: "This sounds like a fantasy, a Camelot-on-the-Nile, and yet it is documented, on the whole, by the kind of evidence that is hard to discredit: contemporary inscriptions, documentary papyri, art, and archaeology."
17 Justin 27.1.9.
18 *OGIS* 54.1–6.
19 *BNJ* 241 (Eratosthenes of Cyrene) T7 = *P.Oxy* 1241 col. 2; Fraser 1972: 1.322–323, 331–333. I am unpersuaded by Murray 2012 that *P.Oxy* 1241 col. 2 is not a list of the librarians.
20 Fraser 1972: 1.323; Clayman 2014: 160. The best discussion of the institutions of the Mouseion and library remains Fraser 1972: 1.312–335; see Woolf 2013 for more recent bibliography.
21 A surviving papyrus account of the Third Syrian War told from the king's perspective may have been written by him; see *BNJ* 160 (Anonymous *Belli Syrii tertii annales*). Since Ptolemy I Soter had written an account of his campaigns with Alexander (*FGrH* 138 (Ptolemaios Lagu)), there is nothing implausible about Euergetes writing a history of his own campaigns.

of the discourse of control and power associated with ancient libraries.²² Yet a history of court science foregrounds the library at Alexandria as an archive, since later authors thematize the fixedness and preservation of authoritative written texts in the library.²³ Later authors tell the famous story about Euergetes' acquisition of the Athenian state copies of the tragedies of Aeschylus, Sophocles, and Euripides: he borrowed them to copy for a deposit of fifteen talents and, upon copying the official versions, forfeited his deposit and returned the copies, keeping the official versions for Alexandria. Whatever the historicity of that story urbanely portraying the transfer of Greek cultural heritage from Athens to Alexandria, the later Greek doctor Galen reports Euergetes' procedures for copying traveler's holdings of books.²⁴ The books were copied, the copies returned to their owners, the originals were placed first in storehouses with an additional subscription noting their origin, and subsequently organized in the libraries. Some scholars have attempted to see a coherent and systematic approach to information science in this account: an accessions department and a cataloguing system, among others.²⁵ Even if much remains unclear in Euergetes' efforts, his motivation at least is not in doubt. The king intended and set up a state-authorized apparatus to acquire, copy, and store books that passed through Egypt, the totalizing claim to universal and encyclopedic cultural heritage.

Given Euergetes' reputation as the great conqueror of the Seleucids and his continuance of a Ptolemaic golden age, it is a surprise that two later authors give Euergetes the nickname Τρύφων "Luxuriant", as if for all his military prowess he indulged in a undisciplined life of excess.²⁶ Still this nickname is also attested for Philopator, Euergetes' son and successor, whom historical sources routinely paint in black colors.²⁷ It is possible to weigh sources and shield the father from the charges that darken the son; yet the nickname was not uncommon among other Hellenistic rulers. Even if the epithet was a pejorative and politically useful for later authors, it nonetheless gestures toward the real glitz and splendor of the Hellenistic courts, irrespective of ruler.

2   Berenice II (PP 14499; TM 6642), the queen, was born in Cyrene, the biological daughter of Magas, self-proclaimed king of the city-state Cyrene nominally under Ptolemaic control.²⁸ Her marriage to Euergetes was her second, after she was said

---

22 Too 2010: 19–24, 31–40. Bagnall 2002 cautions against confusing the Ptolemaic ambition to collect all knowledge with the historical reality of their having done so. For Euergetes' art collecting see Plutarch *Aratus* 12.6.
23 Berrey 2015: 472–475.
24 Galen *In Hippocratis Epidemiarum Librum III* 17A.606–607K = 79.8–22 Wenkebach.
25 Fraser 1972: 1.326–337, opposed by Bagnall 2002.
26 Justin *Prologus* 27; Eusebius *Chronicon* I = 251.26 Schoene.
27 Pliny *Historia Naturalis* 7.57.208.
28 Clayman 2014 is a lengthy biography with discussion of all relevant material.

to have been involved in the murder of her previous, philandering husband in Cyrene.²⁹ Berenice's queenship followed the model from the cult of divinity surrounding Arsinoe II, Berenice's mother-in-law.³⁰ It was Berenice's lock, dedicated to her deified mother-in-law Arsinoe, that **no. 137** Conon catasterized as a new constellation. Poetic lock or not, Berenice's Cyrene was said to have an excellent industry in rose-oil, useful for perfumes of the hair.³¹

Berenice was a prominent symbol of Ptolemaic values and rule. With her husband Berenice embodied the values of the *theoi Euergetai*, the "beneficient gods". Royal decrees such as the Canopus Decree were thanksgivings for the couple's legendary munificence and benevolence in importing grain to Egypt in a time of failed harvest; coins minted with Berenice's image showed her with enlarged eyes, a sign of divinity, and the cornucopia, the overflowing harvest of a liberal *physis*; trading cities along the Red Sea in elephant-hunting country were named "Berenice" in her honor; the court polymath Eratosthenes **no. 136** described her preserving women's property by forbidding dissolution of doweries left to women of Lesbos by their fathers.³² Beyond memorializing her now celestial lock at the end of his *Aetia*, **no. 110** Callimachus celebrated Berenice's chariot victory in the Nemean games through an epinician in the *Aetia* that interweaves a genealogical and syncrestic account of Egyptian's centrality for elements of Greek culture and civilization.³³ Callimachus also wrote epigrams and hymns that referred to her directly or indirectly; scholars have read references to Berenice in other contemporary or near contemporary court poets, a research project that is still on-going.³⁴

When Berenice's son Philopator ascended the throne, he put her under house arrest. She refused to submit "due to her courage", took poison, and died.³⁵ Philopator enshrined her in the family masoleum he had built; once deified, she was worshipped as "Berenice Soter" or Berenice the Savior.³⁶

3–7 There were other members of the immediate royal family at court: Lysimachus (PP 14531, 15054, 16932; TM 10114), brother of the king; Magas (PP 14534; TM 10144), son of the king; Alexander (PP 14479; TM 4613), son of the king; and

---

29 Justin 26.3.3–8.
30 Hazzard 2000: 113–116 cogently argues this point.
31 Athenaeus 15.689a; Clayman 2014: 102–103.
32 Canopus Decree *OGIS* 56.16–20; coins and cities Clayman 2014: 128–133; Lesbian doweries Hyginus *Astronomia* 2.24, 1003–1007 Viré.
33 Callimachus *Aetia* frr. 54–54a; Stephens 2003: 8–9.
34 Clayman 2014: 58, 78–104 on Callimachus. I hesitate to accept the arguments of Clayman 2014: 105–120, 148–158 that the *realia* of Berenice's life ought to be read so strongly as intertexts in the Apollonius' *Argonautica* and that the *Hippika* series of Posidippus' epigrams refers to Berenice II as queen.
35 Zenobius 3.94; Polybius 5.36.1.
36 Clayman 2014: 47 argues that this was in imitation of Arsinoe II, as a protectress of sailors.

an unnamed son of the king (PP 14575; TM 15972). Finally, early in their reign Euergetes and Berenice had a daughter named Berenice (PP 14500; TM 6646) after her mother. The young princess died young, when she was at most twelve. The Canopus Decree promulgated in 238 BCE established a cult in her honor.[37] Euergetes and Berenice had two other children, their successor monarchs.

8 The king Ptolemy IV Philopator (PP 14545, 16712; TM 12910), the successor son of Euergetes, has traditionally been held responsible for the decline of the Ptolemaic kingdom's power during his reign from 222–205/4 BCE.[38] His nickname, Τρύφων "Luxuriant", hints at the atmosphere of riches, dissolution, and excess surrounding his court.[39] Philopator himself so celebrated the god Dionysus that the court poet Euphronius **no. 115** called him *Neos Dionysos* "New Dionysus".[40] Immediately upon his accession he had his mother Berenice II, his uncle Lysimachus, and brother Magas murdered through his minister **no. 15** Sosibius. Worse, he armed 20,000 native Egyptians to supplement his troops in preparation for the battle of Raphia against Antiochus III the Great after spending a year at Memphis instead of Alexandria.[41] The historian Polybius and many modern historians following him attribute the subsequent Egyptian revolt of 204 BCE, the first serious revolt under the Ptolemies, to the development of Egyptian political identity.[42] The historian Plutarch gives the chief charge: Philopator was ruled by women and wasted himself in orgiastic rites.

> The kingdom at once fell into riotousness, drunkenness, and womanish temperament ... for the king himself so ruined his life by women and drink, that, when he was very sober and most serious, he participated in mystery rites and with a drum gathered his participants in the palace, while Agathoclea, mistress of the king, and Oenanthe, her mother, a madame, ran the chief affairs of the kingdom.[43]

---

37 *OGIS* 56.46–73; Pfeiffer 2004: 148–149.
38 Whereas Hölbl 2001 included Euergetes under the Golden Age, Philopator leads off his "Part 2: Change and Decline of the Hellenistic State of the Ptolemies." In this Hölbl is following ancient sources. Polybius 15.34.5 refers to "Philopator's unfitness for kingship"; Polybius 5.34 more generously describes Philopator's assumption to the throne free from external threats.
39 Philostephanus F35 Capel Badino = Pliny *Historia Naturalis* 7.57.208 See Philostephanus **no. 119**.
40 See generally Fraser 1972: 1.203–205.
41 Polybius 5.65.9, 5.62.4.
42 Polybius 5.107; Fraser 1972: 2.143.n178. In a polemic piece Ritner 1992 dissects Fraser's emendation of a famous remark of Polybius, preserved in Strabo, about the respective worth of the Egyptians at Alexandria and accuses Fraser of attributing the decline of Alexandria to the Egyptians.
43 Plutarch *Agis et Cleomenes* 54.1–2: τῆς δὲ βασιλείας εὐθὺς εἰς πολλὴν ἀσέλγειαν καὶ παροινίαν καὶ γυναικοκρασίαν ἐκπεσούσης ... ὁ μὲν γὰρ βασιλεὺς αὐτὸς οὕτω διέφθαρτο τὴν ψυχὴν ὑπὸ γυναικῶν καὶ πότων, ὥσθ', ὁπότε νήφοι μάλιστα καὶ σπουδαιότατος αὐτοῦ γένοιτο, τελετὰς τελεῖν καὶ τύμπανον ἔχων ἐν τοῖς βασιλείοις ἀγείρειν, τὰ δὲ μέγιστα τῆς ἀρχῆς πράγματα διοικεῖν Ἀγαθόκλειαν τὴν ἐρωμένην τοῦ βασιλέως καὶ τὴν ταύτης μητέρα καὶ πορνοβοσκὸν Οἰνάνθην.

Philopator's reputation has never recovered from the problematic image of both his political actions and personal debauchery. Later Greek historians routinely paint Philopator in the black marks of a bad king, including the king's persecution of the Jews in the ahistorical 3 Maccabees.[44] While my aim is simply to underline how difficult it is to draw a picture of life at court under Philopator without falling prey to the moralizing narrative of a hedonistic ruler that surrounds him, his *philoi*, and his leading bureaucrats in primary sources, Philopator's accomplishments should not be minimized. He won the major war of his reign in the Fourth Syrian War, preserved the royal succession of his family despite his disloyal advisors, and patronized elite culture.[45]

Kings were traditionally patrons of elite culture. Euergetes of course had supported poets and possibly written prose himself; other Greek precedents could be named.[46] Philopator respected traditional Ptolemaic kingship: he built the family shrine for his forefathers, including Alexander the Great's body.[47] Philopator was interested in *belles-lettres* as the epigram in support of Aratus' *Phaenomena* described above showed. Philopator also wrote a tragedy *Adonis* in imitation of Euripides, now lost.[48] The death of Adonis, the mortal lover of Aphrodite, at the tusks of a wild boar was a frequent subject of sentimental bucolic poetry in the Hellenistic period.[49] Yet, since the Ptolemies exempted actors from the salt-tax and Euergetes acquired the texts of the Athenians tragedians, Philopator's interest in Euripidean tragedy in particular suggests that tragedy was seen as an art form whose perfected forms had been laid down over a century earlier.[50] Perhaps no poet was writing tragic poetry at the court of Euergetes or Philopator. Sositheus of Alexandria, the only secure Alexandrian member of the so-called Pleiad whose members constituted the Hellenistic canonical tragedians, was active during the first half of the third century; Euphronius **no. 115**, active at the courts of Euergetes and Philopator, may have been a tragedian since he is once called member of the Pleiad, although the evidence is very late and we possess no tragic fragments.[51]

---

44 3 Maccabees is a later piece of historical fiction, perhaps written soon after the Roman conquest of Egypt, about Philopator's attempted persecution of the Jews of Egypt. The writer has taken some pains to provide historically accurate actors: **no. 23** Dositheus the *hypomnematographos*, **no. 141** Andreas physician of Philopator, and unnamed *philoi* are mentioned. Despite these historically accurate details, most historians treat 3 Maccabees as an ahistorical literary construct.
45 Hazzard 2000: 115–122 reasonably argues that Philopator tried to emulate his grandfather Philadelphus in his kingship. Philopator merely tried to keep the empire as he had received it in peace for the benefit of its inhabitants.
46 The literary ambition to write tragedy connects Philopator and Dionysius I of Syracause (Lucian *Adversus Indoctum* 15), if the stories about that monarch are true.
47 Zenobius 3.94.
48 Σ.Aristophanem *Thesmophoriazusae* 1059.
49 Cf. Theocritus 15.100–143. Fraser 1972: 1.198.
50 Fraser 1972: 1.618–619; Fantuzzi and Hunter 2004: 432–437.
51 See nos. **109–116**; Fraser 1.619–621.

Philopator appears as a patron of elite culture with dedications to the archaic Greek poets Homer and Hesiod. We possess the fragmentary ending of a poem commemorating Philopator's dedication of a temple to Homer.

> Long-lived Ptolemy ... established this shrine to Homer on behalf of ... in a dream, Homer who previously wrote the ageless song of the *Odyssey* and *Iliad* from his immortal wits. O blessed euergetists of mortals, who raised the lord best among spears and Muses.[52]

The poem celebrates both Homer and Philopator, who seems to have been prompted to set up the shrine on account of a dream. Perhaps Homer as a god (ἀπ' ἀθανάτων πραπίδων) visited Philopator in sleep and commanded his worshipper to found a shrine in his honor. To encounter a literary divinity in a dream finds a predecessor in **no. 110** Callimachus' dream encounter with Hesiod's Muses: Philopator, Homer's dreamer, is represented in the tradition of a devotee of an inspired author.[53] Furthermore, in line with the Ptolemies' hybrid Greco-Egyptian kingship, the *topos* of a king's dream with divine visitation and subsequent temple building has dynastic Egyptian predecessors: pharaonic temple monuments often record the divine visitation that inspired the building.[54] In the dedicatory poem Homer is lauded for having written his notable epics; Philopator is praised as the son of the Euergetai **nos. 1** and **2** and famed for war and culture (ἐν δορὶ καὶ Μούσαις). We are elsewhere told that Philopator put a statue of Homer in the temple surrounded by all the cities who laid claim to him; by tradition these were cities of the Aegean islands.[55] Philopator is also known to have been a patron of a shrine of the poet Hesiod in the poet's native Ascra in Boeotia.[56]

---

52 *SH* 979; Fraser 1972: 2.862.n423. εὐαίων Πτολεμ[...] τοδ' Ὁμήρῳ / εἴσαθ' ὑπὲρ δι .. [... κ]ατ' ὄναρ τέμενος / τῷ πρὶν Ὀδυσσείας τε καὶ Ἰλιάδος τὸν ἀγήρω / ὕμνον ἀπ' ἀθανάτων γραψαμένῳ πραπίδων. / ὄλβιοι ὦ θνατῶν εὐεργέται, οἳ τὸν ἄριστον / ἐν δορὶ καὶ Μούσαις κοίρανον ἠρόσατε.
53 Callimachus *Aetia* fr. 2. The literary forerunner ought to be preferred to a biographical reading from later historians. Later writers associate Philopator with supersition and the power of dreams in numerous anecdotes, particularly with Philopator's attacks of conscience after various misdeeds: in Plutarch *Agis et Cleomenes* 60.2–3 "superstition and fear fell upon the king" after a great snake coiled itself around Cleomenes' body and protected the head from birds; in Zenobius 3.94 Philopator built a memorial for his mother Berenice and his forefathers after her house imprisonment and suicide "on account of disturbances in his dreams". One cannot but suspect that the Greek moralizing literary picture of a bad king is at work in these texts. It is germane to the Ptolemaic self-conception of a learned kingship patronizing elite culture that Ptolemy VIII Euergetes II also was a devotee of Homer (despite his expulsion of the intellectuals in 145/4 BCE); cf. *BNJ* 234 (Ptolemy VIII) F11 = Athenaeus 2.61c.
54 See Lang 2012: 49–53 on pharoahs and dreams.
55 Aelian *Varia Historia* 13.22: Fraser 1972: 2.862.n423.
56 Fraser 1972: 2.467.n55. Fraser 1972: 1.203 finds dedications to Isis, Harpocrates, and Hathor, in addition to the cultic foundations for Dionysus. It may also be correct to attribute to Philopator the famous dedication of the eroded statuary of eleven Greek poets and philosophers at the Serapieion in Memphis; see Bergmann 2006.

9 Arsinoe III (PP 14492, 14510, 14515; TM 6124), daughter of Euergetes, queen and sister-wife of Ptolemy IV, did not marry her brother immediately upon his accession to the throne. She was present at the battle of Raphia and at a critical juncture she rallied the troops, "crying with let-down locks for them to help themselves, their children, and wives with courage".[57] After the celebratory inscription for the victory at Raphia Arsinoe disappears from the public record of Philopator's court.[58]

But some surviving materials portray Arsinoe within elite court society.[59] By far the most discussed representation of Arsinoe comes from a work called *Arsinoe* by Eratosthenes **no. 136**. Probably a dialogue rather than a biography, it depicts Arsinoe discussing one of Philopator's newly founded Dionysiac festivals called the Lagynophoria, from the carrying of wine-pitchers.

> Since Ptolemy [sc. Philopator] founded all sorts of feasts and festivals, particularly in celebration of Dionysus, Arsinoe asked a man carrying branches what day he was celebrating and what festival it was. He replied: "It is called Lagynophoria. People dine on what they have brought while reclining on straw. Each person drinks from their own *lagynos* (wine-pitcher) they have brought." When he left, she looked at us and said, "These synoecisms are dirty (ῥυπαρά) – for the gathering has to be from a mixed crowd, who bring to the table an out-of-date and ill-fitting meal. But if this sort of festival should be thought right, the kingdom surely wouldn't tire preparing the same things as in the Choes. For there one feasts individually, but the inviter to the festival provides these materials."[60]

---

57 3 Maccabees 1.4: δακρύων τοὺς πλοκάμους λελυμένη βοηθεῖν ἑαυτοῖς τε καὶ τοῖς τέκνοις καὶ γυναιξὶν θαρραλέως.
58 *Prose* nos. 12–14. Hazzard 2000: 118–119 argues that Arsinoe became weak and isolated at court.
59 Fraser 1972: 1.609–611 analyzes a papyrus epigram dedicated possibly to Arsinoe III that describes a Nympheaum, or a semi-circular fountain of the Nymphs composed of Greek marbles and Egyptian granite. Hazzard 2000: 118 finds a number of honorifics dedicated to Arsinoe. Damagetus 1 *HE* shows Arsinoe dedicating a lock of her hair to Artemis: "Artemis who has as her lot bows and strong arrows, to you Arsinoe, *parthenos* of Ptolemy, has left this lock of her own hair in a fragrant shrine, shearing it from her lovely hair." From the term *parthenos* and association with the goddess Artemis Arsinoe is not yet a bride. It remains unclear whether to understand the term *parthenos* as "daughter of Ptolemy" or "virgin bride of Ptolemy", that is, whether Damagetus refers Arsinoe to Euergetes as daughter or to Philopator as her potential brother-husband. A contemporary reader in the Ptolemaic court perhaps saw Arsinoe imitating the famous behavior of her mother Berenice in dedicating a lock to a goddess.
60 *BNJ* 241 (Eratosthenes of Cyrene) F16 = Athenaeus *Deipnosophistae* 7.276a–c: τοῦ Πτολεμαίου κτίζοντος ἑορτῶν καὶ θυσιῶν παντοδαπῶν γένη καὶ μάλιστα περὶ τὸν Διόνυσον, ἠρώτησεν Ἀρσινόη τὸν φέροντα τοὺς θαλλούς, τίνα νῦν ἡμέραν ἄγει καὶ τίς ἐστιν ἑορτή. τοῦ δ' εἰπόντος "καλεῖται μὲν Λαγυνοφόρια, καὶ τὰ κομισθέντα αὐτοῖς δειπνοῦσι κατακλιθέντες ἐπὶ στιβάδων, καὶ ἐξ ἰδίας ἕκαστος λαγύνου παρ' αὐτῶν φέροντες πίνουσιν", ὡς δ' οὗτος ἀπεχώρησεν, ἐμβλέψασα πρὸς ἡμᾶς "συνοίκιά γ'" ἔφη "ταῦτα ῥυπαρά· ἀνάγκη γὰρ τὴν σύνοδον γίνεσθαι παμμιγοῦς ὄχλου, θοίνην ἕωλον καὶ οὐδαμῶς εὐπρεπῆ παρατιθεμένων. εἰ δὲ τὸ γένος τῆς ἑορτῆς ἤρεσκεν, οὐκ ἂν ἐκοπίασε δήπου τὰ αὐτὰ ταῦτα παρασκευάζουσα ἡ βασιλεία [*scripsi*: βασίλεια mss.] καθάπερ ἐν τοῖς Χουσίν· εὐωχοῦνται μὲν γὰρ κατ' ἰδίαν, παρέχει δὲ ταῦτα ὁ καλέσας ἐπὶ τὴν ἑστίασιν."

A biographical reading of the *Arsinoe* sees evidence of Arsinoe's character and Eratosthenes' sympathy with the queen against the vulgarity of Philopator.[61] Yet philological indications suggest the *Arsinoe* was a sort of philosophical dialogue: the interchange of the speakers, multiple characters in the scene, and the philosophical vocabulary.[62] The queen leads a symposiastic discussion about philosophical opinions among several intellectuals and elites, perhaps about the new worship of Dionysus recently introduced by Philopator, a devotee of the god.[63] The discussion puns on ῥυπαρά, both physically dirty (the straw on which symposiasts recline) and socially base (the mixed crowd, who bring their own food), to unite the members of the kingdom in a religious festival. The irony of queen, gently chiding her husband's policies and the non-elite crowd of the festival, reflects the irony in the epigram about Aratus attributed to Philopator. It is difficult to believe that such a dialogue was published without her approval in her own lifetime. The fragment of Eratosthenes' *Arsinoe* cannot be taken as biographical evidence for attitudes of Arsinoe. It is, however, contemporary representational evidence of the Ptolemaic queen's elitist beneficence modeled on previous Greek traditions. The Alexandrians fondly remembered Arsinoe after her death.[64]

10 Ptolemy IV Philopator and Arsinoe III had a son, who succeed his father as Ptolemy V Epiphanes (PP 14546, 17232; TM 12911).

The *Philoi* or Friends of the King

The *philoi* or friends of the king were the outstanding members of the royal entourage, the chief examples of courtiers.[65] The *philoi* represented the king's interests in different ways: some were generals, some were political advisors, all were intermediaries between social groups to which they belonged (native cities, intellectuals, soldiers) and the king. The *philoi* possessed funds, soldiers, and gifts from the

---

61 Fraser 1972: 1.309–310.
62 Fraser 1972: 2.699.n38 objects that Athenaeus terms it a σύγγραμμα instead of a διάλογος. But Athenaeus is simply specifying that the *Arsinoe* is a prose work. While Geus 2002: 61–68 also sees a dialogue, he ends the fragment at ῥυπαρά and believes that the Arsinoe of the fragment is Arsinoe II, the wife of Philadelphus. In accord with his general chronology of Eratosthenes' writings (2002: 54–58) Geus dates the *Arsinoe* to Eratosthenes' life in Athens. At the least, the descriptions of the festivals are especially otiose for a biographical or historical piece, as if either the internal courtly participants of the dialogue were unacquainted with the famous Athenian festival of the Choes or did not know the name of the festival instituted by Philopator.
63 Philopator's affiliation with Dionysus is described below under Euphronius **no. 115**.
64 Polybius 15.25.9.
65 Important papers include Habicht 2006, Herman 1980/81, Strootman 2011, Strootman 2014: 111–184. Chapter 2 considers court friendship at length.

king; the gifts were mostly landed estates and their derived revenue.[66] We do not meet all *philoi* in Alexandria, but it is certain that they returned to court to participate in the king's advisory council on matters of state. Their friendship with the king was probably both personal and political, a mixture of genuine and utilitarian loyalty; it is best described as ritualized friendship after the description in the Introduction. As we will see in chapter 2, the equality and expectation of reciprocity between the king and his *philoi* were the model for the discourse of court friendship that runs throughout Hellenistic court society. Consequently, all other members of court society to follow in this prosopography, insofar as they desired a successful career at court, aspired to the power and privileges of the king's friends.

We know of four *philoi* under Euergetes, involved in foreign diplomacy from Athens, the Aegean islands, Cilicia and Coele Syria (conquests of Euergetes in the Third Syrian War), and the east coast of Africa. We know of four *philoi* under Philopator. Traditionally the Ptolemaic king's *philoi* were from the elite of Greek cities, not from Alexandria or Egypt.[67] The Alexandrian *philoi* of Philopator mark a change from earlier patterns of ritualized friendship. Loyal families had always offered important support to Ptolemaic kings. Proposopographic evidence shows an increasingly closed court society in the later third century BCE with multiple families of hereditary aristocrats drawn from the Macedonian aristocracy and from the families of Greek settlers and colonists who emigrated in the early part of the third century. The court society of these families loyal to the Ptolemaic dynasty was augmented by exiles and individuals who successfully pursued promotion among the priesthood or military, although we meet no native Egyptians among the political or bureaucratic elite under Euergetes and Philopator. (While there were no native Egyptians among the upper political establishment, royal patronage of monuments to native Egyptian ideology and worship continued with the massive temple complexes built at Philae and Edfu and the syncretistic shrine of Serapis in Alexandria.[68]) Families whose members became *philoi* or leading bureaucrats often established other immediate family members such as sons or daughters into positions within the priesthood or military.

11 Simmias (PP 14628, 16304, 16952; TM 13496) – *philos*, elephant-hunter, and explorer – is known only from a passage in the historian Diodorus Siculus who is quoting the earlier Ptolemaic geographer Agatharchides.[69] From Diodorus' descrip-

---

66 Examples in *RC* nos. 11–13.
67 Fraser 1972: 1.101–104; Herman 1987: 153–155.
68 Hölbl 2001: 104–112, 160–161; Clayman 2014: 160–164.
69 Diodorus Siculus 3.18.4 "The third Ptolemy, who was very fond of hunting elephants that lived in that region, sent one of the *philoi* named Simmias to check out the area. He was sent just with appropriate support, as Agatharchides of Cnidus the historian says, and he examined the tribes along the coast." Diodorus is discussing the country and lifestyle of a people who lived at along the coast of the Red Sea at the horn of Africa (today the coast of Sudan and Eritrea). Mooren 1975: no. 16 suggests further identifying Simmias with a geographical writer named Simmeas. This Sim-

tion it is likely that Simmias was responsible for capturing elephants and returning them to Euergetes for hunting; hence Simmias was Euergetes' elephant-hunter. Simmias' expedition was financially supported and equipped by the king. Soter and Philadelphus had acquired elephants from the Red Sea coast, but Euergetes' expedition went further south than theirs.[70] The stated purpose of exploring this new and little known area was a combination of elephant hunting and ethnography, probably with a particular focus on political and financial considerations, since Egyptian merchants traded along the Red Sea coast.[71] Here were the foundations of the cities "Berenice", at least some named for **no. 2** Berenice II. Inscriptions from Euergetes are known from the ancient coastal site of Adulis, north of the area explored by Simmias, so Simmias' expedition to parts south along the coast was a demonstration of Ptolemaic power. A central passage from a partially lost inscription portrays Euergetes as a victorious pharaoh who campaigned into Asia in the Third Syrian War with elephants.[72] The purpose of the expedition will therefore have called for a manager of military, political, and potential economic interests loyal to Euergetes, so it is probable that Simmias also possessed some military rank. Two elephant-hunters of a generation later under Ptolemy IV Philopator are known from inscriptions; they were both *strategoi* or generals turned political governors.[73] Since the Third Syrian War broke out soon after Euergetes' accession to the throne, Simmias' expedition could not have been responsible for the acquisition of the elephants used in that war. It is probably best to see Simias' expedition as a consequence of Euergetes' success in the Third Syrian War, intended to scout for new sources of elephants.

12 The date of Athenion (PP 14578; TM 4506), an ambassador to Jerusalem to demand tax payment, depends on the status of Coele Syria: dates under Ptolemy III Euergetes, Ptolemy IV Philopator, or Ptolemy V Epiphanes are possible.[74] There is a *kanephoros* or eponymous priestess of Arsinoe Philadelphus in 237/6 named Aspasia, daughter of an Athenion.[75] Since the *kanephoros* priestesshood was conferred on elite families close to the king, I suggest that Athenion of the ambassador-

---

meas is known from the late antique geographer Marcian of Heraclea Pontica 1.565.30–31 "Simmeas who wrote a *periplus* of the *oikoumene*".
70 See Weber's 1993: 147.n6 prosopography of Philadelphus' elephant-hunting expeditions. Sidebotham 2011: 44 argues that there had been a systematic policy of elephant acquisition since at least Philadelphus.
71 Sidebotham 2011: 39–53.
72 *OGIS* 54.9–13 "He campaigned into Asia with his forces – foot, horse, navy, and Troglydite and Aethiopian elephants, which both his father and himself were the first to hunt from those regions and, upon bringing them back to Egypt, made them ready for the purposes of war."
73 *OGIS* 82, 86.
74 Josephus *Antiquitates Iudaicae* 12.157–171; Mooren 1975: no. 17.
75 P.Petrie 3.58c; Clarysse and Van Derveken 1983: no. 54.

ship is to be identified with Athenion, father of Aspasia. Such a date would mark Athenion as a *philos* under Euergetes.

The story of Athenion's embassy to Jerusalem is worth retelling, since it illustrates the diplomatic task of the king's *philoi* and the potential for court friendship.[76] Since Euergetes controlled Coele Syria he also controlled ancient Judea and hence demanded tribute from the high priest of Jerusalem. But the high priest failed to meet the usual tax of twenty talents. Euergetes, angered, sent his *philos* Athenion as an ambassador to demand the required tribute. Josephus, nephew of the high priest, declared he would go as ambassador to Egypt to seek remission of the tribute after welcoming Athenion with gifts and banquets. Athenion independently returned to Egypt and abundantly praised Josephus to Euergetes so that the king became friendly toward him. Josephus meanwhile prepared for his journey and undertook it, meeting tax-collectors on the way who mocked his poverty and simplicity.

> When [Josephus] came to Alexandria and he heard that Ptolemy was in Memphis, he went to meet him and conversed with him. The king was seated in a carriage with his wife and his *philos* Athenion. Athenion was the one who had been ambassador to Jersualem and been hosted by Josephus. Athenion, upon seeing him, at once made him acquainted with the king, saying that this was the man about whom he had spoken when he returned from Jerusalem, and that he was a good young man with friendship-of-honor.[77]

Josephus asked the king to forgive his uncle the high priest. Ptolemy was pleased by Josephus' "grace and wit" and hosted him for several days, seating him in banquets at the palace among the leaders of Syria until the matter of the taxes was resolved: Josephus bought the tax-farming contract and extorted huge sums from Coele Syria.[78]

While we do not need to believe the historicity of this story as Flavius Josephus tells it, the story nonetheless provides good representational evidence of ritualized friendship in diplomacy. Friendship is the discourse for social courtesy and interchange among elites. The king sent his *philos* to conduct foreign diplomacy on his behalf. The king was urbane and approved elite graces. Access to the king, a model monarch of sophistication and able to recognize social graces in foreigners, is con-

---

[76] In what follows I retell Josephus *Antiquitates Iudaicae* 12.157–171; cf. Strootman's 2014: 151–152 interpretation of 'brokerage'.

[77] Josephus *Antiquitates Iudaicae* 12.170–171: ὡς δ' εἰς τὴν Ἀλεξάνδρειαν ἀφικόμενος ἐν Μέμφει τὸν Πτολεμαῖον ἤκουσεν ὄντα, ὑπαντησάμενος συνέβαλεν αὐτῷ. καθεζομένου δὲ τοῦ βασιλέως ἐπ' ὀχήματος μετὰ τῆς γυναικὸς καὶ μετὰ Ἀθηνίωνος φίλου, οὗτος δ' ἦν ὁ πρεσβεύσας εἰς Ἱεροσόλυμα καὶ παρὰ Ἰωσήπῳ ξενισθείς, θεασάμενος αὐτὸν ὁ Ἀθηνίων εὐθὺς ἐποίει τῷ βασιλεῖ γνώριμον, τοῦτον εἶναι λέγων, περὶ οὗ παραγενόμενος ἐξ Ἱεροσολύμων ἀπήγγειλεν, ὡς ἀγαθός τε εἴη καὶ φιλότιμος νεανίσκος.

[78] The matter of Josephus and the taxes is not finally resolved until Josephus *Antiquitates Iudaicae* 12.185, when Josephus has taught the Syrians a lesson.

trolled by the king's *philoi*. Friendship is established by gifts repaid at a later time, whether physical objects or social recognition. The king entertains his audiences at sympotic banquets, a large-scale imitation of the intimate symposia.

13 Castor (PP 14608; TM 9652) is known from a poorly preserved inscription in Athens, praising him for his goodwill toward the people and stating that he was chosen by Euergetes for his task.[79] The inscription can be dedicated precisely to 226/5 because of its preserved dating formula but Castor's relation to the king is damaged. The restored text gives "Castor, friend and intimate of King Ptolemy."[80] There are comparable "intimates" of monarchs so it is possible that Castor was of some relation to Euergetes. Nonetheless we should remember that court rankings often traded on the language of friendship, suggesting that "intimate" here indicates a friend rather than a blood relation.[81] Castor seems to have been engaged in the same sort of diplomacy with Athens as other *philoi*.

14 Antiochos (PP 14584; TM 5054) was a *philos* and *strategos* of Euergetes.[82] In the aftermath of Euergetes' victory in the Third Syrian War, Antiochos was given Cilicia to govern; he was probably also one of generals successful in the war. It is possible that this Antiochus is to be identified with Antiochus the eponymous priest under Ptolemy II Philadelphus in 248/7, or that he is to be identified as the father Antiochus of Seleucus the eponymous priest under Ptolemy III Euergetes in 237/6.[83]

15 Sosibius of Alexandria (PP 48, 2179, 5272, 10100, 14631, 17239; TM 3630) is of the two major figures in the court politics of Euergetes and Philopator. As a young man he probably competed and won a victory contest celebrated in elegiacs by **no. 110** Callimachus.[84] Under Euergetes he was *dioiketes* or chief minister in at least 243/2 and eponymous priest of Alexander in 235/4.[85] His daughter Arsinoe was the eponymous priestess *kanephoros* of Arsinoe Philadelphus in 215/4 under

---

79 See Mooren 1975: no. 15; *IG* 2².838.
80 *IG* 2².838: Κάστωρ φίλο|ς| ὢν καὶ οἰκεῖος τοῦ βασιλέως] Πτολεμαίου.
81 See *OGIS* 9 for a comparable *oikeios*. While Herman 1987: 18.n28 cites the blood relations of father and brother for terms of ritualized friendship, see Introduction footnote 55 for the documentation that kings adopted the language of blood relationship only for political equals.
82 He is called *amicus* and *dux* in Jerome *In Danielem* 11.7 = 905.982–984 Glorie.
83 Clarysse and Van Derveken 1983: nos. 43, 54. The first identification is Mooren's 1975: no. 14 suggestion.
84 I accept with Acosta-Hughes and Stephens 2011: 2–3 that Callimachus frr. 384, 384a Pfeiffer refers to the Sosibius under discussion here, the chief minister of Philopator. The ascription "Sosibius, Alexandrian, son of Dioscorides" seems assured from Callimachus fr.384a Pfeiffer with Pfeiffer's 1949 commentary *ad loc.* with reference to *OGIS* 79, 80.
85 Weber 1993: 149.n6; Clarysse and Van Derveken 1983: no. 56.

Philopator.[86] Later his son also named Sosibius was a *somatophylax* or bodyguard to Ptolemy V Epiphanes.[87] That he and his family stood in prominent official positions for over thirty years under three monarchs is evidence of his immense power as well as political skill.[88] In recognition of his service to the Ptolemaic royal house later antiquity even credited him with having forged a treatise on kingship under the earlier philosopher Theophrastus' name.[89]

But it was under Philopator when the vast extent of Sosibius' power became evident. He was the most powerful of Philopator's *philoi*; his initials were marked on coins minted with the image of Euergetes; he was called *proxenos* or foreign representative of two Boeotian towns; he was awarded a *dorea* or landed estate by the king.[90] At the accession of Philopator Sosibius took charge of murdering certain members of the royal family: **no. 3** Lysimachus, Euergetes' brother and Philopator's uncle; **no. 4** Magas, son of Euergetes and brother of Philopator; and even **no. 2** Berenice II, wife of Euergetes and mother of Philopator.[91]

For all his goodwill and preservation of Philopator's interests, Sosibius turned against the royal house after the king's death. Sosibius took charge of murdering the Spartan king exiled to the Ptolemaic court, Cleomenes III of Sparta, and **no. 9** Arsinoe III, daughter of Euergetes and wife of Philopator, upon Philopator's death.[92] At the death of Philopator Sosibius and **no. 16** Agathocles plotted to establish themselves in charge of the kingdom. The historian Polybius states that they forged a will and established themselves *epitropoi* or guardians of the ruling child Ptolemy V Epiphanes. While Sosibius' end is not narrated, perhaps he was murdered by his co-conspirator Agathocles.

16 With Sosibius, his co-conspirator, Agathocles of Samos (PP 4986, 10061, 10078, 14047, 14576, 16813; TM 4386) was one of the two major political figures at Philopator's court. He came from a family long associated with the Ptolemies.[93]

---

86 Clarysse and Van Derveken 1983: no. 76.
87 Polybius 15.32.6–8; Mooren 1975: no. 35.
88 Even Polybius 15.34.3–4 grudgingly expresses admiration for Sosibius' political skills in his evaluation of Agathocles: "Nor did I intend to use this treatment [sc. that of assigning causes or fortune] regarding what has been said, on account of the fact that there was neither martial bravery nor notable ability in Agathocles, nor a lucky and enviable handling of affairs, nor finally the political shrewdness and special immorality in which Sosibius and many others spent their lives when managing king after king, but that opposite qualities attached to the aforesaid man."
89 Athenaeus 4.144e refers to our Sosibius (with Murray 2007: 18 against Haake 2013: 169.n23), since Callimachus frr. 384, 384a Pfeiffer refers to the same individual.
90 Plutarch *Agis et Cleomenes* 54.5: Σωσιβίου δὲ τοῦ πλεῖστον ἐν τοῖς φίλοις δυναμένου; coins Brett 1952: 8 no. 29, although this reading is hard to credit; IG 7.3166 (*proxenos* and *euergetes* of Orchomenos) and OGIS 80 (*proxenos* and *euergetes* of Tanagra); *dorea* in his name P.Teub. 3.860.
91 Polybius 15.25–34.
92 Polybius 15.25–34. More precisely Polybius 15.33.11 blames Philammon for Arsinoe's murder.
93 Clayman 2014: 197.n4.

Agathocles held official titles of *philos* and eponymous priest of Alexander in 216/215;⁹⁴ he was possibly *dioiketes* at some point. But his influence extended beyond those roles. When younger Agathocles had been a cup-bearer for Philopator's symposia; when older Agathocles gathered among his unofficial titles flatterer and beloved of Philopator.⁹⁵ Philopator's affections for Agathocles extended to the latter's family, with the king sleeping with Agathocles' sister. After Philopator's death, Agathocles' conspiracy for control of the Ptolemaic kingdom saw him eliminate rivals and rule alone as regent. His brief time in power is a collage of hedonism, violence, depravity.⁹⁶ He has traditionally (and in my view, deservedly) been held responsible for the worst excesses of Philopator's reign.

Agathocles enjoyed the vaunted luxury of Philopator's court, including his pursuits of *paideia*. Agathocles wrote a commentary on Philopator's tragedy *Adonis*.⁹⁷ If Agathocles was probably imitating the behavior of the learned poet-scholars of Alexandria in writing commentaries on literary works, it may be no less a stretch of the imagination to conjecture that the commentary was considered an example of scholarship by a leading courtier. Since Philopator's reign was marked by excesses worse than bad poetry, Philopator's *Adonis* and Agathocles' commentary on it escaped the flaying criticism of later grammarians to disappear from the historical record.

After Philopator's death Agathocles and Sosibius also murdered Arsinoe III and had themselves made guardians of the child-king Ptolemy V. But the conspiracy caught up with Agathocles. A military commander in Pelusium, Egypt's eastern border, gathered forces and became the leader of opposition. The treason ultimately ended badly for the conspirators: Polybius recounts the infamous end of Agathocles, and **nos. 104–105** his sister and mother Agathoclea and Oenathne at the hands of the Alexandrian mob, a tale so embellished by previous historians that Polybius complains of their sensationalism.⁹⁸

17 Ptolemaeus (PP 5238, 14624; TM 12936) was a *philos* of Philopator known from the later historian Plutarch's *Agis and Cleomenes*.⁹⁹ Plutarch's text heroizes Cleomenes III, a king of Sparta who briefly restored Spartan control over the Peloponnesus but was defeated by Antigonus III and consequently lived in exile at the

---

94 Friend Athenaeus 6.251d; eponymous priest IJSweijn 1961 no. 71.
95 Cup-bearer Polybius 15.25.32; flatterer Athenaeus 6.251d; beloved Σ.Aristophanem *Thesmorphoriazusae* 1059.
96 Polybius 15.25.21–22 remarks that Agathocles filled up the vacant places of the *philoi* but gives no names.
97 Σ.Aristophanem *Thesmorphoriazusae* 1059: ἐζήλωσε δὲ αὐτὸν [sc. Euripides] Πτολεμαῖος ὁ Φιλοπάτωρ ἐν ᾗ πεποίηκε τραγῳδίᾳ Ἀδώνιδι, περὶ ἧς ὁ ἐρώμενος αὐτῷ Ἀγαθοκλῆς γέγραφεν.
98 Polybius 15.31–34. The sensationalism is discussed below under **no. 121** Ptolemy of Megalopolis, a historian and military official at Philopator's court.
99 Mooren 1975: no. 19 contains the relevant passages.

court of Euergetes and then Philopator. Plutarch picturesquely juxtaposes Cleomenes' royal life – martial, courageous, without luxury, loyal to friends and generous to defeated enemies – with that of the slothful, luxurious, and disaffected Philopator. Within this moral dichotomy of royal lives Ptolemaeus appears as a supporting actor of Philopator: he acted politely and with free speech toward Cleomenes but, thinking he was out of earshot, reproached the guards for not keeping a tighter leash on him.[100] Ptolemaeus was the first to die in Cleomenes' vain bid for freedom.[101]

Ptolemaeus belonged to a distinguished Ptolemaic family. His father Chrysermos (PP 2022, 10107; TM 3972) won some athletic contests as a youth, held an unspecified high office under Euergetes, and possessed a landed-estate.[102] His son Ptolemaeus (PP 5241, 14946; TM 12881) and grandson Galestes (PP 1870, 14904; TM 6723) were listed as *proxenoi* of Delphi in 188/7 and his other son Glaucon (PP 14905; TM 6749) was *proxenos* of Delphi in 185/4.[103] The family took Alexandrian citizenship sometime over these three generations: Chrysermos, the father, has a Thracian ethnic but the sons Ptolemaeus and Glaucon are Alexandrians; Ptolemaeus' own ethnic is unknown. Ptolemaeus himself was eponymous priest under Euergetes in 224/3.[104]

18 Phormion of Byzantium (PP 14635; TM 14784), *philos* of Philopator, is known from an honorific inscription in Oropus in Attica.[105] The city names him *philos* of Philopator and says that he has acted on behalf of the city both privately and with the king. They grant Phormion the right of being the representative of the Oropes in Alexandria, with all its attendant rights in Attica, and copy their decree upon the base of a statue of King Philopator and Queen Arsinoe.

Bureaucrats, Military Officials, Judiciary, Minor Functionaries

There were several bureaucratic functionaries at court. The two most powerful offices were the *dioiketes*, the interior minister responsible for immediate governance of Egypt,[106] and the *hypomnematographos*, chief secretary of the king, who drafted

---

100 Plutarch *Agis et Cleomenes* 57.2–3.
101 Plutarch *Agis et Cleomenes* 58.9.
102 The texts are SEG 27.1114, *P.Cairo.Zen.* 3.39355, and *P.Enteux* 60. In addition to PP, see Fraser 1972: 2.191.n87 (although I do not agree with Fraser's identification of Chrysermos of *OGIS* 104 as a member of this family).
103 *Syll.*³ 2.585.136–137; *Syll.*³ 2.585.190–191.
104 Clarysse and Van Derveken 1983: no. 66.
105 *OGIS* 81.
106 Bagnall 1976: 3–10 provides a general overview to the administrative system governing land, taxes, soldiers, and the law below the immediate level of the court. Since my work is only concerned with court actors, none of these connections will be pursued.

his orders and corresponded with all important officers of the kingdom. We know the names of three *dioiketeis* at the court of Euergetes and one *hypomnematographos*; we know three *dioiketeis* and no *hypomnematographos* at the court of Philopator.

19 Apollonius (PP 16; TM 937) is well-known to modern scholarship thanks to an enormous cache of about two thousand papyri written to and from Zenon, his secretary, called the Zenon archive. He was *dioiketes* under Ptolemy II Philadelphus for over a decade and conducted Philadelphus' daughter in marriage to Antiochus II as a conclusion to the hostilities between the Ptolemaic and Seleucid kingdoms. He had a number of estates in the Fayum, financial gifts for his service to Philadelphus. Apollonius ceased to be *dioiketes* in 245 under Euergetes; surviving documents do not explain why.

20 Chrysippus (PP 52; TM 14952) is attested as *dioiketes* in at least the years 229–225 under Euergetes; he was also *archisomatophylax*, chief of the body-guard of the king.[107] A letter from 225 BCE shows preparations for hosting Chrysippus and his retinue for a trip to the Fayum to inspect road-work being done in the region.[108] This letter and other surviving correspondence shows that the *dioiketes* traveled fairly widely across the country, was concerned with building projects and above all with agriculture, the central industry in Egypt's economy.[109] Although the titles *somatophylax* "guard" and *archisomatophylax* "chief-guard" are equally attested in the third century, the *archisomatophylax* was the more powerful position.[110] Given the influence that the king's bodyguards had in later Ptolemaic history, presumably the *archisomatophylax* was political position. Chrysippus' double titles of *dioiketes* and *archisomatophylax* formally consolidate power and established him head of Egypt's civilian apparatus and among the friends as the head of the lower ranks.[111]

21–22 Four other *dioiketes* are known. Eutychus (PP 30; TM 8205) was *dioiketes* in 238/7 under Euergetes. Theogenes (PP 32; TM 8836) is attested as *dioiketes* in 217 and after 216/5 under Philopator. Sosibius of Alexandria (PP 48, 2179, 5272, 10100, 14631, 17239; TM 3630) and Agathocles (PP 4986, 10061, 10078, 14047,

---

107 *P.Petrie* 3.53l–m; Mooren 1975: no. 40.
108 *P.Grenf.* 2.14b = *Chest.Wilck.* no.411. "Amenneus to Asclepiades, greetings. Just as you wrote, we have made ready for the presence of Chrysippus, *archisomatophylax* and *dioiketes*, ten *leukometopoi* birds, five tame geese, fifty birds, fifty [wild?] geese, two hundred birds, and one hundred pigeons. We have borrowed five riding donkeys and kits for these, and we have made ready forty pack donkeys. We are at the road-work. Farewell. Year 22, Choiak 4." I have translated the *lacunae* based on Wilcken's suggestions.
109 Fraser 1972: 1.147.
110 Compare Mooren 1975's lists: 3 *archisomatophylakes* in the third century nos. 37–41 (two of these are fictional persons) vs. 4 *somatophylakes* in the third century nos. 33–36.
111 Mooren 1975: 1–5, 14–15.

14576, 16813; TM 4386) were probably the *dioiketes* under Philopator at some point; I discussed them both above as **nos. 15–16** among the *philoi* of Philopator.

23 Dositheus (PP 8, 5100; TM 1583) was *hypomnematographos* in 240 and eponymous priest in 223/2 under Euergetes.[112] As *hypomnematographos* he was responsible for drafting Euergetes' correspondence, orders, laws, and so on. Later Ptolemaic bureaucracy had an *epistolographos* who was responsible for correspondence with foreign dignitaries; but, since we have no evidence for such a position under Euergetes or Philopator, we do not know whether Dositheus was also responsible for Euergetes' foreign correspodence. Dositheus at least must have had extensive contacts with all the leading individuals in the kingdom, both officials and leading civilians. We know disappointingly little about such a powerful bureaucrat at the central node of official communications.

24–29 There were military officials who governed the provinces also at court. During their governorships they were not in Alexandria but of course returned to the palace from time to time. Since some of their family members became eponymous priests, it is a solid inference that the military officials formed an important base of courtiers for the Ptolemies. We know the names of three provincial governors under Euergetes[113] and we know the names of five military officials under Philopator, including **no. 31** Pelops described below.[114]

30 A military official coming from outside the Ptolemaic kingdom was Hippomedon of Sparta (PP 14605, 15048, 16115; TM 9405), a native Spartan who took voluntary exile. Hippomedon was appointed *strategos* by Euergetes to "The Hellenspont and Regions in Thrace", celebrated the mysteries on Samothrace, and was there the subject of an honorary inscription praising him for his care toward the public and for having acted in conformity with the wishes of the king.[115] A contemporary Cynic philosopher uses Hippomedon's success as a notable example that exile does not deprive one of worldly goods, nor boldness of speech: in addition to Hippome-

---

112 *Hypomnematographos* from *P.Mich.Zen.* 55; eponymous priest in 223/222 in Clarysse and Van Derveken 1983: no. 68.
113 Philinus of Athens (PP 15083; TM 14589) was *strategos* of Salamis; Thraseas (lacks PP no.) was *strategos* of Cilicia (see Weber 1993: 150.n5); Xanthippus (PP 15060; TM 10922) governed the provinces across the Euphrates.
114 This second group comes primarily from Polybius' account (15.25.12–27) of the machinations of Sosibius and Agathocles after the deaths of Philopator and Arsinoe with the overt purpose of preserving alliances, while in fact removing possible notable opponents from Alexandria. Philammon (PP 15082; TM 14572) who was *Libyarch*; Pelops (PP 15064; TM 11854) discussed as **no. 31**; Ptolemy of Megalopolis (PP 15068, 16944; TM 12779) discussed as **no. 121**; Theodotus (PP 15045; TM 8857), the former governor of Coele Syria who betrayed his province to Antiochus III the Great; Tleopolemos (PP 50, 2180, 14634; TM 14273), *strategos* of Pelusium.
115 Hippomedon was apparently stationed in Thrace c.240/39 according to *IG* 12.8.156. Bagnall 1976: 160–161 argues that he continued in this role until at least 219.

don's governing of Thrace as *strategos*, he was a councilor and advisor of the king.[116] For Hippomedon to receive landed estates and participate in the planning councils of state implies that he was among the political elite of the kingdom.

31 The best recorded career of a military official under Euergetes and Philopator is that of Pelops (PP 15064; TM 11854).[117] He was from an established and loyal family under the Ptolemies: his father, also named Pelops (PP 5227, 14618; TM 11854) had been a *philos* of Ptolemy II Philadelphus; his mother Iamneia was *kanephoros* of Arsinoe Philadelphus in 243/2.[118] A fragmentary honorific inscription attests that Pelops was stationed in Cyrene during the reign of Euergetes, possibly as *strategos* or *Libyarch*. Later, under Philopator, he was *strategos* on Cyprus. Finally, in the aftermath of Philopator's death as **nos. 15–16** Sosibius and Agathocles plotted to establish themselves as regents over the young Ptolemy V Epiphanes and consequently sent the leading men out of Egypt, Pelops was sent to the Seleucid monarch Antiochus III "to council him to observe the friendship and not trespass his agreements with the child's father."[119] Pelops had evidently been a loyal officer and military governor to Euergetes and Philopator through his service in Cyrene and Cyprus.

32–41 In addition to the grandees of the royal bureaucracy and leading military officials we know the names of several more minor figures,[120] local Alexandrian judiciary officials,[121] and several stewards running the daily business of the court. The names of the stewards come from funerary vases recovered from an archaeological context. A series of similar painted vases were recovered from the cemeteries around Alexandria; they are now known as Hadra vases from the Alexandrian cemetery of that name.[122] Many entered antiquities collections in the late nineteenth century and consequently lack a well-defined archaeological context. None-

---

116 Teles *De Fuga* 23.4–12 "'They do not command,' say [the objectors], 'they are not trusted, they do not have boldness of speech (παρρησίαν).' Some [sc. exiles] at least command the garisons of cities for the kings, command the trust of peoples, and receive great estates (δωρεὰς) and commissions ... Are not Hippomedon the Spartan, now established by Ptolemy in charge of Thrace, and are not the Athenians Chremonides and Glaucon councilors and advisors (πάρεδροι καὶ σύμβουλοι)? To speak not of ancient matters, but things in our own time."
117 Bagnall 1976: 252–253 discusses his tenure on Cyprus.
118 On the family of Pelops in general see Fraser 1972: 1.104. On Iamneia see Clarysse and Van Derveken 1983: no. 48, *ABSA* 56: 15 no. 39.
119 Polybius 15.25.13: Πέλοπα μὲν ἐξέπεμψε τὸν Πέλοπος εἰς τὴν Ἀσίαν πρὸς Ἀντίοχον τὸν βασιλέα, παρακαλέσοντα συντηρεῖν τὴν φιλίαν καὶ μὴ παραβαίνειν τὰς πρὸς τὸν τοῦ παιδὸς πατέρα συνθήκας.
120 Nicon in Polybius 15.27.37; Nicostratus in Polybius 15.27.7; Scopas in Polybius 15.25.16.
121 Local judiciary officers included Chrysermos (PP 8040a; TM 17591), Diodoros (PP 8033c; TM 1443), and Zenis (PP 8033d; TM 1748).
122 Cook 1966 is fundamental.

theless, they seem to have been used as funerary urns to contain cremated remains. About thirty are inscribed with names of the deceased and dates of death or burial. The vases give the regnal year but never the name of the reigning king. In spite of this difficulty, chronological studies of the vases have yielded a dating sequence from 271 BCE to 210/209 BCE. The deceased appear largely to have been ambassadors (πρεσβεύτης, θεωρός) to the Ptolemaic court on political or religious missions. Often the inscription is accompanied by the Ptolemaic official who buried the ambassador, here always called ἀγοραστής: the term means "purveyor, buyer" but is probably used elastically for a steward or concierge in charge of provisions and general problem-solving in the daily running of the court. I give one example.

> Year 9. Sotion son of Cleon, of Delphi, *theoros* announcing the festival of the *Soteres*. [Buried] by Theodotus *agorastes*.[123]

This particular vessel dates from 212 BCE in the reign of Philopator and contains the name of the ambassador, his city-ethnic, and (uncharacteristically for the inscribed Hadra vases) the purposes of his mission, and finally the name of the Ptolemaic steward who buried him, Theodotus (PP 14672; TM 8876). Perhaps the fact that Sotion announced a festival in honor of Philopator's great-grandparents, Ptolemy I Soter and his wife, explains why the inscription lists the purposes of Sotion's mission. From these Hadra vases we gain the names of four stewards at the courts of Euergetes and Philopator and the names of numerous foreign ambassadors to the court, all from Greek cities.[124]

Priests

42–102 Classical Athens and other Greek city-states dated their years by the one-year term of important local officials; hence the term "eponymous" official.[125] In Ptolemaic Egypt a double dating system arose: years were numbered both by the regnal year of the monarch as well as by certain priesthoods.[126] The eponymous priesthood of the deified Alexander and the deified Ptolemies was introduced sometime during the reign of Ptolemy I Soter; the priesthood is earliest attested in 284 BCE. The eponymous male priest was joined eventually by three eponymous female priestesses: the *kanephoros* or basket-carrier of Arsinoe Philadelphus (intro-

---

123 Cook 1966 no. 10 = Hadra vase Metropolitan Museum accession number 90.9.37: θ' Σωτίων | Κλέωνος | Δελφὸς | θεωρὸς τὰ | Σωτήρια | ἐπαγγέλλων | διὰ Θεοδότου | ἀγοραστοῦ.
124 Stewards (*agorastes*) included Kratides (PP 14678; TM 9896), Sarapion (PP 14690; TM 13297), Philon (PP 14694; TM 14727), and of course Theodotus (PP 14672; TM 8876). I have not included the visiting ambassadors in my count of court actors.
125 The standard reference for Ptolemaic priests is Clarysse and Van Derveken 1983.
126 The eponymous priest was based at Alexandria although later an eponymous priesthood was instituted at the important Greek city of Ptolemais in Egypt. I will not be concerned with the priesthood at Ptolemais, since it is unimportant for the period of this book.

duced 269 BCE), the *athlophoros* or prize-bearer of **no. 2** Berenice Euergetis (introduced 238 BCE), and the priestess of **no. 9** Arsinoe Philopator (introduced 199 BCE). The full dating system appears on the Rosetta Stone (an inscribed stone decree of Ptolemy V Epiphanes from 196 BCE) after the king's name and his titles:

> ... In the ninth year [of Ptolemy V Epiphanes], when Aetus son of Aetus was priest of Alexander, *theoi Soteres, theoi Adelphoi, theoi Euergetai, theoi Philopatores*, and the god *Epiphanes Eucharistos*; when Pyrra daughter of Philinus was *athlophoros* of Berenice Euergetis, when Areia daughter of Diogenes was *kanephoros* of Arsinoe Philadelphus, when Eirene daughter of Ptolemy was priestess of Arsinoe Philopator ...[127]

As here, the priest and priestesses are named and their fathers are also standardly named. The dated decrees, laws, and legal contracts preserved as primary documents have yielded a wealth of prosopographical information about important families under the Ptolemies. Historians have been able to reconstruct a nearly complete list of eponymous priests and priestesses for the reigns of Euergetes and Philopator from 246–212 BCE, although we lack names for all the *althophoroi* of Berenice Euergetis during this time. A new priest of Alexander and the Ptolemies and a new *kanephoros* were instituted each year; only rarely does the selected individual repeat the office for a second time. During the third century BCE the eponymous priesthoods were conferred on elite families close to the king; these individuals were probably all members of families associated with the court in Alexandria as *philoi*, bureaucrats, or military offices. I have counted 31 uniquely named eponymous priests and 29 eponymous *kanephoroi* for the years 246–204 BCE.

There were also local priests in Alexandria.[128]

Flatterers, Entertainers, Lovers

We might expect that life at the court of two monarchs surnamed "Luxuriant" was lived to excess. Ancient sources do not disappoint. "I wish you had rather brought the harp-playing girls and cinaedi; for these at the moment most rouse the king."[129] "Symposiasts were brought together for the king from all the city and were called merry-makers."[130] "When the king arrived in Egypt he enlarged those habits of

---

127 OGIS 90.4-6: ἔτους ἐνάτου ἐφ' ἱερέως Ἀέτου τοῦ Ἀέτου Ἀλεξάνδρου καὶ θεῶν Σωτήρων καὶ θεῶν Ἀδελφῶν καὶ θεῶν Εὐεργετῶν καὶ θεῶν Φιλοπατόρων | καὶ θεοῦ Ἐπιφανοῦς Εὐχαρίστου, ἀθλοφόρου Βερενίκης Εὐεργέτιδος Πύρρας τῆς Φιλίνου, κανηφόρου Ἀρσινόης Φιλαδέλφου Ἀρείας τῆς Διογένους, ἱερείας Ἀρσινόης Φιλοπάτορος Εἰρήνης | τῆς Πτολεμαίου.
128 OGIS 65.
129 Plutarch *Agis et Cleomenes* 56.3: ἐβουλόμην ἄν, ἔφη, σε μᾶλλον ἥκειν ἄγοντα σαμβυκιστρίας καὶ κιναίδους· ταῦτα γὰρ νῦν μάλιστα κατεπείγει τὸν βασιλέα.
130 BNJ 161 (Ptolemy of Megalopolis) F2 = Athenaeus 6.246c: συμπότας φησὶ τῷ βασιλεῖ συνάγεσθαι ἐξ ἁπάσης τῆς πόλεως, οὓς προαγορεύεσθαι γελοιαστάς.

wickedness through the aforementioned symposiasts and companions bereft of justice; he not only satisfied himself in innumerable acts of violence but gained so much in confidence that ill-repute arose in the districts and many of the *philoi* who kept their eyes on the king's example also followed his wishes."[131] As tempting as it is to embrace the ancient gossip and innuendo, accepting this evidence fails to separate fact from fiction – since only the second quotation derives from a contemporary historian of Philopator.

If a prosopography is the most empirically accurate means of ascertaining the truth of court luxury, it is also limiting. We know very few names of flatterers, entertainers, and lovers at the courts of Euergetes and Philopator. In fact, we know the names of only one flatterer at the courts of Euergetes and Philopator. Flatterer (κόλαξ) was not a social rank at the Ptolemaic courts. Rather, as we will see in chapter 2, the flatterer was a debased mode of court friendship. The best example of a courtier labeled both friend and flatterer is **no. 16** Agathocles son of Oenanthe, recounted above among the *philoi* of Philopator.

103 Tryphon (PP 14710; TM 14369) was a jester at the court of Philopator. He appears in a story told about Hyrcanus, son of Josephus, whose ritualized friendship with **no. 12** Athenion (PP 14578; TM 4506) and mission to Euergetes were recounted above. As Josephus tells the story, the courtiers and guests were hierarchically arranged at the banquet.[132] Tryphon as the jester had the comic license to move among the established social order. The pile of bones in front of the young Hyrcanus suggested a joke at his expense: Tryphon alluded to the tax extortion of Hyrcanus' father Josephus in Syria a generation earlier. Euergetes laughed at the joke. As in the earlier tale of Josephus, we may reasonably doubt the historicity of the story of Tryphon and Hyrcanus.[133] Nevertheless, this tale provides good representational evidence of jesters at the court of Euergetes and Philopator. Comedic entertainment happened at banquets and symposia; jesters used their comedic license to punch above their social station; wit relied on knowledge about social norms as well as knowledge about the history of the courtiers. The contemporary historian **no. 121** Ptolemy of Megalopolis remarked that Philopator's merry-makers became his symposiasts;[134] yet we know no names.

---

131 3 Maccabees 2.25–26: διακομισθεὶς δὲ εἰς Αἴγυπτον καὶ τὰ τῆς κακίας ἐπαύξων διά τε τῶν προαποδεδειγμένων συμποτῶν καὶ ἑταίρων τοῦ παντὸς δικαίου κεχωρισμένων οὐ μόνον ταῖς ἀναριθμήτοις ἀσελγείαις διηρκέσθη, ἀλλὰ καὶ ἐπὶ τοσοῦτον θράσους προσῆλθεν ὥστε δυσφημίας ἐν τοῖς τόποις συνίστασθαι καὶ πολλοὺς τῶν φίλων ἀτενίζοντας εἰς τὴν τοῦ βασιλέως πρόθεσιν καὶ αὐτοὺς ἕπεσθαι τῇ ἐκείνου θελήσει.
132 Josephus *Antiquitates Iudaicae* 12.211–213.
133 Tryphon is also attested as a jester at the court of Ptolemy in *Suda* π3040 but the story appears to have been copied from Josephus. The *Suda*'s attestation provides no independent evidence that Tryphon was a historical person.
134 *BNJ* 161 (Ptolemy of Megalopolis) F2 = Athenaeus 6.246c.

104–106 The list of named lovers at the courts of Euergetes and Philopator is thin. We know of no mistresses of Euergetes and only two lovers of Philopator. One was the flatterer and *philos* Agathocles, recounted above as **no. 16**. The other was his sister Agathoclea (PP 14714; TM 4382). Plutarch says that Agathoclea, her mother Oenanthe (PP 14731; TM 10957), and a third woman named Aristonica (PP 14715; TM 5841) were all Samians, flute-players, and dancers "who walked over the diadems of kings", a reference to their control over Philopator.[135] Agathoclea and her mother Oenanthe merit no descriptions in ancient sources for their wit, beauty, or intelligence. Plutarch and Atheneaus blame the evils of Philopator's rule on their manipulations.[136] Polybius' account of Agathoclea's mutilation at the hands of the Alexandrian mob emphasizes her breasts and nakedness, as if it were poetic justice for a body that had seduced the king to have no effect on the common people.[137]

107–108 We hear of two actors who recited the works of Homer and Hesiod in the great theater of Dionysus near the royal quarter in Alexandria, probably in the second half of the third century BCE;[138] we hear of a tragic actress singing tales of the destruction of Troy.[139] Performances in theaters and symposia, along with plays, mimes, and dances required entertainers, actors, musicians, and the like; but, although certain traveling actors' guilds such as the *technitae* of Dionysus performed around the Mediterranean world, we have no names remaining in the historical record.[140]

Poets

We know of eight writers at the courts of Euergetes and Philopator whose primary literary output was poetry. Of some, we hear nothing more than an anecdote involving the king. Others became spokesmen for royal ideologies of antiquarianism, the protean richness of the natural world, and the preeminence of Greco-Egyptian

---

135 Plutarch *Amatorius* 753D: διαδήμασι βασιλέων ἐπέβησαν. Justin 30.1.9 and Jerome *In Danielem* 11.13 = 907.1034–1035 Glorie add different instruments.
136 Plutarch *Agis et Cleomenes* 54.1-2; Athenaeus 13.577a.
137 Polybius 15.31.12; 15.33.8–9.
138 *BNJ* 632 (Jason) F1 = Athenaeus 14.620d: Hegesias (PP 17001; TM 8424) and Hermophantus (PP 16997; TM 7950). See Fraser 1972: 1.23 on the site of the theater.
139 Dioscorides of Nicopolis 2 *HE*. The singer is called Athenion (PP 16977; TM 4497) "an actress in tragedy", although I have little confidence she is an historical individual and so have not included her in the bolded enumeration.
140 Named individuals known from contemporary poetry in Alexandria are Aristagoras (PP 16983; TM 5742), a dancer, and Damomenes (PP 16989; TM 6793), a dance leader. I see no reason to believe that these are historical individuals. There was a genre of burlesque tragedies and comedies, spread out in the Mediterranean and the Egyptian *chora*, but Alexandrian evidence seems to be limited to the first half of the third century: Fraser 1972: 1.620–621; Esposito 2010: 279–281.

culture. Ptolemaic aesthetics presented the Mediterranean civilizations as a province of Egyptian developments, embracing a sort of cosmopolitanism. The geography, demography, and mythology of Egypt appear in nearly all poets. Euergetes and Philopator emerge as a champions of Greco-Egyptian culture: Egyptian gods are assimilated to their Greek counterparts; Egyptian cities and landscape are praised; Egypt, always recognized as an older civilization than Greece, is celebrated as the mother land that gave civilization and colonies to other Mediterranean places.[141] Antiquarianism, itself a reflex of literary belatedness, emerges as a prominent theme. Greek literature became canonized, with court writers referencing the earlier writers Homer, Hesiod, Sappho, Sophocles, Euripides, as well as their own contemporaries.[142] Writers celebrated ancient Athens and picked out for learned elaboration recherché facts involving forgotten lore and obscure social customs.[143] Book learning was prominent; the performance in traditional Greek poetry made for an occasion was made to inhabit textual features, so that writers emphasized movement, dance, and song in their writing. The scientific was not absent in these themes. Poets celebrated the wonders of nature and its strangeness. They invested their readers in the power of the natural world to change its appearances, as well as instructing readers on the advancement of human knowledge in understanding that natural world. Human civilization too was celebrated, either in the foundation of cities, the progress of technology, or the written word that made possible scholarship itself.

I have accorded the rank of place to poetry among writing at court because Ptolemaic court actors valued poetry more highly. Yet it is essential to realize that Ptolemaic court writing consisted largely of prose. We know three times as many prose authors as poets and at least seven authors (Callimachus **no. 110**, Neoptolemus **no. 112**, Archelaus **no. 113**, Istrus **no. 117**, Hermippus **no. 118**, Philostephanus **no. 119**, and Eratosthenes **no. 136**) wrote both prose and poetry. The ability of a poet to write in multiple genres had been declared an aesthetic principle by Callimachus;[144] the notion evidently embraced literary prose as well.

109 The two most important poetic figures at Philadelphus' court were Apollonius of Rhodes and Callimachus of Cyrene. Apollonius (PP 16510; TM 5407) has been partially recounted above as the tutor of Euergetes. Apollonius is known now primarily as the author of a surviving four book epic, *Argonautica*, and a possible

---

141 For literary readings in Greco-Egyptian culture see chapter 4 footnote 1; overview in Dieleman and Moyer 2010.
142 On canonizing of poetry see Fantuzzi and Hunter 2004: 444–461.
143 On historical antiquarianism see Bravo 2007, whose definition (p.517) is worth highlighting: "works that (1) refer to a period in the past their authors considered as 'ancient' ... (2) have as their object not what could be constructed and represented by a complex and dramatic narrative but rather a series of fact to be assembled and described."
144 Callimachus fr. 203 Pfeiffer; Fantuzzi and Hunter 2004: 15–26.

literary opponent of Callimachus, although some notion of a literary quarrel between them is now generally discounted.¹⁴⁵ The *Argnonautica* is a highly literary retelling of the myths of Jason, his Argonauts, and Medea, written in complex archaizing language. Since the characters move over the Mediterranean and neighboring areas to places that will become parts of the Ptolemaic Empire, it is possible to read the poem on one level as a piece of court propaganda: the epic offers mythic exempla and justification for Ptolemaic ideology.¹⁴⁶ The poem incidentally includes much scientific material writ into myth, including the optics of attraction and geographical material about the wider Mediterranean. Recent scholarship has suggested that the poem's composition dates to Alexandria under Euergetes.¹⁴⁷ Apollonius composed other poems that have not survived, including several works that dealt with Egyptian themes, such as the catasterism of the mythological pilot Canobus and the foundation of the Greek cities Alexandria and Naucritus in Egypt.¹⁴⁸

110 Callimachus of Cyrene (PP 14676, 16517, 16863; TM 9559) remained at the court as it transitioned from Philadelphus to Euergetes. He wrote praise poetry for the new ruling couple, as in his celebration of the chariot victory of **no. 2** Berenice in the Nemean games.¹⁴⁹ Callimachus' poetic output included surviving hymns and epigrams, and fragments from an epyllion about archaic Attica, a victory ode for a Ptolemaic courtier, a wedding song and later an apotheosis of Arsinoe II, iambic satire, and an influential complex episodic poem called *Aetia* or "Causes".¹⁵⁰ Here the poet posed questions to the Muses about recherché cultic practices told at least partially within the frame of an inspired dream. The *Aetia* ended with the establishment of cultic practice for Arsinoe-Aphrodite-Zephyritis, told by the casterized lock of Berenice discovered in the heavens by **no. 137** Conon of Samos, a passage already recounted in the Introduction. Callimachus was also a proflic prose author. His prose works, which have almost entirely vanished apart from titles, were apparently structured as lists (πίνακες). He wrote in prose on geographical features, city foundations, illustrious authors and their writings.¹⁵¹

Callimachus' most extensive surviving prose comes from the genre of paradoxography.¹⁵² This genre was popular among Ptolemaic prose authors with Callima-

---

145 See Harder 2012: 2.88–89 with further references.
146 Stephens 2003: 171–237; Clayman 2014: 105–120.
147 Murray 2014.
148 See Apollonius of Rhodes pp. 4–8 *CA* for the surviving fragments.
149 Callimachus *Aetia* frr. 54–60j with Harder's 2012 acccompanying commentary.
150 While Callimachus clearly had connections to the royal house, scholars have also credited him with affiliations to other members of the court: fr. 384 Pfeiffer was probably written for **no. 15** Sosibius, epigram 3 *HE* may refer to the father of **no. 132** Caphisophon, *Aetia* fr. 110 concerns **no. 137** Conon. Callimachus' relation to **no. 109** Apollonius has already been discussed.
151 Geography: Callimachus frr. 404, 412, 413, 414–428, 457–459 Pfeiffer; city foundations: *Suda* κ227; authors: Callimachus frr. 429–456 Pfeiffer.
152 Callimachus frr. 407–411 Pfeiffer. For scholarship, see *RE* 18.3: 1137–1166 *s.v.* 'paradoxographoi'; Schepens and Delcroix 1996.

chus, Archelaus **no. 113**, and Philostephanus **no. 119** each writing a paradoxography. Hellenistic paradoxography described and catalogued odd natural features, such as red rivers, whistling caves, diving birds, and so on. The name "paradoxography" is a medieval invention; ancient authors simply called it θαυμασία or παράδοξα, "Wonders" or "Paradoxes". The genre surveyed a select aspect of the known world and presented an image of natural regular complexity. The authors relied not only on autopsy but on the reports of previous authorities. A paradoxology achieved multiple aims: it offered authors an opportunity to display book learning and the fruits of wide reading; it invested readers in the curiosities of the natural world. Whether Callimachus originated this genre, the literary features of paradoxography fit well with a court culture invested in antiquarianism, cosmopolitanism, and turning the natural world into a source of spectacle and entertainment.

Callimachus' combination of poetic prowess as an articulate spokesman for Ptolemaic power and as an organizer of the book world captured in the Alexandrian library was a powerful model for the next generation of Ptolemaic court writers. Antiquity recognized at least four, primarily prose authors – **nos. 117–120** Istrus, Hermippus, Philostephanus, Apollas – as "Callimacheans", presumably his direct students.

111 The poet Dioscorides (PP 16683, 16694; TM 7435), known from forty surviving epigrams, is perhaps the most influential of the poets after Apollonius and Callimachus. He is called a citizen of Nicopolis in manuscript and, since several epigrams refer to an Egyptian or Alexandrian context, he is thought to come from the Alexandrian suburb Nicopolis.[153] His funerary epigram on Machon, an Alexandrian comic poet under Ptolemy II Philadelphus, suggests he flourished in the second half of the third century BCE, during the reign of Euergetes and Philopator.

> May you, light dust, carry the ivy, friend-of-the-contest, growing above the tomb of the comedic poet Machon. For you do not have a rewashed dress but are wrapped in a shroud worthy of ancient art. This the old man will say: "City of Cecrops, sometimes even by the Nile a pungent thyme flourished among the Muses."[154]

The epigrams fall into recognizable series: erotic epigrams on both sexes, and funerary epigrams, and several on literary figures who date from the Archaic period

---

[153] Dioscorides of Nicopolis 38 *HE* (see *apparatus criticus*) for the lemmatist's ascription of the ethnic; epigrams 14, 24, 33, 34, 37, 39 *HE* refer to some Egyptian or Alexandrian context.
[154] Dioscorides of Nicopolis 24 *HE*: τῷ κωμῳδογράφῳ, κούφη κόνι, τὸν φιλάγωνα / κισσὸν ὑπὲρ τύμβου ζῶντα Μάχωνι φέροις· / οὐ γὰρ ἔχεις κύφωνα παλίμπλυτον ἀλλά τι τέχνης / ἄξιον ἀρχαίης λείψανον ἠμφίεσας. / τοῦτο δ' ὁ πρέσβυς ἐρεῖ· 'Κέκροπος πόλι, καὶ παρὰ Νείλῳ / ἔστιν ὅτ' ἐν Μούσαις δριμὺ πέφυκε θύμον'.

down to near-contemporary Ptolemaic dramatists and comedians. I pass by the epigrams on literary history, important in their own right as Alexandrian markers melding aesthetic reception with biographical readings. Among the best of the remaining are an epigram describing how Atys, a devotee of the goddess Cybele, frightened a lion; an epigram advising how to avoid the swollen belly of a pregnant wife during intercourse; and an epigram on a slave *tropheus* buried by his master in a free man's tomb.[155] All together his epigrams show frequent reversals in narratological perspective, a fascination with physical movement in performance, interest in the experience of colonization, and a sense of literary belatedness.

Consider the funerary epigram which plays on the reversal of fortune of Philocritus, evidently a Greek colonist to Egypt, who had left a mercantile livelihood for an Egyptian farm. The epigram shows colonists' awareness of distance from their homeland, a sense of literary belatedness, and a playful comic touch of black humor.

> Memphis covers Philocritus in a foreign grave; he had stopped his merchant's life and just had a taste of the plow. There the great flood of the Nile running with furious water stripped the small clod from its master. While alive he fled the salt sea, but now, buried by waves, the wretch has a sailor's tomb.[156]

The Nile was famous for the fertility its annual flood left Egyptian soil.[157] Here the riches of the flood, normally a boon to farmers, cover Philocritus in the watery grave of a sailor he thought to have left when settling down to farm. Philocritus' career from trader to farmer echoes the literary life of the father of Hesiod, the archaic poet, who settled from a trading city in Asia Minor to a farming hamlet in Boeotia. Dioscorides' final comment on Philocritus, "wretch", evokes the note of misery in Hesiod's description of his father's farming village: "in a miserable hamlet, Ascra, bad in winter, terrible in summer, not ever good".[158] Dioscorides' literary respect for his archaic model extends to castigating the life of village farmers. The epigram on Philocritus pictures an Egyptian colonist's life far away from the glories of Egyptian immigrant life in celebrated other Ptolemy court poetry, such as Theocritus *Idyll* 17 and **no. 114** Parmenon.

---

155 Dioscorides of Nicopolis 16, 7, 38 *HE*.
156 Dioscorides of Nicopolis 33 *HE*: ἐμπορίης λήξαντα Φιλόκριτον ἄρτι δ' ἀρότρου / γευόμενον ξείνῳ Μέμφις ἔκρυψε τάφῳ, / ἔνθα δραμὼν Νείλοιο πολὺς ῥόος ὕδατι λάβρῳ / τἀνδρὸς τὴν ὀλίγην βῶλον ἀπημφίασε· / καὶ ζωὸς μὲν ἔφευγε πικρὴν ἅλα, νῦν δὲ καλυφθεὶς / κύμασι ναυηγὸν σχέτλιος ἔσχε τάφον.
157 Cf. Hunter 2003: 154–157 on Theocritus 17.79–80.
158 Hesiod *Opera et Dies* 639–640: οἰζυρῇ ἐνὶ κώμῃ, / Ἄσκρῃ, χεῖμα κακῇ, θέρει ἀργαλέῃ, οὐδέ ποτ' ἐσθλῇ.

112 Neoptolemus of Parium (PP 16706, 16874; TM 10604) was a literary critic and poet.[159] He wrote works titled *Witticisms*, *On Epigrams*, as well as glosses on rare words in Homer, a poem called *Dionysus* that credited the god with discoveries, and a poem called *Three Lands*. The most important poetic fragment from this poem is attributed in manuscript both to Neoptolemus and another third-century poet.

> Ocean, in which all land is bound and washed all around[160]

Three Lands is presumably in reference to a theory of three continents: Europe, Africa, and Asia. These three continents were together one landmass, called the *oikoumene* or "known world" in Hellenistic geography. Two other lines have been attributed to Neoptolemus from the same poem.

> For a chain of land does not surround the ocean but it is poured into boundlessness: so the ocean is in no way stained.[161]

Yet these lines are not ascribed to Neoptolemos nor anyone else. The geographer Strabo, in the midst of his review of a previous treatise on the question of the extent of the ocean and the *oikoumene* or known world, quotes them without attribution. Some scholars have thought them to come from **no. 136** Eratosthenes' *Hermes*.[162] While it is not possible to attribute the verses to either Neoptolemus or Eratosthenes with certainty, they are the work of someone who believed in an ocean circling the known world. Since to follow the discussion of Hellenistic geography down to its conclusion in Strabo would take us far from the present prosopography, I will sketch the primary contributions of Eratosthenes. While the extent of the known world had long been a question, Eratosthenes brilliantly reversed the issue and posed the question about the extent of the ocean: were in fact the basins of the Mediterranean sea, the Atlantic ocean, the Red Sea, and the Indian ocean one confluent and continuous body of water?[163] Eratosthenes thought so and Neoptolemus followed him.[164]

---

159 Fragment collection in Mette 1980, scholarship in Asmis 1992, Gutzwiller 2010. Mette 1980: 15 provides the slim evidence for thinking Neoptolemus an Alexandrian working in the second half of the third century BCE.
160 Neoptolemus fr. 2a Mette = Achilles Tatius *In Arati Phaenomena* 22 (52.1 Maass): Ὠκεανός, τῷ πᾶσα περίρρυτος ἐνδέδεται χθών.
161 Neoptolemos fr. 2 CA = Strabo 2.3.5 = 100C.22–23 (244.22–23) Radt: οὐ γάρ μιν δεσμὸς περιβάλλεται ἠπείροιο, / ἀλλ' ἐς ἀπειρεσίην κέχυται· τό μιν οὔτι μιαίνει. Mette 1980 does not attribute these lines to Neoptolemus.
162 Kidd 1989: 2.250, referring to earlier scholarship, takes no position.
163 On the circumambient ocean see Strabo 1.1.8 = 5C.13 (10.13 ff) Radt. For the controversy see Kidd 1989: 2.249–263 with reference to previous scholarship.
164 I choose an interpretation that foregrounds the scientific knowledge of court society over that offered by the ancient scholars and scholiasts who quote this fragment; see Mette 1980: 1–2. They

To see Neoptolemus following Eratosthenes' scientific learning in his poetry enhances our understanding of Neoptolemus' literary criticism. An ancient scholar reports that the Roman poet Horace's *Ars poetica*, an important prescriptive poem of literary criticism, "gathered Neoptolemus' precepts on poetry, not all, but the most eminent".[165] Neoptolemus evidently gave a three-fold structure to the purposes of literature, in which the poet's technical capacities, his style, and subject matter were the central axes of criticism. In this work Neoptolemus may have argued against Eratosthenes' literary criticism. Eratosthenes had made the remark in his *Geography* "every poet aims at entertainment, not instruction".[166] Neoptolemus, on the other hand, insisted that the poet has as his object instruction as well as entertainment.[167] In light of Neoptolemus' literary criticism, his poetry endorsing a continuous and confluent ocean surrounding the *oikoumene* probably sought to present enlightened scientific knowledge as well as entertain the reader.

113 Archelaus from Egyptian Chersones (PP 16674; TM 6280) was an epigrammatist and prose writer.[168] Both his poems and prose concerned wondrous happenings in nature, either the transformation of one animal into another through the cycle of life or the special natures of particular animals. He read some of his epigrams possibly in a sympotic context before a Ptolemy (probably Euergetes) on the supposed birth of wasps from the corpses of horses.[169] His prose work circulated under

---

place it in defense of Homer's statement that the ocean is a river, surrounding the known world. Thus, they claim, the later poets allegorically understood *okeanos* "ocean" as the horizon of the astronomical objects like the fixed stars that rise and set into the *okeanos*. Such an interpretation in reference to Homeric geography remains possible even when we set the dispute with Eratosthenes over the world ocean first. Eratosthenes' most stinging rebuke to treating Homeric geography seriously is his famous remark in Eratosthenes *Geographica* fr. 6 Roller = Strabo 1.2.15 = 24C.10–12 (56.10–12) Radt "one would find where Odysseus wandered when one finds the leather-worker who stiched up the bags of winds" that Aeolus gave Odysseus.

165 Neoptolemus fr. 5 Mette = Porphyrio *Commentum in Horatii Flacci Artem Poeticam* 1: *congessit praecepta Neoptolemi* τοῦ Παριανοῦ *De arte poetica, non quidem omnia, sed eminentissima*.

166 Eratosthenes *Geographica* fr. 2 Roller = Strabo 1.1.10 = 7C.2 (14.2) Radt: ὅτι ποιητὴς πᾶς στοχάζεται ψυχαγωγίας, οὐ διδασκαλίας. Gutzwiller 2010: 340–342, aligning Eratosthenes with Polybius' criticism of rococo historians, suggests that Eratosthenes meant to separate the moral ends of poetry and prose. But Eratosthenes' technical prose did not lack entertainment and his poetic fragments are packed with technical learning. Geus 2002: 265–267 denies that Neoptolemus' criticism has any relevance for Eratosthenes. I would set Eratosthenes' authorial intent by this remark within the narrow context in which it appeared: within his *Geographica*, as a criticism motivated against Homer's readers. Eratosthenes echoes the criticism of the uses of Homer articulated in Plato's *Ion*. The debate, known from other third century poets, is part of the discussion of poetic inspiration and proper mimesis continued from Plato; cf. Fantuzzi and Hunter 2004: 1–26.

167 Neoptolemus fr. 6a.39–41 Mette = Philodemus *On Poems* 5, *PHerc*. 1425 col.XVI, 143–144 Mangoni.

168 For the poetic fragments see *SH* 125–129; for prose testimonia see *PGR* 24–28. More generally Fraser 1972: 1.778–780.

169 I follow Fraser's 1972: 2.1086.n443 dating, who includes alternative views.

the title Ἰδιοφυῆ "Unique Natures" which, as we have seen, is also attested for the poetic work of Philopator. Archelaus writes of *physis* as an awesome, changing force. Change through continuity and natural wonder in its Egyptian zoology, such as crocodiles, vipers, scorpions, appear as central themes. Archelaus' poetry stands on the border between hybridity and natural wonders, themes also found in other court writers of a hybrid Greco-Egyptian culture and the writers of geographical paradoxography.

114 Parmenon of Byzantium (PP 16710; TM 11475) was a poet of the late third century. The longest extant fragment appears to be a hymn in praise of the Nile or rather its divine manifestation, called Egyptian Zeus. The extant fragment opens with a survey of towns in the Nile delta.

> Nile, Egyptian Zeus, the Canopians ... citizens, and those who possess Butus, Mende, the fortress of the Goat, and those who posses the wall of Phacussae, who live in Letopolis and Cynopolis ...[170]

Earlier court praise poetry incorporated a demographic survey of Ptolemaic possessions.[171] Perhaps Parmenon figured his work as a belated response to earlier court poetry. Furthermore, the ethnographic survey of a geographical territory finds parallels in the work of the paradoxographers and geographers **nos. 110, 113, 117–120, 136.**

115 Euphronius from Egyptian Chersones (PP 16686, 16853; TM 8222) was a poet and grammarian.[172] His grammatical work marked an important advance: he was the first to write ὑπομνήματα or a running commentary, a form which became stan-

---

170 *SH* 604a: Αἰγύπτιε Ζεῦ Νεῖλ', ἀτὰρ Κανωπῖται / ⟨...⟩ πολῖται, Βοῦτον οἵ τε Μένδην τε, / Αἰγὸς πολίχνην, καὶ Φακούσιον τεῖχος, / οἵ τ' ἄστυ Λητοῦς καὶ Κυνὸς περὶ κλῆρον / φοιτῶσι ...
171 Cf. Theocritus 17.77–94; Hunter 2003: 154–155 notes it is an old theme but offers Ptolemaic exemplars.
172 For discussions see Susemihl 1891: 280–281; *RE* 6.1: 1220–1221 *s.v.* 'Euphronius (7)'; Fraser 1972: 2.347.n117, 2.663.n100; Montana 2015: 126–128. There are difficulties with the works and associations of this individual. Scholarship has long credited Euphronius as a possible member of the tragic Pleiad under Ptolemy II Philadelphus on the strength of remark by the grammarian Choeroboscus (cited in Susemihl 1891: 280.n58). Yet on the strength of an emended remark in the *Suda* and a comment elsewhere in Choeroboschus scholarship has also credited him as the teacher of **no. 126** Aristophanes of Byzantium (cited in Montana 2015: 126.n311). Bergk's emendation of Choeroboscus that Aristophanes was a student of Euphronius must be accepted for the chronology to work, a fact already seen by Susemihl 1891: 281.n60. If we accept that Euphronius was a member of the tragic Pleiad, then he alone of all poets at the courts of Euergetes and Philopator composed tragedies. The evidence of Choeroboscus is admittedly late but it does appear to be good evidence. Strecker's 1884 discussion of Euphronius' grammatical study shows that his work overlapped considerably in topics and sources with **no. 136** Eratosthenes' *On Ancient Comedy*. Yet neither cites the other; cf. Geus 2002: 297.

dard among later grammarians.[173] His grammatical studies of classical Athenian comedians parallel the literary interests of **no. 111** Dioscorides, **no. 126** Aristophanes, and **no. 136** Eratosthenes in antiquarian literature.[174] He was also the first to write a metrical form called Priapeia, of which one fragment remains.

> Not initiated, o mystagoges of the new Dionysus, ... but I too come after doing service from benefaction, crossing by the Pelusian marsh at dusk.[175]

Here Euphronius addresses the courtiers and adherents of Philopator as the "mystagoges of the new Dionysus". Philopator associated himself with Dionysus and held him to be a founder of the Ptolemaic dynasty.[176] **No. 136** Eratosthenes' *Arsinoe* portrays Queen Arsinoe in conversation about a wine-drinking ritual enacted probably for Dionysus; a papyrus contains Philopator's order for regulating worship of Dionysus; the oarage of Philopator's famous forty-banked ship was covered with the imagery of the ivy-leaf and *thyrsus* of Dionysus.[177] The king even portrayed himself sometimes as the god, with ritualized tattoos of ivy-leaves, and some terracotta portraits identified as Philopator are crowned with an ivy-leaf of Dionysus.[178] Since Dionysus is identified as "the god of wine and revelry, of drama and of the luxury of nature and of man", Euphronius' poem thus plays on Philopator's desire to celebrate those luxuries and extend his generosity to cultured figures.[179] Perhaps Euphronius suggests that his Priapeia poetry participates in Dionysiac rituals and is descended from worthy literary stock (ἐξ εὐεργεσίης ὠργιασμένος).

116 A nephew of the famous **no. 110** Callimachus of Cyrene, Callimachus the Younger (PP 16700; TM 9562) wrote hexameter poetry about islands, possibly foundation stories, a *topos* of **no. 109** Apollonius of Rhodes and Callimachus himself.[180] None of his work survives.

---

173 Σ.Aristophanem *Aves* 1403b.
174 Strecker 1884: 6–7 points to comments on Aristophanes, Cratinus, Phrynichus, and Aristagoras.
175 Euphronius pp. 176–177 *CA*: οὐ βέβηλος, ὦ τελεταὶ τοῦ νέου Διονύσου, / .... / κἀγὼ δ' ἐξ εὐεργεσίης ὠργιασμένος ἥκω, / ὁδεύων Πηλουσιακὸν κνεφαῖος παρὰ τέλμα.
176 *BNJ* 631 (Satyros of Alexandria) F1 = Theophilus *Ad Autolycum* 2.7. Gambetti's discussion (*BNJ* 631) traces the evolution of the mythological genealogy of the Ptolemaic dynasty, from Euergetes equally honoring Heracles and Dionysus in *OGIS* 54, to Philopator's embrace of Dionysus alone in *BNJ* 631 (Satyros of Alexandria) F1.
177 See Fraser 1972: 1.203–4, 2.344–48.n112–119. Eratosthenes' *Arsinoe* is discussed in **no. 9**. The papyrus *prostagma* is *BGU* 6.1211. The forty-banked ship is described under **no. 119** Philostephanus.
178 On Philopator's tattoos see *Etymologicum Magnum* 220.19–20 s. v. γάλλος· ὁ Φιλοπάτωρ Πτολεμαῖος διὰ τὸ φύλλοις κισσοῦ κατεστίχθαι ὡς οἱ γάλλοι. Fraser 1972: 2.347.n118 compares 3 Maccabees 2.29, where the registered Jews are to be branded with the mark of Dionysus. On the terracotta portraits see Fraser 1972: 2.348.n119.
179 Fraser 1972: 1.205.
180 *SH* 309.

Other poets writing about the court included the epigrammatist Damagetus (PP 16678; TM 6779) and Herodas (PP 16691; TM 8719) whose *Mimiambi* have been referenced in the Introduction. There is no reason to believe that these poets resided at court.

Prose Authors: Antiquarians, Geographers, Historians

The prose authors in history and geography at court were greatly influenced by **no. 110** Callimachus. The Callimacheans – **nos. 117–120** Istrus, Hermippus, Philostephanus, Apollas – continued in prose the themes of a hybrid Greco-Egyptian culture, antiquarian learning, and celebration of the wonders of the natural world. Historians at court who wrote contemporary history celebrated Ptolemaic prowess, military success, and also the luxury of the court. Our modern words distinguishing the authorship of special kinds of prose genres – political history, social history, mythology, travel writing, geography, antiquarianism – are largely subsumed under the ancient Greek words ἱστορικός "historian" and συγγραφεύς "prose author" that refer indistinctly to prose authorship involving historical researches in book learning. Polybius seems to refer to the prose under discussion here: "The genealogical mode [of historical writing]," he said, "moves the friend-of-lectures, the inquisitive, and the curious with its account of colonies, foundations, and social relations."[181]

117 Istrus of Alexandria (PP 14384, 16698, 16927; TM 9474), antiquarian, historian, geographer, and poet (although we possess no poetry) was one of the four direct pupils of **no. 110** Callimachus called Callimacheans.[182] Although he is called a slave of Callimachus, this may be part of the discourse of dependency that marks relations between teachers and impoverished and thus dependent students in the ancient world.[183] He wrote an influential treatise at least fourteen books long on ancient Athens and its customs (various titles are transmitted).[184] Istrus also wrote

---

181 Polybius 9.1.4: τὸν μὲν γὰρ φιλήκοον ὁ γενεαλογικὸς τρόπος ἐπισπᾶται, τὸν δὲ πολυπράγμονα καὶ περιττὸν ὁ περὶ τὰς ἀποικίας καὶ κτίσεις καὶ συγγενείας. Chapter 2 discusses the discourse of friendship behind the labored translation "friend-of-lectures".
182 The most important fragment collection is *BNJ* 334 (Istros); scholarship includes Berti 2009a, Berti 2009b and the bibliographic essay to *BNJ* 334. Berti (*BNJ* 334 (Istros) "Biographical Essay") has argued that Istrus dates to the early part of the third century as a contemporary of Callimachus in an argument that I cannot share. I return to the traditional dates of Callimachus' students as active in the second half of the third century BCE.
183 Berrey 2014b: 434–436.
184 A particular fragment that Istrus took from other prose authors (*BNJ* 334 (Istros) F6) discusses a spring on the Acropolis called the *Clespydra*, whose flows correspond with the etesian trade winds that affect the Nile. A contemporary Egyptian reader may have been reminded of Callimachus' paradoxographical records of rivers, waters, and fountains (frr. 407, 410, 457–459 Pfeiffer) with

several works that drew from Homer, the Archaic lyric poets, and the Athenian tragedians; local histories of regions in the Peloponnesus; several different works of religious histories; two different works on the history of athletic contests; and a work of criticism directed against the late fourth-century historian Timaeus. His interests varied over ancient Greek places, religious rites, culture, and literature in a manner that depends on Callimachean learning and Alexandrian book culture.

Although the natural world appears little in his work, Istrus was a prominent propagandist for a hybrid Greco-Egyptian culture. In addition to those previously listed Istrus' other works were called *On the Egyptian City of Ptolemais* and *Egyptian Colonies*.[185] Since Ptolemais was a large city founded in upper Egypt near Memphis by Ptolemy I Soter, Istrus' work was undoubtedly a work combining geography, demography, and religious history with Ptolemaic court propaganda. In the *Egyptian Colonies* Istrus appears to have continued the earlier thesis of the historian and propagandist Hecataeus of Abdera, probably active at the court of Ptolemy I Soter, "that the Egyptians were the oldest people and that the valley of the Nile was the birth-place of mankind and that, consequently, all other peoples were emigrants and colonies of Egypt."[186] His propogandizing for Egyptian origins and thus Ptolemaic patronage of Greek cities and customs may also have extended to his works on athletics as cultic foundations for Egyptian colonizers.

118 Hermippus of Smyrna (PP 16918; TM 7904), called both a Peripatetic and a Callimachean, was also among the direct students of **no. 110** Callimachus.[187] His works were the product of the book-culture of ancient Alexandria. He wrote a work on the Persian Magi, a book called *On Sayings Taken from Homer*, a *Phaenomena* about constellations mentioned in the epigram attributed to Philopator recounted in the opening of the chapter (for which he earned the epithet "astronomer"), and possibly a work on paradoxography including a story about how the river Nile got its name from an ancient Egyptian king.[188] But Hermippus is most famous for his biographies that have sometimes assumed a central role in modern discussions of the development of the ancient biography. He wrote the lives of a variety of persons: lawgivers, philosophers, sages, poets, and rhetoricians. These figures come from all prior periods of Greek history, from the mythological age to the Archaic and Classical down to the end of the fourth century. The fragments from later authors show an eye for detail, antiquarian research, catalogues of writings, descrip-

---

potential interest for an Egyptian audience: the Clepsydra's filling coincides with the Nile's flood at the etesian winds (cf. Herodotus 2.20).
185 *BNJ* 334 (Istros) F43–47.
186 Berti "Biographical Essay" *BNJ* 334 (Istros).
187 For the following paragraph Bollansée 1999: 1–18, 117–186 and *FGrH* 1026 (Hermippos of Smyrna) is central.
188 As "astronomer" Athenaeus 11.478a; for paradoxography see *FGrH* 1026 (Hermippos of Smyrna) F90–91.

tive narratives, and seemingly an eye for reconstructing an author's life from his writings. In displaying the products of his vast reading in such a way Hermippus shared the same interests in antiquarianism and book-learning as his fellow Callimacheans.

119 Philostephanus of Cyrene (PP 16961; TM 16961), a student of **no. 110** Callimachus, was a poet, antiquarian, geographer, and paradoxographer. His one surviving poetic fragment concerns the geography of Sicily, a topic covered in his paradoxology. His works included *Cities in Asia*, a paradoxographical work *On Rivers*,[189] *On Islands*, including a discussion of Cyprus as a colony from Egypt with **no. 117** Istrus.[190] His work bears the hallmarks of Callimachus, with a fascination for aetiological accounts of both human action and the natural world with book learning.[191]

In addition to his geography and paradoxology, most interesting for a history of court science is Philostephanus' work *On Discoveries*.[192] Philostephanus was not the first to write an account of inventions. It was already a subject of discussion among the Peripetatics. Strato of Lampsacus, one of the heads of Aristotle's school in the first half of the third century and teacher of Ptolemy II Philadelphus, had written a work in two books called *Refutations of Discoveries*.[193] Prose treatises treating discoveries seem to have covered technological artifacts and proverbs in a kind of cultural history.[194] Philostephanus' *On Discoveries* attributed early inventions to mythical Greek heroes, including the discovery of the plow to the Egyptian king Osiris, called Dionysus in *interpretatio Graeca*.[195] In this way Philostephanus may have sought to supplement and expand Istrus' argument for the chronological priority of the Nile valley in human history. But he also brought the story of inventions down to his own time.

---

189 Philostephanus F29 Capel Badino = Σ.Dionysium Periegetem 289 "Philostephanus says that at that time [the Po river (Ἠριδανός)] was named the Rhône (Ῥοδανόν) by the natives", a fragment that derives from both book learning and an interest in dialectical questions, similar to the grammarian Hellanicus **no. 131** described below.
190 Philostephanus F4 Capel Badino = Stephanus of Byzantium s. v. Κύπρος. Billerbeck 2014: 156 shows that, although the epitomized manuscripts of Stephanus do not transmit this fragment of Philostephanus, the indirect tradition represented by Constantine Porphyrogenitus transmits it. I accept the authenticity of the fragment.
191 Philostephanus F18 Capel Badino = Σ.Iliadem 2.145 (*Scholia Didymi*), where Callimachus' *Aetia* is also cited.
192 Capel Badino 2010: 40–41, 101–103.
193 Diogenes Laertius 5.60. The fragments of Strato's work are collected as frr. 84–86 Sharpe in Desclos and Fortenbaugh 2011. See the discussion of predecessors of Strato at fr. 85 Sharpe.
194 *On Discoveries* from Theophrastus on the technology of the potter's wheel see Σ.Pindarum O.13 27c; *On Discoveries* from Strato on proverbs see Clement of Alexandria *Stromateis* 1.14.61.1.
195 Philostephanus T3; F11, 35–37 Capel Badino. Might we also see a comparison with **no. 112** Neoptolemus of Parium's *Dionysus*?

Philostephanus says that Jason was the first to sail a warship ... and Philostephanus says that Ptolemy Soter made [a warship] with twelve banks of oars, Demetrius [Poliorcetes] son of Antigonus made them with fifteen, Ptolemy Philadelphus with 30 and Ptolemy Philopator surnamed Tryphon with 40.[196]

This important testimonium shows that Philostephanus credited contemporaries with discoveries. Perhaps surprisingly from a modern perspective, Philostephanus does not credit the engineers or designers who built the warships but the monarchs who commissioned them and sailed on them. Just as the ancient festival games celebrated the owners' of chariot-teams for their victories (including **no. 2** Berenice at Nemea), not the drivers, so in the ideology of aristocracy the paymaster claims some due for the honor of discovery. In this way the Ptolemies might claim the honor associated with the discourse of friendship and patronage as "friends-of-the-*technai*".[197] Philostephanus probably construed discoveries within a framework of ascending progress directed toward the present day, just as did the Peripatetics who preceeded him in this genre.[198] *On Discoveries* may have been in implicit praise of Philopator's patronage, invoking in its teleological antiquarianism a hybridized Greco-Egyptian culture populated by ancient Greek heroes as well as Egyptian ones.

120 Apollas of Pontus (PP 16821; TM 5171) was a historian and geographer who wrote in prose *On Peloponnesian Cities* and *On Delphi*, each one book long. These works seem to have focused on local history and geography, embracing social customs and human artifacts.[199] Apollas is called a 'Callimachean', a term reserved for the direct students of **no. 110** Callimachus, in a discussion of the locale of famous banquet of Scopas discussed in Callimachus' *Aetia*.[200] Although no geographical information about Apollas is known beyond his toponym, the epithet 'Callimachean' locates him in Alexandria during the second half of the third century. Apollas' prosaic interests in geography, cultic practices, and antiquarianism follow those of the other students of Callimachus.

---

196 Philostephanus F35 Capel Badino = Pliny *Historia Naturalis* 7.57.207–208: *longe nave Iasonem primum navigasse Philostephanus auctor est ... ad XII ordines Philostephanus Ptolomaeum Soterem, ad quindecim Demetrium Antigoni, ad triginta Ptolomaeum Philadelphum, ad XL Ptolomaeum Philopatorem qui Tryphon cognominatus est*. Philopator's forty-banked ship appears also in Plutarch *Demetrius* 43.5 and Athenaeus 5.203e–204d = *BNJ* 627 (Kallixeinos of Rhodes) F1, where the description is credited to Callixeinus of Rhodes. Keyser (*BNJ* 627) argues that Callixeinus' *On Alexandria* dates to 190–170 BCE and is reliant upon earlier writers. Philostephanus is not a likely source for Callixeinus' description but both might be drawing upon a common technical source.
197 Philon *Belopoeica* 50.26 Th: φιλοδόξων καὶ φιλοτέχνων βασιλέων, described under **no. 139**.
198 Zhmud 2006: 117–165.
199 *BNJ* 266 (Apollas of Pontus) F1, F9.
200 *BNJ* 266 (Apollas of Pontos) F6 = Quintilian *Institutio Oratoria* 11.2.14. I follow the emended text *Apollas Callimach(i)us*, which Harder 2012: 1.228 Callimachus *Aetia* fr. 64.13 gives in the *apparatus*.

121 Ptolemy of Megalopolis (PP 15068, 16944; TM 12779) had a career first as a military official then as a memorist.²⁰¹ Since he was one of the men sent abroad upon **nos. 15–16** Sosibius and Agathocles' assumption of the guardianship of Ptolemy V Epiphanes, Ptolemy was probably a leading personality in the kingdom and a potential opponent to the conspirators. His father may have been a general under Euergetes and Ptolemy himself later became the *strategos* of Cyprus under Ptolemy V Epiphanes. Ptolemy wrote a book called *Histories of Philopator*, some of it perhaps drawn from his personal experience. The surviving fragments of the memoir conform to the later cultural expectation of Ptolemaic debauchery, with discussion of the merry-makers, wine-pourers, and mistresses populating the royal symposia. Yet it remains unclear whether those fragments more accurately capture the interests of Ptolemy's own book or the interests of his excerptors. Polybius complains that certain writers used "prodigies and elaborations for the amazement of the readers" when narrating the scandalous reign and end of **no. 16** Agathocles.²⁰² Polybius' criticism of his predecessors in this passage, among others, used to be thought directed against a genre of so-called tragic history; yet recent scholarship has exposed such a generic category as a modern scholarly invention.²⁰³ Whether Ptolemy of Megalopolis' historical accounts were salacious and exaggerated, whether the *Histories of Philopator* might been the very book Polybius complains of, we do not know; but we should at least recognize that the aesthetics of the fantastic and amazing, represented in the paradoxographers, were popular in Philopator's court.

Polybius complains about Ptolemy's own dissolute later life.²⁰⁴ Given Polybius' own historiographical demands that authors have personal experience in their subject matter, one might have thought that he would welcome Ptolemy's special insight into Philopator's debauchery. Despite Polybius' criticisms of his lifestyle, Ptolemy of Megalopolis' family remained in faithful service to the Ptolemaic house. Under Ptolemy V Epiphanes his daughter became the eponymous priestess of Arsinoe Philopator memorialized on the Rosetta Stone.²⁰⁵

122 The historian Demetrius of Byzantium (PP 5082, 16910; TM 6936) wrote two works covering the same historical period, circa 280–271 BCE: one about the migration of the Galatians to Asia Minor, the second about war between Antiochus I and Ptolemy II Philadelphus over Libya.²⁰⁶ No fragments survive from either work. If

---

201 *BNJ* 161 (Ptolemy of Megalopolis) is fundamental for the following paragraph.
202 Polybius 15.34.1: τὰς τερατείας καὶ διασκευάς, αἷς κέχρηνται πρὸς ἔκπληξιν τῶν ἀκουόντων.
203 Polybius' criticism of tragic history occurs at 2.56 against the historian Phylarchus. Bollansée 2005 gives a helpful account of so-called 'tragic history' and argues against the thesis Ptolemy of Megalopolis was Polybius' source.
204 *BNJ* 161 (Ptolemy of Megalopolis) T2 = Polybius 18.55.7.
205 *BNJ* 161 (Ptolemy of Megalopolis) T3 = *OGIS* 90.5–6.
206 *BNJ* 162 (Demetrios of Byzantion) is fundamental for the following paragraph.

Demetrius the historian may be identified as Demetrius the eponymous priest in 220/19 and 219/18, he was the son of Apelles.[207] Apelles (PP 2151) was an immigrant from Byzantium to Egypt and became a military commander (ἡγούμενος στρατιωτῶν), probably under Philadelphus and Euergetes.[208] Demetrius' historical interests in the Galatian wars in Asia Minor may have resulted from his family connections to the area. At the least, his interest in Ptolemaic history is more evidence of a broader court interest under Philopator for the earlier history of the Ptolemies themselves. Furthermore, if Demetrius the historian is rightly identified as Apelles' son, he was the brother of **no. 126** Aristophanes of Byzantium (PP 16513; TM 5858). It is possible to see Demetrius the historian as part of a Ptolemaic family involved at life at court over two generations: from the father Apelles, an immigrant and commander under Philadelphus and Euergetes, to his sons Demetrius and Aristophanes, intellectuals under Philopator and Epiphanes who were rewarded with formal religious and intellectual offices.

Prose Authors: Philosophers, Grammarians

Prose authors apart from antiquarians, geographers, and historians existed at court. Philosophers and grammarians had long been recognized as authors of a distinctive kind of prose writing, although at least three of the following authors wrote prose works that were closer to antiquarian history than philosophy or literary criticism.

123–125 We hear of three philosophers at court. The one with the longest residence appears to be an obscure philosopher called Panaretus (PP 16776; TM 11344), student of the sceptical academic philosopher Arcesilaus; he received twelve talents annually from Euergetes.[209] Perhaps the payment was a retainer fee for tutoring Philopator. We are better informed about the Stoic philosopher Sphaerus (PP 16788; TM 13750). Sphaerus had lived and lectured in Sparta at the court of Cleomenes III, even writing two works on the Spartan constitution.[210] He lived for some years at Alexandria after following Cleomenes into exile. An anecdote about Sphaerus' audience before Philopator reveals Sphaerus' wit and Philo-

---

[207] Clarysse and Van Derveken 1983: nos. 71–72.
[208] *Suda* α3933.
[209] On Panaretus see Atheneaus 12.552c. This Panaretus is included in a list of scholars aspiring to *leptotēs* so well that their bodies too became skinny; is the joke that twelve talents weighed him down sufficiently?
[210] *FGrH* 585 (Sphairos der Borysthenite) T3a = Plutarch *Agis et Cleomenes* 23.2; *FGrH* 585 (Sphairos der Borysthenite) T1 = Diogenes Laertius 7.178.

pator's intelligence.²¹¹ The king attempted to make Sphaerus assent to a false opinion and thus reveal himself not to be a Stoic sage. The king placed two pomegranates made of wax before Sphaerus, who assented; caught out, Sphaerus replied that he had assented not to the fact that they were pomegranates but to the reasonableness of the fact that they were pomegranates. The sagacious Stoic was unprepared to deal with the represented realism afforded by Ptolemaic luxury, whereas Philopator was well educated about particular details of Stoicism. Despite, or perhaps, even because of the intellectual ambush Sphaerus respected Philopator. Sphaerus wrote a treatise on kingship, a prescriptive genre about the appropriate behavior of kings and courtiers. Another court philosopher Mnesistratus (PP 16772; TM 10479), otherwise almost unknown,²¹² accused Sphaerus of denying that Philopator was a king. No, replied Sphaerus, "Ptolemy is such a man and is a king."²¹³ Although Cleomenes attempted a revolt and an escape from his Alexandrian house-arrest recounted above under **no. 17** Ptolemaeus, we have no knowledge whether Sphaerus too died among Cleomenes' followers.

Although we find no Peripatetics at the courts of Euergetes and Philopator, this brief account of three transitory philosophers does not obscure the fact that Aristotelian themes ran through the scholarship of Ptolemaic court society. From antiquarian researches to cultural and technological histories to geographic and literary investigations, scholarship ran along the lines sketched out and practiced over the hundred previous years by the Peripatetics.

126–131 Aristophanes of Byzantium (PP 16513; TM 5858), the most famous of the six grammarians at court, perhaps exemplifies Peripatetic scholarship best.²¹⁴ (The Greek γραμματικός indicates a literary critic, not an elementary school teacher.) Aristophanes became head of the library after **no. 136** Eratosthenes' death and may have been *tropheus* or tutor to **no. 10** Ptolemy V Epiphanes. He was probably the brother to **no. 122** Demetrius of Byzantium. The ancient tradition offers him as a pupil of the leading grammarians, poets, and other librarians in Alexandria: the librarian Zenodotus of Ephesus, the poet Machaon, the grammarian Dionysius the Iambos, **no. 110** Callimachus, **no. 115** Euphronius, and **no. 136** Eratosthenes. I re-

---

211 Diogenes Laertius 7.177. Brouwer 2002: 203–205 argues that Sphaerus is not presenting himself as a sage. Sage or not, the point of the anecdote is premised upon Sphaerus' high intellectual standing as a worthy representative of Stoicism.
212 Diogenes Laertius 7.177; Diogenes Laertius 3.46; Athenaeus 7.279d.
213 Diogenes Laertius 7.177 = 583.13–15 Dorandi: πρὸς δὲ Μνησίστρατον κατηγοροῦντα αὐτοῦ ὅτι Πτολεμαῖον οὔ φησιν εἶναι βασιλέα ⟨...⟩ εἶναι, 'τοιοῦτον δ' ὄντα τὸν Πτολεμαῖον καὶ βασιλέα εἶναι.' Dorandi 2013: 583 follows Causabon in marking a lacuna here, such that almost any supplement to fill it will involve an interpretation that Sphaerus criticized Philopator's kingship. I reject supplements of that character.
214 For life and work see *RE* 2.2: 994–1005 *s.v.* 'Aristophanes (14)', Pfeiffer 1968: 171–209, Fraser 1972: 1.459–461, 2.662–666.

gard the tradition as hopelessly confused. But these relationships suggest that Aristophanes was well-known to numerous court actors under Philopator. Aristophanes' work is fundamental to the surviving texts and transmission of many ancient Greek authors. He published editions of or treatises on Homer, Hesiod, the lyric poets, the tragedians, Aristophanes and Menander. He was interested in lexicography, biography, character studies, and proverbs. The only surviving work attributed to him is an *epitome* of Aristotle's zoological studies, perhaps representing an interest in book learning and antiquarianism rather than a particular curiosity about the natural world.[215] His interest in grammatical studies mirrors the work of his contemporaries the Callimacheans, all following in the footsteps of the researches of the Peripatos.

The remaining grammarians evidently shared the aesthetic interests of the poets and antiquarian prose authors. Three were probably active under Euergetes. Agathocles (PP 16812, 16893; TM 4398) is the name ascribed to the authorship of two different sets of writings, grammatical materials and a local history, but the author is probably the same individual.[216] This Agathocles was the student of the first librarian at Alexandria, Zenodotus of Ephesus, and hence probably active under Euergetes. The surviving fragments showcase learned astronomical, botanical, and geographical knowledge, among other topics. Two other grammarians were active as students of Zenodotus, also probably active under Euergetes.[217] Two grammarians were probably active under Philopator. Asclepiades of Nicaea or Alexandria (16837; TM 6390) was a grammarian and antiquarian who wrote at least one securely attested title, *On the Axones of Solon*, a treatise about the laws and political organization of ancient Athens. It is unclear when he lived, although one admittedly confused testimonium suggests he was active in the later years of Philopator's reign.[218] Finally, Hellanicus (PP 16851; TM 7676) was a grammarian and student of Agathocles.[219] Most famous as the teacher of the leading grammarian of the next generation, Aristarchus of Samothrace, he was probably active under Philopator and his successor Epiphanes. His fragments show a particular interest in dialectical questions; he was also part of a group of ancient grammarians who maintained that the *Iliad* and *Odyssey* had separate authors.

---

[215] Hellmann 2006 suggests that it was intended as a kind of handbook.
[216] *BNJ* 472 (Agathokles); Montanari 1988: 15–42.
[217] Theophilus (PP 16859; TM 8978) and Anaxagoras (lacks PP and TM nos.); for both see Montanari 1988: 113–118.
[218] *BNJ* 339 (Asklepiades of Nikaia).
[219] Montanari 1988: 45–73.

Scientific Prose Authors

Finally there were scientists at court, including physicians, astronomers, mathematicians, mechanicians, and geographers. Treating these individuals as a distinct group creates a partial historical illusion. After all, we have seen among the court poets and prose authors an interest in celebrating the natural world and technological innovation. Geography, in particular, is divided between the paradoxographers, historians, and geographers below. Yet the Greek terms for authorship of prose genres distinguished by scientific discipline were well established by the Hellenistic period. Some of the ten scientists listed below were authors; others were not. The scientists who did write prose wrote both scientific works and historical or antiquarian treatises alongside the influential Callimacheans. The scientists' association with court figures of both political power and literary repute often contextualizes the generic forms and scientific aspirations of their written work.

132 The doctor Caphisophon (PP 14990, 16615, 16650; TM 2164) may have been a royal physician under Euergetes.[220] A papyrus letter preserved among the Zenon archive, dated to the spring of 240 BCE, shows Caphisophon's stature in Alexandria.[221] Zenon received a letter from a correspondent in Alexandria worried about a mutual friend. The mutual friend had been arrested on some charges but his friends exerted their influence to free him. Among those friends was Caphisophon, son of Philippus the doctor (PP 16640).[222] There was good reason Caphisophon might have had success. We possess a stone decree from Cos honoring him, since Euergetes had sent him as a religious ambassador to the shrine of the healing god Asclepius on the island of Cos with an introductory letter.[223] In addition to his diplomacy Caphisophon practiced medicine. Ancient physicians credit Caphisophon with a compounded emollient to be applied to the body. The emollient was so well-known by name that it was unnecessary to write out the recipe, but was useful for removing warts, moles, or other pustules of the body.[224] It is a reasonable inference that Caphisophon wrote a recipe book.

---

[220] On Caphisophon see Fraser 1972: 1.369; Masser 2005: 55–56; *EANS* 168. I agree with Masser 2005: 56.n147 that *SEG* 33.672 ought not be referred to Caphisophon.
[221] *P.Mich.Zen.* 55; Edgar 1931: 126.
[222] I am ambivalent whether Callimachus epigram 3 *HE* refers to this Philippus.
[223] *OGIS* 42 = *SEG* 33.671 = Samama no. 132.
[224] Caelius Aurelianus *Celerum Passionum* 2.136 = 236.2 Bendz; *Tardum Passionum* 2.34 = 564.10 Bendz; *Tardum Passionum* 3.55 = 710.22 Bendz. Scholars have restored the name Caphisophon in two places in the text on the strength of this third, main citation. Soranus *Gynaecia* 3.37 = 117.15–17 Ilberg credits Caphisophon with a composition useful for a mole or cyst on female genitalia; Oribasius *Eclogae Medicae* 53.5 = 215.8 Raeder attributes a pitch-plaster out of bitumen, sulphur, and salt-marsh; Oribasius *Eclogae Medicae* 59.3 = 224.27 Raeder attributes an emollient without specifying the type.

Since Ptolemy introduced and commended Caphisophon to the Coans by letter, Caphisophon was probably personally unknown to the political class of Cos and had not previously undertaken political or religious missions outside of Egypt. It is a plausible speculation that Caphisophon's position as a religious ambassador in the embassy to Asclepius on Cos was the social reward for his political connections.[225] The honorific inscription shows that Caphisophon was not simply an attendant physician of some courtier. Instead Caphisophon possessed honor and social esteem in his own right. The evidence strongly suggests that Caphisophon was the physician of the king, since physicians who secured royal approval for certain actions, acted as authorized political agents of the monarch, gained independent honors, and developed innovations in medicine were almost certainly royal physicians of the court.

133–134 From a papyrus letter dated 243 BCE we hear of another doctor under Euergetes, Neon (PP 16623; TM 2594), buying rugs and carpets.[226] The unnamed letter writer adds a final bit of information for his recipient: "Know that Neon is in favor with the king."[227] Pursuit of political influence seems to have run in the family. We possess an honorary inscription from Neon's son, Agathoboulos (PP 15784; TM 4378), on the dedication of a statute of **no. 15** Sosibius, the *philos* of Philopator.[228] Whether or not Agathoboulos was a doctor like his father, it seems clear that both Neon and his Agathoboulos were Alexandrian courtiers under Euergetes and Philopator.

135 Finally, we hear of a third physician under Euergetes, Xenophantus (PP 16624; TM 10938). Strikingly, we have an inscription where Euergetes dedicates a statute of his own doctor.

> King Ptolemy son of Ptolemy and Arsinoe, *theoi Adelphoi*, [dedicates a statue of] Xe[nophan]tus, son of Sosicrates, of Heracleion, his own doctor.[229]

The stone is damaged at the physician's name and Fraser has offered three possible restorations of the name.[230] Subsequent editors have followed Fraser's reading of Xenophantus, which I too adopt.[231] Xenophantus' ethnic, Heracleion, is the name

---

225 That argument implicitly dates the stone inscription after 240 BCE; see Samama no. 132.
226 *P.Cair.Zen.* 4.59571.
227 *P.Cair.Zen.* 4.59571.12–14: γίνωσκε δὲ καὶ Νέωνα εὐημεροῦντα παρὰ τῶι βασιλεῖ.
228 *OGIS* 79.1–6.
229 Samama no. 393: [βασιλεὺς Πτολεμαῖος Πτολε]μαίου | [καὶ Ἀρσινόης θεῶν] ἀδελφῶν | Ξε[νόφαν]τον Σωσικράτους Ἡράκλειον | τὸν αὑτοῦ ἰατρόν. Nutton 1988: 193–194, puts the earliest instance of the later term for the king's physician, ἀρχιατρός, in the late third century; Samama no. 233 objects to the restoration of *archiatros* in that inscription.
230 Fraser 1972: 2.546.n287.
231 *Prose* no. 15; Samama no. 393.

of one suburbs of Alexandria.²³² Heracleion was a small town (called Thonis in Egyptian) between the town of Canopus and the Canopic branch of the Nile; it has since fallen into the sea 6.5 kilometers from the modern coast due to earthquakes and division of the Canopic branch.²³³ Heracleion was a significant port and had an important temple to Egyptian gods.²³⁴ The inscription was discovered in modern Alexandria, not Heracleion, indicating that Euergetes set up the honorific statue where Xenophantus worked. It was a great honor to have a statue for a living individual dedicated by a monarch; parallels are infrequent.²³⁵ Whether Xenophantus' contribution was political, medical, or personal, Euergetes thought it worth memorializing the individual.

136 Eratosthenes of Cyrene (PP 14645, 16515; TM 7752), poet, philosopher, mathematician, astronomer, geographer, and historian was perhaps the leading intellectual of his age; he is a central actor in this book.²³⁶ His polymathy earned him the sobriquet "Beta" as if like the letter he were second best in everything and Archimedes' flattery as "learned, remarkably preeminent in philosophy, and honoring any science that passes under your eyes".²³⁷ Again and again we find him as the central node in the intellectual social network at court. He wrote intellectual pieces for the monarchs Euergetes, **no. 2** Berenice II, and **no. 9** Arsinoe III; he corresponded with Archimedes; he probably exchanged literary polemics with **no. 138** the astronomer Dositheus, **no. 141** the doctor Andreas, and **no. 112** the poet Neoptolemus of Parium; he was associated with **no. 126** Aristophanes of Byzantium.²³⁸ Eratosthenes perhaps best represents the intellectual and social potential of a court scientist.

Born in Cyrene, he went to Athens for philosophical education; he was called to Alexandria by Euergetes to be librarian probably soon after his accession in

---

232 BNJ 631 (Satyros of Alexandria) has an updated list of the known demes of Alexandria from Fraser 1972: 1.38–47.
233 Goddio 2011: 1–26, 69–130.
234 The Canopus decree of Euergetes discusses the annual sailing of the ritual bark of Osiris from Heracleion to Canopus: OGIS 56.50–54, Pfeiffer 2004: 154–157.
235 Fraser 1972: 2.546.n287 compares other dedications to individuals made by Ptolemaic rulers: in an early dedication a Ptolemy (probably Philadelphus) honors a named builder of a thirty-banked warship; Physcon and the Cleopatras honor a named *epitropos* and member of the first friends.
236 There is still no standard collection of Eratosthenes' fragments and surviving materials are divided between various collections. Geus 2002 is the best introduction to the surviving materials although there is great room for disagreement. I have not added any additional testimonia or fragments to Geus' lists, although I divide up the surviving material of the scientific works differently.
237 Geus 2002: 34–39 disputes the interpretation of *Suda* ε2898; Archimedes *Methodus* 71.col1.33–71.col2.8 Netz et al. = 428.18–20 Heiberg: σπουδαῖον καὶ φιλοσοφίας προεστῶτα ἀξιολόγως καὶ τὴν ἐν τοῖς μαθήμασιν κατὰ τὸ ὑποπίπτον θεωρίαν τετιμηκότα.
238 *Suda* ε2898 states that Eratosthenes was a student of Callimachus, a notion now discounted.

246 BCE. While no ancient source specifies that he was *tropheus* or tutor to Philopator, some modern scholars have inferred so from a note in Eratosthenes' own poetry and the ancient tradition that the head of the library often served as tutor to the monarch's children.[239] Although the reasons for his call to Alexandria have been much discussed, it seems fairly certain that Eratosthenes had already gained fame for his poetic and philosophical work: the librarian was always a philosopher or poet.[240] He lived in the remainder of his life in Alexandria until sometime in reign of **no. 10** Ptolemy V Ephiphanes (reigned 205/4–181 BCE). Geus infers from the reasons of his call to Alexandria that Eratosthenes' literary life divided neatly between poetry and philosophy in Athens and scientific, historical, and literary-critical works in Alexandria.[241] While this thesis is too strong, undoubtedly the majority of Eratosthenes' scientific writing happened in Alexandria. Geus has assembled the evidence for ten philosophical works, six poetic works, six literary-critical works, and four historical works;[242] I pass over discussion of these to focus on the scientific material.

Eratosthenes' scientific works included mathematics, astronomy, and geography. Of his mathematical works,[243] we have testimonia and a text called the μεσόλαβος "mean-taker" on a certain instrument,[244] testimonia on an instrument called κόσκινον "sieve" for finding prime numbers,[245] testimonia on the calculation of harmonics,[246] and testimonia about a treatise called περὶ μεσοτήτων "*On Means*".[247] The testimonia show a thematic interest in proportionality and mechanical processes to achieve a mathematical solution. Of his astronomical works, we possess testimonia about a treatise called περὶ τῆς ὀκταετηρίδος "*On the Eight-Year Cycle*", a title also attested for **no. 138** the court astronomer Dositheus,[248] and later revisions of a work called either ἀστρονομία or καταστηριγμοί "*Astronomy*" or "*Cat-*

---

[239] Geus 2002: 27–28 seems not to see that PP is repeating a suggestion of Wilamowitz; see Fraser 1972: 2.477.n127.
[240] Geus 2002: 28–30; Fraser 1972: 1.322–323.
[241] Geus 2002: 54–58.
[242] Geus 2002: 49–53. See Fraser 1972: 2.487.n182 and *BNJ* 241 (Eratosthenes of Cyrene) for disagreements about Geus' discussion of Eratosthenes' historical works.
[243] Geus 2002: 139–205 incorporates almost all Eratosthenes' surviving material relating to mathematics, with the exception of the *mesolabos*, into the philosophical dialogue *Platonikos*.
[244] Vitruvius 9.*pr*.13–14; Plutarch *Marcellus* 14.9–11; Pappus *Collectio* 3.21–23; Eutocius *In Archimedis de Sphaera et Cylindro Libros II* 88.3–96.27. Chapters 3 and 4 analyze this text.
[245] Nicomachus *Arithmetica* 1.13.2–19 (compare the three diagrams in the *apparatus* of Hoche 1866: 30–31); Asclepius of Tralles *In Nicomachi Arithmetica* 1.86.
[246] Nicomachus *Harmonica* 11.6; Ptolemy *Harmonica* 2.14 (three tables); see further testimonia in Geus 2002: 171.n131.
[247] Pappus *Collectio* 7.3; Pappus *Collectio* 7.22; Pappus *Collectio* 7.29. Pappus *Collectio* 7.3 clearly indicates that this was a separate treatise in two books.
[248] Geminus *Isagoge* 8.24; Achilles Tatius *In Arati Phaenomena* 19 (47.22–24 Maass).

asterisms". This work, which survives in a complex tradition of many revisions of uncertain date, evidently contained a description of constellations and the stars that composed them: due to its subsequent revisions, it is unclear what of the surviving material ought to be attributed to Eratosthenes.[249] Further astronomical testimonia of Eratosthenes concern the relative distance of certain planetary orbits and a wind rose, material which may belong to either to Eratosthenes' astronomical or geographical works.[250] Finally, Eratosthenes' geographical works were titled περὶ τῆς ἀναμετρήσεως τῆς γῆς "*On the Measurement of the Earth*" and γεωγραφικά "*Geography*".[251] These were major scientific treatises. They determined the shape and circumference of the earth, the division of earth's surface into water and land, the consequences for zonal theory, the extension and limits of the *oikoumene* "the known world", the establishment of parallels and meridians, a geographic and ethnological description of each of the sectors of the *oikoumene*, and critically reviewed previous contributions to geography. It remains a question to what extent Eratosthenes engaged, appropriated, or critiqued the geographical knowledge in the paradoxographical writings of the other court authors.[252] Later ancient geographers took Eratosthenes' geographical works as the standard for much technical information, although Roman conquests in the western Mediterranean and Europe soon superseded his limited knowledge of those areas.

Although scholars have held different views about Eratosthenes' relation to his Egyptian surroundings and royal patrons, I am in agreement with Geus that Eratosthenes was supportive of his royal patrons and knowledgeable about his Egyptian surroundings.[253] While an ancient report that Eratosthenes translated Egyptian kingships lists has been challenged,[254] earlier scholars have credited him

---

[249] On this complex question, see Geus 2002: 211–223 and Pàmias and Geus 2007.

[250] On planetary distances see Geus 2002: 239–43 on Aetius 2.31.3 (*Doxographi Graeci* p.362.25–363.4 Diels); [Galen] *Historia Philosophia* 72, Stobaeus 1.26.5, Eusebius *Preparatio Evangelica* 15.53.3; on the wind rose see Geus 2002: 251–256 on Achilles Tatius *In Arati Phaenomena* 33 (68.25–26 Maass) and Vitruvius 1.6.9–11. Geus' other source is not credible for Eratosthenes: [Galen] *In Hippocratis De Humoribus* as printed in Kühn is a Renaissance forgery that skillfully excised large portions from genuine ancient sources; von Staden 1989: 65.n49.

[251] Geus 2002: 223–261 and Roller 2010 have both made collections of the fragments and testimonia of these works.

[252] Eratosthenes criticizes Callimachus in his geography at Eratosthenes *Geographica* fr. 8 Roller = Strabo 7.3.6 = 299C.21–22 (260.21–22) Radt "[Eratosthenes says] that others are forgiven but not Callimachus who lays claims to grammatical knowledge". I cannot find named references to other paradoxographers in Eratosthenes' work.

[253] On the Egyptian context see Geus 2002: 336–337; Pàmias 2004 takes contrapositions on some issues.

[254] Georgius Syncellus 103.9–10 *Ecloga Chronographica* says, on the authority of Apollodorus the chronographer, that Eratosthenes translated from the Egyptian a list of kings in Egypt "by means of Egyptian memoranda" (Αἰγυπτιακοῖς ὑπομνήμασι) into Greek. This would imply he read demotic or hieroglyphs. But the attribution ought to be dismissed: Eratosthenes' *On Chronologies* did not

with telling of Thoth's invention of letters in his poem *Hermes*, inspiring the Egyptian calendar reforms of Euergetes' Canopus Decree, incorporating Egyptian mythology into his *Astronomy*, and changing the hieroglyphic sign for Elephantine because of the measurement of the earth's circumference; I credit him further with focusing on Egyptian examples in his *Geography* and directing his engineering accomplishments to the native village economy of Greco-Egyptian scribes under the direction of the *dioiketes*.[255] In chapter 4 I will more precisely describe Eratosthenes' scientific work in relation to its Egyptian context. More generally, Eratosthenes' call to Alexandria, his numerous literary dedications to Ptolemaic monarchs, and his position as head of the library and possible *tropheus* to Philopator suggest the mutual exchange of expert knowledge and social respect paradigmatic for gift-exchange within the framework of court science. For his scientific success, his centrality to social networks of power and other intellectuals, and his embeddedness in his Greco-Egyptian context in Alexandria, Eratosthenes was probably the most influential court scientist of the Ptolemaic period.

137 Conon of Samos (PP 16454; lacks TM no.) was a mathematician and astronomer living in Alexandria. His correspondence with Archimedes and patronage dedication of the constellation Coma Berenices to Euergetes have been described in the Introduction.

138 Dositheus of Pelusium (PP 16537; TM 7503) was an astronomer and mathematician. The name Dositheus means "god-given" and is typically Jewish; consequently Dositheus has been called the first Jewish scientist.[256] He became a mathematical correspondent of Archimedes when **no. 137** Conon died. The earliest correspondence between them is Archimedes' opening letter of his *Quadrature of the Parabola*.[257] The introduction from Archimedes to Dositheus supposes that Do-

---

have any beginning before the Trojan War. See Fraser 1972: 2.487.n182. Geus 2002: 311–312 argues weakly against Fraser.

255 On *Hermes* see *SH* 397; on the Canopus Decree see Geus 2002: 207–210; on Eratosthenes' sky catalogue of mythological figures, where "ship of Osiris?" (πλοῖον) is preserved, see Geus 2002: 219.n50, 221.n62; on the change in hieroglyphs see Winter 1987, Geus 2002: 259.n237; for Egyptian examples in his geography see Eratosthenes *Geographica* fr. 15 Roller = Strabo 1.3.4 = 49C.10–50C.20 (122.1–124.20) Radt; for Eratosthenes in conjunction with the village economy see below chapter 4. The last two examples are my own.

256 Netz 1998. Dositheus' ethnic indicates that he was from Pelusium, the fortified border town on the easternmost branch of the Nile Delta. Yet Ptolemy' *Phaseis* 67.4 (see Heiberg's wish to change his text 1907: CLIII.n1) attributes Dositheus with astronomical observations made in Colonia. This Colonia, wherever it was (Ptolemy has given its Roman name), was on a similar line of latitude to Phocis, since the Ptolemy goes on to equate the hours of sunlight at spring equinox in those places. At some point in his career Dositheus traveled outside of the Ptolemic empire.

257 Archimedes *Quadratura Parabolae* 262.2–13 "Archimedes to Dositheus, greetings. When I heard that Konon had died, who was never wanting in friendship toward me, and that you were a friend of Konon and a friend of geometry, I grieved for his death because he was a good man and something special in mathematics. I have undertaken to send to you in writing (just as we used to

sitheus resided in Alexandria and was a well-enough known member of Conon's circle that his name was transmitted to Archimedes with Conon's death: the inference seems assured that Dositheus was Conon's student. Whatever Archimedes' evaluation of Dositheus,[258] he continued to send him mathematical correspondence. Four treatises, including *Quadrature of the Parabola*, addressed to Dositheus from Archimedes survive.

Dositheus must have been a mathematician of some quality. The mathematician Diocles' *On Burning Mirrors*, preserved only in Arabic, shows that Dositheus solved a variant of a problem about focalized parabolas which was initially posed to Conon.[259] Dositheus was the first mathematician to recognize the fundamental property of the parabola, that there is a point (now called the focus) at which all rays reflected from the parabola meet. Diocles says that Dositheus solved the problem practically: this seems to mean only that Dositheus did not give a formal Euclidean proof, not that he constructed a parabolic mirror.[260] It seems plausible that Dositheus offered some sort of mathematical argument, possibly derived from mechanical models, analogous to the non-Euclidean proofs offered by his contemporaries Archimedes, **no. 136** Eratosthenes, and Nicomedes.[261]

Dositheus is known to have written at least four works: *Appearances of the Fixed Stars*, *Weather-signs*, *On the Octaeteris of Eudoxus*, and *To Diodorus*.[262] The first and second were astronomical data about the stars and seasons; parts of Dositheus' measurements are preserved in the later astronomers Ptolemy's *Phaseis* and Geminus' *Introductio*. The third work, on Eudoxus' so-called *octaetereis* or eight-year calendar cycle, is also a subject of a work by Eratosthenes; whosever book was published first, the second book was probably an agonistic treatment of the first. Mention of Dositheus' work *To Diodorus* is preserved among the lives of Ara-

---

write Konon) some geometric theorems, which earlier were not investigated but now have been investigated by me; they were discovered first by mechanics then later proven through geometry."
258 Netz 2004: 34 has argued Archimedes' forms of address to Dositheus elsewhere show that Dositheus was not the same caliber mathematician as Conon.
259 Diocles *On Burning Mirrors* 3–6 "Python the Thasian geometer wrote a letter to Conon in which he asked him how to find a mirror surface such that when it is placed facing the sun the rays reflected from it meet the circumference of a circle. And when Zenodorus the astronomer came down to Arcadia and was introduced to us, he asked us how to find a mirror surface such that when it is placed facing the sun the rays reflected from it meet a point and thus cause burning … One of those two problems, namely the one requiring the construction of a mirror which makes all the rays meet in one point, is the one which was solved practically by Dositheus."
260 Toomer 1976: 140 comments: "My translation 'solved practically' is meant to imply, not necessarily that Dositheus actually constructed a parabolic mirror, but that he stated the focal property of the parabola, perhaps without giving a formal proof."
261 I discuss Eratosthenes' and Archimedes' mechanical mathematics in chapters 4 and 5. On Nicomedes see Netz 2004: 298–306.
262 *RE* 5.2: 1607–1608 *s. v.* 'Dositheus (9)'; *EANS* 277.

tus and seems to have at least partially treated the court life of Aratus, the astronomical poet.

> Dositheus the Pelusian says in *To Diodorus* that Aratus went to Antiochus son of Seleucus and resided with him (διατρῖψαι παρ' αὐτῷ) a sufficient time.²⁶³

The formulation transmitted of Dositheus, διατρῖψαι παρ' αὐτῷ, is expressed in the technical language (διατρῖψαι παρά τινι βασιλεῖ) for a *philos* or honored guest to reside for an unspecified but extended period of time at the court of a monarch.²⁶⁴

Given Dositheus' contacts with Archimedes, his association with Conon, and his apparent agonism with Eratosthenes, it is reasonable to infer that he served the court of Euergetes and Philopator in the same capacity as Conon.²⁶⁵

139 Philon of Byzantium (PP 16561; TM 14708) is one of the major surviving writers of Greek mechanics.²⁶⁶ He wrote a nine book treatise called the *Mechanical Collection*, of which only four books survive: *Artillery Constructions, Pneumatics, Siege Preparations*, and *Siege-Crafts*. The missing parts seemingly treated levers, harbors, and more complicated machines. The treatise focuses on practical devices useful for war. Philon spent extensive time spent with engineers in both Alexandria and Rhodes and expresses a direct appreciation for the work of the Alexandrian mechanician Ctesibius.²⁶⁷ Ctesibius was a master mechanician of the previous generation who was active under Philadelphus and dedicated a mechanical musical drinking horn in the temple of the deified Arsinoe-Aphrodite-Zephyritis.²⁶⁸ It is likely that Philon was active in Alexandria under Euergetes and Philopator.

Philon's opening to the *Artillery Constructions* recognizes the importance of royal patronage in building complicated mechanical devices. After a historical discussion about the proportional size of the hole that holds the tension-spring, the central concept in the building of effective ancient artillery, Philon discusses the methodological steps that lead from trial-and-error to established mathematical principles.

---

263 *Vita Arati* 3 = 16 Martin: Δωσίθεος δὲ ὁ Πηλουσιακὸς* ἐν τῷ πρὸς Διόδωρον ἐλθεῖν φησιν αὐτὸν [sc. Ἄρατον] καὶ πρὸς Ἀντίοχον τὸν Σελεύκου καὶ διατρῖψαι παρ' αὐτῷ χρόνον ἱκανόν. At the sign (*) I have written Πηλουσιακὸς for the transmitted πολιτικὸς, in accord with *Pelusinus* of the Latin translation. Martin's 1974 *apparatus* accepts this interpretation of the Greek.
264 Herman 1980/1: 106.
265 No historical evidence names him court astronomer, although no evidence names Conon court astronomer either. The inference that **no. 137** Conon was court astronomer depends on **no. 110** Callimachus' recognition of his catasterism of Berenice's lock.
266 See *EANS* 654–656; Roby 2016.
267 Philon in Alexandria and Rhodes *Belopoeica* 51.10–13 Th; appreciation for Ctesibius *Belopoeica* 56.14 Th, 67.28–68.3 Th.
268 For Ctesibius' dedication of a musical *rhyton* see Hedylus 4 *HE*.

Some of the ancient engineers discovered that the element, beginning, and measure of the machines' construction was the diameter of the hole [sc. that receives the tension-spring]. ... It was not possible to know this except by experimentally increasing and decreasing the perimeter of the hole. ... Later engineers, reflecting on former mistakes and looking for a standard element from subsequent trials, brought on the principle and basis of construction, I mean the diameter of the perimeter that holds the tension-spring. Those craftsmen in Alexandria accomplished this first, having gotten large financial support on account of the fact that they secured the patronage of kings who were friends-of-renown and friends-of-the-*technai*.²⁶⁹

Philon discusses the need for practical considerations alongside theoretical methods to advance ballistics knowledge. The attitude of generic crossing between "pure" and "applied" science reflects the court aesthetic of hybridization seen in the Ptolemaic propagandists of a Greco-Egyptian culture. Further, Philon's use of the language of patronage – finances, honor, friendship – parallels the patronage of other known court actors.²⁷⁰

140 An unnamed Phoenician mechanician (PP 16562; TM 14777) designed a dry-dock with which to launch Philopator's forty-banked ship described by **no. 119** Philostephanus.²⁷¹ This dry-dock was evidently located in Alexandria's harbor on the Mediterranean, perhaps on the preferred Great Harbor on the eastern side of the Heptastadion that attached the island of Pharos to the mainland.²⁷²

141 Andreas of Carystus (PP 16574; TM 4909), doctor and literary polemicist, is a central figure in this book.²⁷³ Personal physician to Philopator, he was murdered just before the battle of Raphia in 217 BCE during a failed assassination attempt on the king.²⁷⁴ The assassination ended a life of close association with the intellectual and political elite of Alexandria.

---

269 Philon *Belopoeica* 50.14–28 Th: ἐπὶ γὰρ τῶν ἀρχαίων τινὲς ηὕρισκον στοιχεῖον ὑπάρχον καὶ ἀρχὴν καὶ μέτρον τῆς τῶν ὀργάνων κατασκευῆς τὴν τοῦ τρήματος διάμετρον ... οὐκ ἄλλως δὲ ἦν ταύτην λαβεῖν, ἀλλὰ ἐκ πείρας αὔξαντάς τε καὶ συναιροῦντας τὸν τοῦ τρήματος κύκλον ... τοὺς δὲ ὕστερον ἔκ τε τῶν πρότερον ἡμαρτημένων θεωροῦντας καὶ ἐκ τῶν μετὰ ταῦτα πειραζομένων ἐπιβλέποντας εἰς ἑστηκὸς στοιχεῖον ἀγαγεῖν τὴν ἀρχὴν καὶ ἐπίστασιν τῆς κατασκευῆς, λέγω δὲ τοῦ κύκλου τὴν διάμετρον τοῦ τὸν τόνον δεχομένου. τοῦτο δὲ συμβαίνει ποιῆσαι τοὺς ἐν Ἀλεξανδρείᾳ τεχνίτας πρώτους μεγάλην ἐσχηκότας χορηγίαν διὰ τὸ φιλοδόξων καὶ φιλοτέχνων ἐπειλῆφθαι βασιλέων. Schiefsky 2015 discusses the issue at length in Philon's work.
270 χορηγία cf. Diodorus Siculus 3.18.4 on **no. 11** above; on *philos*-compounds see chapter 2.
271 Athenaeus 5.204c–d; see the discussion in *BNJ* 627 (Kallixeinos of Rhodes) F1.37.
272 Fraser 1972: 1.21–22 shows that the eastern harbor was the preferred ancient harbor.
273 I cite the fragments and testimonia of Andreas as vS according to the list of von Staden 1989: 472–477. I have located two additional testimonia: Galen *Subfiguratio Empirica* 8 = Empiricist fr. 10b, p.69.10–20 Deichgräber, which I call *An.*47b, and Oribasius *Collectiones Medicae* 49.4.53–55 = 4.9–10 Raeder, which I call *An.*16b. I agree with von Staden that medical fragments and testimonia referring to "Andreas the doctor", "Andreas of Carystus", and "Andreas the Herophilean" all point to the same individual, to be distinguished from Andreas ὁ κόμης; see Calà 2012.
274 *An.*1 vS = Polybius 5.81. The later title *archiatros* still did not exist at this time.

He was from Carystus, on the island of Euboea, son of Chrysareus.[275] Like most immigrant intellectuals to Egypt Andreas did not take Alexandrian citizenship.[276] He was a member of the Herophilean medical sect.[277] Since the early Herophileans were entirely based in Alexandria, it is likely that Andreas received his medical education in the city. Herophilus gathered a group of medical students around him who lived in his house; he died sometime between 260–250 BCE.[278] It is unclear whether Andreas was a direct student of Herophilus or a student of Herophilus' students. The Hellenistic medical sects were loose ideological affiliations who agreed broadly on the appropriate areas of medical activity and the principles to study them.[279] Andreas' Herophilean leanings manifested themselves in pharmacology, gynecology, surgery, and medical doxography, typical areas of medical activity for the Hellenistic Herophileans.[280] We know five of Andreas' titles all only one book long: *To Sosibius, On Animal Bites, Casket* (on plants and pharmacology), *On Things Falsely Believed, On Medical Genealogy.*[281] In later medical tradition Andreas was most renowned for his work on pharmacology and a surgical instrument he designed for resetting bones: we possess 22 testimonia about Andreas' pharmacology and 9 testimonia relating to Andreas' machine.[282]

During his own lifetime Andreas was certainly socially embedded and prominent at court. He named a particular drug after the god Serapis, the hybrid Greek-Egyptian god promoted by the Ptolemies as a unitary object of worship for different ethnic groups in the kingdom;[283] he wrote about the deliberate dilution of black poppy at Alexandria;[284] he quoted the rose-compound of his contemporary, the Alexandrian doctor Nileus, who had improved upon Hippocrates' machine for resetting bones.[285] Andreas also wrote a treatise dedicated to **no. 15** Sosibius, Philopator's most powerful *philos*.

---

275 *An.*5 vS = Cassius Iatrosophista *Problemata* 58; *An.*38 vS = Galen *Glossarium s.v.* Ἰνδικόν 19.105K.
276 Fraser 1972: 1.307.
277 *An.*9 vS = Soranus *Gynaecia* 4.1 = 131.4–7 Ilberg; *An.*22 vS = Celsus *De Medicina* 5.pr.1.
278 See von Staden 1989: 43–50 for Herophilus' dates.
279 Berrey 2014a, von Staden 1982.
280 von Staden 1989: 448–57, von Staden 1999a.
281 *An.*9, *An.*18, *An.*25, *An.*45, *An.*47 vS.
282 Pharmacological testimonia: *An.*18–19, 21–24, 26–40, 48–49 vS; testimonia about the surgical machine: *An.*11–17 vS. Chapter 4 discusses Andreas' machine.
283 *An.*37 vS = Dioscorides *De Materia Medica* 3.126–7 "There is another *orchis*, which some people call Serapis' just as Andreas on account of the usefulness of its root".
284 *An.*34 vS = Pliny *Historia Naturalis* 20.76.200: "Andreas adds, that one is not so directly blinded by [the poppy], since its strength is adulterated at Alexandria".
285 *An.*32 vS = Asclepiades Iunior *apud* Galen *De Compositione Medicamentorum secundum Locos* 12.765K: "The rose compound of Neileus: just as Andreas says it is useful against great pain, a discharge that is a lot but weak, pustules, and [uterine] prolapses".

> Andreas in his *To Sosibius* (it's a letter) agrees with the Herophileans adding only that the disabled embryo is also thin: "Those embryos that do not have weight produce difficult births."[286]

Two other testimonia of Andreas concern procreation;[287] perhaps they too belonged to *To Sosibius*. The medical content shows that the letter was not social pleasantry to Sosibius but rather some sort of display of Andreas' learning in a dedication. Since Sosibius had a prominent daughter who was later the eponymous *kanephoros* of Arsinoe Philadelphus in 215/4, it is possible to give Andreas' *To Sosibius* a biographical reading as an advisory letter about the best times and means of procreation for Sosibius' daughter. Nonetheless, I would see Andreas' *To Sosibius* within the context of gift-exchange. The epistolary genre matches surviving court science treatises. Further, Soranus' comment that Andreas agrees with the Herophileans suggests that Andreas positioned his medical advice within an intellectual tradition, a literary technique found in all court science treatises.[288] So I suggest that Andreas addressed a court science treatise to Sosibius within a social context of gift-exchange.

Andreas' literary polemics took as their objects both contemporaries and prominent beliefs. Andreas wrote a treatise called *Things Falsely Believed* in which he attacked **no. 113** the poet Archelaus' "Unique Natures".[289]

> Archelaos says in *Idiophye* that sea-eels, coming up onto land, mate with vipers and have teeth similar to vipers, but Andreas says that this is false: that the sea-eel does not live on the beach nor does the viper.[290]

The rationalism of Andreas' text refutes Archelaus' paradoxography and pointedly contradicts the court poet of the earlier generation: category boundaries firmly resist the aesthetics of mixing. Andreas also refuted the claim that the sight of a particular bird cures jaundice.[291] Despite the appearance of rationalism in his fragments Andreas also propagated some dubious animal lore. In one instance he claimed that the blood of the salamander (reputed to extinguish fire) made clothes

---

286 *An*.9 vS = Soranus *Gynaecia* 4.1 = 131.4–7 Ilberg: ὁ δὲ Ἀνδρέας ἐν τῷ Πρὸς Σωσίβιον (ἐστὶν δὲ ἐπιστολικόν) τοῖς ἀπὸ Ἡροφίλου συντίθεται μόνον προσθεὶς τὸ παραλελυμένον ἔμβρυον καὶ ἰσχνόν· Τὰ γὰρ τοιαῦτα, φησί, βάρος μὴ ἔχοντα δυστοκίαν ἐργάζονται.
287 *An*.7, 8 vS.
288 van der Eijk 1999 notes fewer doxographies in Soranus than usually thought. Chapter 3 discusses generic features of court science treatises.
289 *An*.45, 46 vS.
290 *An*.46 vS = Σ.Nicandrum *Theriaca* 823: Ἀρχέλαός φησιν ἐν τοῖς Ἰδιοφυέσι προιούσας τὰς μυραίνας τοῖς ἔχεσι μίγνυσθαι, ἔχειν δὲ ὀδόντας ὁμοίους ἔχεσιν. φησὶ δὲ ψευδὲς εἶναι καὶ Ἀνδρέας καὶ μήτε τὴν μύραιναν προιέναι μήτε τὸν ἔχιν παρ' αἰγιαλὸν διατρίβειν.
291 *An*.44 vS.

or hands impervious to fire when smeared on them;[292] in another case he said that the bites of sea-eels annul those of vipers.[293] Since poisonous animals were a frequent topic of Hellenistic medical authors' pens, it is possible that Andreas was drawing on either of two popular sources of the previous generation. Apollodorus the *Theriakos* ("Bestiary"), who was active 280–240 BCE, may have written a book for a king Ptolemy on wines and certainly wrote on salamanders and other poisonous animals;[294] Numenius of Heraclea, who also flourished 280–240 BCE, wrote two works in verse on materia medica.[295] At the least Andreas' mix of rationalism and animal-lore evokes the natural world as the same source of entertainment and spectacle known from the contemporary court paradoxographers.

A further example of Andreas' appropriation of contemporary literary aesthetics comes from *On the Genealogy of Physicians*. Of this work only one testimonium is preserved.

> Having trained in medicine and liberal arts, once his parents died, Hippocrates emigrated from his native land, as the wicked Andreas says in *On Medical Genealogy*, on account of having set fire to the archives in Cnidus.[296]

The surviving ancient biography of Hippocrates attributed to Soranus of Cos records this malacious slander from Andreas. Andreas' work seems to have entered common circulation, since the Roman antiquarian Varro also claimed that Hippocrates established bedside medicine after burning down the books kept in the temple on Cos.[297] It is unclear whether Andreas' mention of Cnidus is a fault in transmission (perhaps we should emend to 'Cos') or the remnant of a larger narrative *aition* in which Hippocrates' scorn of Cnidus explains the supposed medical dispute between Cos and Cnidus found in the Hippocratic writings. It is also unclear whether Andreas discussed physicians other than Hippocrates. While the younger contemporary of Andreas, Sotion of Alexandria had written a book called *Succession of the Philosophers*, Andreas' title *On Medical Genealogy* relates to an older literature dedicated to unraveling family lineages.[298] To be sure, earlier physicians had writ-

---

[292] *An.*21 vS.
[293] *An.*18 vS.
[294] *EANS* 106. Jacques 2002: 285–92 contains the extant fragments. Apollodorus fr. 16 Jacques = Pliny *Historia Naturalis* 22.13.31 concerns the medical use of salamanders, mentioned in *An.*21 vS.
[295] *EANS* 583; *SH* 568–596 preserve the extant fragments. *SH* 583 mentions the σκολόπενδραν, also mentioned in *An.*25 vS.
[296] *An.*47 vS = Soranus *Vita Hippocratis* 4: συνασκηθεὶς δὲ ἐν τῇ ἰατρικῇ καὶ τοῖς ἐγκυκλίοις μαθήμασι τῶν γονέων αὐτοῦ τελευτησάντων μετέστη τῆς [ἰδίας] πατρίδος, ὡς μὲν κακοήθως Ἀνδρέας φησὶν ἐν τῷ Περὶ τῆς ἰατρικῆς γενεαλογίας, διὰ τὸ ἐμπρῆσαι τὸ ἐν Κνίδῳ γραμματοφυλακεῖον.
[297] Pliny *Historia Naturalis* 29.1.4.
[298] The most obvious example is Pherecydes of Athens; *BNJ* 3 (Pherekydes of Athens) F59 is Pherecydes' genealogy of Hippocrates. von Staden 1999a: 149–156 has an excellent discussion of Andreas' work and its predecessors, including Pherecydes.

ten biographies of their contemporaries.²⁹⁹ The antiquarian task of writing a biography of Hippocrates fell to Andreas. In this light, Andreas' biographical lineage of physicians seems like his contemporary **no. 118** Hermippus of Smyrna's antiquarian biographies.³⁰⁰ Massar has argued that these works served a public hungry for a popular history of a *technē* told through the lives of its inventors.³⁰¹ Andreas and Hermippus' works were certainly different histories of a field than the Peripatetic histories of discoveries from a century earlier, such as Eudemus' history of mathematics or Meno's history of medicine. Andreas' polemical antiquarianism deflated any claims that Hippocrates' medical knowledge was the product of long clinical observations rather than book learning; whether it correspondingly deflated Hippocrates' status as a hero of Greek wisdom is unclear. Certainly Andreas' impugning of Hippocrates earned him the later ire of the physician Galen. He called Andreas hybristic and compared him to a herald describing a run-away slave without ever seeing him, as if knowledge acquired by empirical observation was not Andreas' *forte*.³⁰²

Finally, the most intriguing example of Andreas' centrality in Alexandrian social circles and his literary agonism concerns **no. 136** Eratosthenes. A late Byzantine dictionary and its derivations preserve evidence of a literary quarrel between Andreas and Eratosthenes:

> Andreas the physician was surnamed Book-Aegisthus (βιβλιαίγισθος) by Eratosthenes, because he secretly rewrote his books.³⁰³

---

299 Dionysius of Ephesus had written *A Record of Physicians*, including his contemporaries Erasistratus of Ceos and Nicias of Miletus; *EANS* 263. Bacchius, Herophilus' direct student, wrote a book *Memoirs of Herophilus and Those of his House*; Galen *In Hippocratis Epidemiarum Librum VI* 17B.145K = 203.18–26 Wenkebach. But in spite of the fact that he had written commentaries' on Hippocrates' works, Bacchius did not apparently write a biography of Hippocrates.
300 I make the comparison without trying to establish a chronological priority. Certainly Hermippus' researches were possible because of the Alexandrian library; but then Hermippus included no medical figures – Pythagoras (*FGrH* 1026 F1, 21–27) and Thales (*FGrH* 1026 F10, 13, 17) and Eudoxus (*FGrH* 1026 F9), scientific figures yes, but no medical figures such as Philistion or Hippocrates. Eudoxus was a physician as well as astronomer; importantly *FGrH* 1026 (Hermippus of Smyrna) F9 = Diogenes Laertius 8.88 does not recognize his medical works. Yet Hippocrates' works were gathered in the Alexandrian library; Berrey 2015: 472–473.
301 While Massar 2005: 215–220 sees the parallel of contemporary grammarians writing biographies, she would place Andreas' and other writers' genealogy of Hippocrates alongside the Hippocratic pseudepigrapha written during the Hellenistic period, such as the *Letters* and *Embassy*.
302 *An*.47b vS = Galen *Subfiguratio Empirica* 8 = Empiricist fr. 10b, p.69.10–20 Deichgräber; *An*.24 vS = Galen *De Simplicium Medicamentorum Temperamentis ac Facultatibus* 11.795–796K.
303 *An*.2a vS = *Etymologicum Magnum* 198.12–21 s. v. βιβλιαίγισθος· Ἀνδρέας ὁ ἰατρὸς ἐπεκλήθη ὑπὸ Ἐρατοσθένους, ὅτι λάθρα αὐτοῦ τὰ βιβλία μετέγραψε. Compare *An*.2b vS for other paraphrases of this statement derived from other Byzantine lexica associated with the *Etymologicum Magnum*.

The picturesque coinage βιβλιαίγισθος "Book-Aegisthus" is typical of Eratosthenes' style.[304] Aegisthus was the mythological seducer of Clytemnestra, wife of Agamemnon, the leader of the Greek forces in the Trojan War. Andreas apparently seduced Eratosthenes' books, i.e. he plagiarized Eratosthenes' content and claimed it as his own. There is only one strange testimonium about the overlap in content between Andreas and Eratosthenes: Hippocrates' genealogy. Eratosthenes, like Andreas, wrote a genealogy of Hippocrates – but who imitated who?[305] A whole host of biographical speculation could be ill-founded on this testimonium; for our purposes it suffices to show once again the agonistic conflict between court scientists and confirms that court scientists read each other's work. Eratosthenes' polemic suggests that Andreas' authorial technique integrated prior discourses and knowledge into his own work while disputing previous authorities.

While we know of other physicians active in Alexandria during the reigns of Euergetes and Philopator, none appears to have been a royal physician to the king.

People missing from the historical record

A descriptive prosopography of historical actors is the most accurate and detailed way of presenting an image of court society but it possesses important limitations as well. There were undoubtedly many other people at the Ptolemaic court in the years between 246–205 BCE than those mentioned above. It is safe to assume that the kings had more *philoi*, more *strategoi*, more priests, bureaucrats, and intellectuals at their courts than the names we possess. In addition to the known functionaries named above at the court, there were people whose social function but not names are known from the historical record. They include people with titles like οἱ περὶ τὴν αὐλὴν νεανίσκοι "the royal pages", a social group of elite youths probably drawn from established Ptolemaic families; the σωματοφύλακες "the royal bodyguard", the elite heads of the larger household troops, the θεραπεία; and the individuals in charge of certain domestic duties in the court, such as the ἀρχεδέατρος "chief steward", the ἀρχιθύρωρος "chief door-keeper", the ἀρχικύνηγος "chief

---

[304] Geus 2002: 298.n72 notes previous attempts to emend ὑπὸ Ἐρατοσθένους into ὑπὸ Ἐρασιστράτου, but the emendation is chronologically impossible since Erasistratus will have died by 230 BCE. Erasistratus will have lived long enough (into the late 240s or 230s BCE) to controvert the emerging Empiricist sect; Erasistratus fr. 35 Garofalo = Ps.-Dioscorides *De Iis Quae Virus Eiaculantur Animalibus* 26.49K.

[305] On Eratosthenes' genealogy for Hippocrates see Soranus *Vita Hippocratis* 1. I cannot share Geus' 2002: 298 speculation that Andreas plagiarized parts of Eratosthenes' *On Ancient Comedy*. The overlap concerning Hippocrates makes even less sense, although von Staden 1999a: 156 maintains that Andreas plagiarized Eratosthenes' historical work *Chronology*.

huntsman", and the ὁ πρὸς ταῖς ἡνίαις "stable-master".[306] While these particular titles are better established in later periods than the third century BCE, it seems reasonable that various social groups existed during 246–205 BCE, perhaps without formal title. From other sources we hear of different ranks of soldiers and guards, unnamed jesters and professional flatterers (Diphilus' hungry men, perhaps?), heralds, cooks, and other free people at the court, all undoubtedly below the elite heads of the private staff and official bureaucracy.

Finally, there were enslaved people at the court; their names are completely absent from the historical record. **No. 41** Theodotus the steward alone of all the people recounted in this chapter does not belong to the political or cultural elite. His name, preserved on funerary vases, comes from historical knowledge gained through an archaeological context. By contrast, our knowledge of almost everyone else recounted in this chapter comes from literary records in papyrological, epigraphic, and manuscript tradition. Literary records were produced outside of Alexandria and circulate in space and time; people who had the ability to travel and write in Ptolemaic Alexandria were mostly bureaucrats or elite. Given the circumstances of historical preservation, it is not surprising that the names and activities of elite individuals are better preserved in literary remains than the names and activities of lower classes are preserved in archaeological remains. The vast number of lower class people responsible for the details of quotidian life in the usual palace complex – slaves, food purveyors and cooks, carpenters and potters, builders and contractors, weavers and washers, fire-stokers and smiths, and other people both free and enslaved – have disappeared unnamed from the historical record. They were not the social focus of the court scientists and so will not figure in the present book, but life at the court of Ptolemy 246–205 BCE would have been impossible without them providing the human labor underlying a cultured society of gift-exchange.

Appendix: Doubtful and Excluded persons

142 Callicrates (lacks PP no.), flatterer of a king Ptolemy, carried around a signet ring with the image of Odysseus and gave the names of Odysseus' more obscure children to his own children.[307] The transmitted text of this passage says that Callicrates was the flatterer of Ptolemy III Euergetes. But the historian Euphantus of

---

[306] See Fraser 1972: 1.102 for this list with further references. Bikerman 1938: 36–40 assembles the evidence from the Seleucid court; Strootman 2014: 165–172 has an expansive notion of titulature, with particular responsibilities associated with titled groups. On the "royal pages" see Strootman 2014: 136–144.
[307] *BNJ* 74 (Euphantus of Olynthus) F1A = Athenaeus 6.251d.

Olynthus, from whom information about Callicrates was derived, was already dead by the middle of the third century. Hence most scholars have emended the text so that Callicrates was the flatterer of Ptolemy I Soter.

143 Abdaraxos of Alexandria (PP 16522; TM 3114) appears in a papyrus listing of famous mechanicians; said to have "completed the machines in Alexandria", he cannot be dated more precisely than the third century BCE.[308]

144 Hippe (PP 14725; TM 9392), a courtesan named in the poems of the poet Machon, probably was a courtesan under Ptolemy II Philadelphus rather than Ptolemy IV Philopator.[309]

145 Chares (PP 16891; TM 14860) was a grammarian, said to be a student of Apollonius (of Rhodes?), who wrote *On the Histories of Apollonius*.[310] We know too little to place him securely in Alexandria.

146–150 Five individuals – Dalion, Aristocreon, Bion, Basilis, and Simonides the younger (all lack PP nos.) – are Hellenistic writers who recorded the distance along the Nile between Syene in Upper Egypt and Meroe in Ethiopia.[311] The common suggestion is that all traveled before the Egyptian revolt of 205 BCE closed off Greek access to the upper Nile. While they all seem to have been active during the third century BCE, they do not appear to have been connected to the court.

151–152 Aristocrates of Alexandria (PP 14591, TM 5813) and Ptolemaeus of Alexandria (PP 14623, TM 12846) are brothers known from an inscription on the island of Thera. They are proclaimed *philoi* of the king and honored by him. The stone cannot be dated any more precisely than the forty years covering the reigns of Ptolemy IV Philopator and Ptolemy V Epiphanes. Since the evidence is ambiguous, I exclude Aristocrates and Ptolemaeus as *philoi* of Philopator.

153 Mnaseas of Patara (PP 16937; TM 10461), a rationalizing geographer and mythographer, was said to be the student of **no. 136** Eratosthenes.[312] There is no reason or evidence to connect Mnaseas with the court of Philopator, although nothing excludes a stay in Alexandria.

154 Menander (PP 16870; TM 10337) was said to be a student of **no. 136** Eratosthenes.[313] This Menander might be identified with the Menander used by Jose-

---

[308] *EANS* 29.
[309] Machon fr. 18 Gow = Athenaeus 13.583a–b.
[310] Σ.Apollonium 2.1052–1057. *RE* 3.2: 2130 *s. v.* 'Chares (14)'.
[311] Pliny *Historia Naturalis* 6.29.183 Mayhoff. Their fragments are collected as *BNJ* 666 (Dalion), *BNJ* 667 (Aristokreon), *BNJ* 668 (Bion Soleus), *BNJ* 669 (Simonides, de Aethiopia), and *BNJ* 718 (Basilis).
[312] *BNJ* 241 (Eratosthenes of Cyrene) T1 = *Suda* ε2898; cf. Geus 2002: 45–46.
[313] *BNJ* 241 (Eratosthenes of Cyrene) T1 = *Suda* ε2898; cf. Geus 2002: 45.

phus.³¹⁴ Josephus' Menander was a Greek writer on Phoenician antiquities. If these two sets of testimonia belong to the same individual, there is no reason to suppose a connection to the Ptolemaic court.

155 Aristis of Cyrene (PP 16830; TM 5776) is said to be the student of **no. 136** Eratosthenes by the *Suda*. Nothing is known of him nor is there any reason to connect him with the court of Philopator.³¹⁵

156 The historian Satyros of Alexandria (PP 16948; TM 13311) has been thought to be a court historian, writing on the demes of Alexandria with a *flourit* sometime under Philopator. But Gambetti has challenged part of this view.³¹⁶ Satyros did write on the genealogy of Philopator but treated the demes of Alexandria only insofar as they related to Philopator's revised genealogy, not as a thematic principle of his work. Satyros cannot be dated more precisely than after Philopator.

157 Satyrus of Callatis (PP 16948; TM 13311), a biographer, appears not to have lived or been associated with Alexandria. Despite sharing the same PP and TM numbers as Satyrus of Alexandria, Satyrus of Callatis is to be distinguished from him.

158 Bolos of Mendes (PP 16740; TM 6711), a native Egyptian writing in Greek, presents a particularly difficult case.³¹⁷ He wrote a series of works on geography, astronomy, and medicine that he ascribed to Democritus, the famous fifth-century philosopher. Bolos' work bears close resemblance to the paradoxographers and antiquarian biographers. Thus he seems to follow in the footsteps of **nos. 117–120** the Callimacheans in Alexandria. In this respect he probably dates to the early second century BCE, after the end of Philopator's reign.

---

314 *BNJ* 783 (Menander).
315 *BNJ* 241 (Eratosthenes of Cyrene) T1 = *Suda* ε2898; cf. Geus 2002: 46.
316 *BNJ* 631 (Satyros of Alexandria).
317 *BNJ* 263 (Bolos of Mendes); Fraser 1972: 1.440–444.

2 Kingship, Symposia, Gift-Exchange: Parameters of Friendship

> From Zeus let us begin and you, Muses, end at Zeus, when we sing the best of the gods in song. Of men moreover let Ptolemy be said foremost, and last and middle: for he is the best of men. Heroes, who came earlier from the demigods, attained wise songs for doing noble deeds. Then I, knowing how to say noble things, would praise Ptolemy – for praise songs are also a gift of the immortals themselves.[1]

The poet Theocritus opens his praise poem to the pharaoh Ptolemy II Philadelphus, written between 279–270 BCE, structured in a priamel of descending levels of divinity – gods, heroes, men – that echoes archaic Greek praise poetry.[2] Gods and heroes have been praised in song before; so Ptolemy, best of men, is a suitable subject for praise poetry. Greek poetry made into a literary *topos* that men's enduring fame achieved in song is analogous to praise given to gods.[3] Literary fame is a gift. Theocritus' opening praise of Ptolemy skillfully evokes the language of gift-exchange, especially playing on whether Ptolemy's gift is suitable for a mortal, hero, or god. Later in the poem Theocritus praises Ptolemy for liberality with his wealth, for Ptolemy does not pile up his gold:

> The famed houses of the gods have much gold, since Ptolemy always offers first-fruits with other gifts, and he has given much gold to powerful kings, much to cities, much to his good friends. No man, who knows how to strike up a resonant song, leaves the holy contests of Dionysus without receiving a gift worthy of his art. The interpreters of the Muses sing of Ptolemy in exchange for his benefaction (ἀντ' εὐεργεσίης).[4]

Ptolemy's generous distribution of his wealth to the gods, political allies, cultural grandees, and friends fulfills the ideal of a generous, pious, and friendly king. In promise for this good turn (ἀντ' εὐεργεσίης) the poets sing in praise of Ptolemy.

---

1 Theocritus 17.3–8: Ἐκ Διὸς ἀρχώμεσθα καὶ ἐς Δία λήγετε Μοῖσαι, / ἀθανάτων τὸν ἄριστον, ἐπὴν †ἀείδωμεν ἀοιδαῖς· ἀνδρῶν δ' αὖ Πτολεμαῖος ἐνὶ πρώτοισι λεγέσθω / καὶ πύματος καὶ μέσσος· ὃ γὰρ προφερέστατος ἀνδρῶν. / ἥρωες, τοὶ πρόσθεν ἀφ' ἡμιθέων ἐγένοντο, / ῥέξαντες καλὰ ἔργα σοφῶν ἐκύρησαν ἀοιδῶν· / αὐτὰρ ἐγὼ Πτολεμαῖον ἐπιστάμενος καλὰ εἰπεῖν / ὑμνήσαιμ'· ὕμνοι δὲ καὶ ἀθανάτων γέρας αὐτῶν.
2 Theocritus alludes to Pindar *Olympia* 2.1–2. More generally on the opening of the poem, see Hunter 2003: 93 ff.
3 Cf. Theocritus 16.1–2: "It is always a concern for the daughters of Zeus to hymn the immortals in songs, to hymn the deeds of good men."
4 Theocritus 17.108–116: ἀλλὰ πολὺν μὲν ἔχοντι θεῶν ἐρικυδέες οἶκοι, / αἰὲν ἀπαρχομένοιο σὺν ἄλλοισιν γεράεσσι, / πολλὸν δ' ἰφθίμοισι δεδώρηται βασιλεῦσι, / πολλὸν δὲ πόλιεσσι, πολὺν δ' ἀγαθοῖσιν ἑταίροις. / οὐδὲ Διωνύσου τις ἀνὴρ ἱεροὺς κατ' ἀγῶνας / ἵκετ' ἐπιστάμενος λιγυρὰν ἀναμέλψαι ἀοιδάν, / ᾧ οὐ δωτίναν ἀντάξιον ὤπασε τέχνας. / Μουσάων δ' ὑποφῆται ἀείδοντι Πτολεμαῖον / ἀντ' εὐεργεσίης.

The social mechanism governing the interchange between kings and poets is clearly a gift-exchange.

Theocritus' poem throws open a window to gift-exchange at Hellenistic courts.[5] On the one side was a poet who controls enduring literary fame; on the other, a socially superior king, who (Theocritus claims) "could outweigh all kings with his wealth" and possessed 33,333 villages.[6] The idealized king is generous with his wealth to equals and subordinates. What but cultural services could a poet have offered to such a rich, powerful, and ideal king? Yet the asymmetry between poet and king is not only social but in types of services as well.[7] The potential exchange of literary praise and worldly goods between social unequals offers a reason why monarchs were eager to sponsor intellectuals at court.

As the evidence from Theocritus shows, gift-exchange was the mechanism of distribution between the king as a patron and producers of culture as clients. The mechanism of gift-exchange was embedded within a complex negotiation of social roles and cultural resources. To move immediately to the analysis of gift-exchange ignores the parameters important in establishing the gift-giving market. Important questions would remain unclear: How did kings choose which clients to sponsor and which gifts to accept? In what social setting did gift-exchange occur? What other forms did gift-exchange from intellectuals take? How did intellectuals use their knowledge of the natural world as a cultural resource to achieve political aims at court? In order to answer these questions, I am concerned to articulate the idealized interests of king and courtier in the gift-economy, their self-presentation, and the place of their interactions. Building on Massar's argument that Hellenistic physicians were agents of *paideia* or culture, my strategy is to implicate scientists who worked at court, including physicians, in the wider social codes, discourses, and cultural spaces of Hellenistic courtiers.[8] Since the argument implicates court scientists in the broader practices of courtiership, my chronological focus stretches from the end of the third century to the end of the first century BCE. This time

---

5 Hunter 2003: 24–45 carefully discusses literary patronage in Theocritus, particularly from *Idylls* 16 and 17, always underscoring the thin evidence. Hunter's 2003: 177–185 reading of the quoted passage picks up on the themes of royal *philanthropia* and friendship that will be highlighted in this chapter. For scholarship on gift-exchange see Fraser 1972: 1.305–314 (with discussion in chapter 1), Herman 1987, Mauss 1990, Weber 1993: 1, 87–95, Murray 2008, Strootman 2014: 145–184 and the discussion in Introduction footnote 15.
6 Theocritus 17.82–84, 95. See Hunter 2003: 158, Netz 2009a: 207.n51, who might also have used this example among his 2009a: 17–62 "carnival of calculation".
7 I emphasize this point against the otherwise perceptive remarks of Crook 2013, who regards non-material gifts as "fictive" for the asymmetric expectations of reciprocity they engender. I treat the ambiguity of expected reciprocity as central to a gift-economy within a hierarchical society.
8 Massar 2005: 103–122, 171–201 shows that physicians served Hellenistic kings as diplomats, ambassadors, and advisors of state alongside other intellectuals such as philosophers, and that medicine was accepted as a domain of *paideia*.

period includes our best evidence for the practices, sites, persons, and discourses of courtiership. We have epigraphic and papyrological evidence for titles and relative ranks of courtiers; the historian Polybius includes numerous examples of court deliberations; and Ps.-Aristeas' fictional *Letter to Philocrates* recounting the translation of the Septuagint under Ptolemaic patronage that best portrays court society probably dates to the second half of the second century BCE. So this chapter takes up the themes of kingship, symposia, and finally gift-exchange itself in turn, but I foreground the role of friendship before all else.

Scientific Selves, Scientific *Personae*

Historians of science have become familiar with the ways that scientific authors endow their work with moral authority. The author's ethical self-presentation – not a confession of creed but rather a discursive construction of characteristic habits, such as truth-telling, discrimination between alternatives, observational accuracy, honorific forms of address, credit to other participants, the recognition of subjectivity and personal bias – is the scientific self or scientific *persona*.[9] The scientific self endows believability in his or her embodied writing person to the observations, arguments, and conclusions in the text before the reader. The ethical self of the scientist is an historicized being, historically-situated in particular societies, with a historically-discursive construction given to the above facets of their moral self.

I call the *personae* of Hellenistic scientists "courtiers".[10] The terms "courtier" and "scientist" are anachronistic of course. The term scientist did not come into English usage until relatively late, only in 1834.[11] The term courtier passed into English from Norman French: it bespeaks a medieval world of nobles and attendants.[12] The realization that our chief words applied to ancient practitioners of investigations of the natural world at the court of Ptolemy are modern or medieval, not

---

9 I adopt the terms "scientific selves" and "scientific *personae*" from Daston and Galison 2007: 191–251.
10 There is a substantial argument whether Hellenistic court society fits the image of early modern court society drawn in Norbert Elias' *Court Society*: see Herman 1997 and Strootman 2014 for opposing views. With due allowance for the specific situations of individual monarchies, a reading of the primary evidence has firmly convinced me that the perspective of the *longue durée* is applicable to the social institutions of monarchy and kingship. It is not anachronistic to use terms such as 'courtiers', 'levée', and others drawn from the world of the early modern courts in reference to their ancient Hellenistic counterparts. The specifically Greek version of the communal meal at which the monarch dines with his subordinates, the symposium, emerges in my analysis as the specific cultural space for interaction between king and ancient scientist.
11 *OED s.v.* 'scientist' 1.
12 *OED s.v.* 'courtier' *n.1*, 1.

ancient, threatens to open up a gap between ancient and modern, as if the subject of our study exists only in our own minds, not in the world of our historical actors. Such reservations are necessary historical cautions but are effectively quibbles about nominalism. To call scientists other premodern designations – virtuoso scientist, Gelehrter, natural philosopher – is equally anachronistic by the standards of historicized language. Ancient scientists at court such as Eratosthenes and Archimedes did not describe themselves as courtiers or scientists, but they practiced investigations of the natural world at the court. "Scientists" and "courtiers" are their *personae*, even as we hasten to qualify and historicize these terms with regard to investigations of natural world at the court of Ptolemy.

Personae are not individual selves but cultural presentations. The issue of cultural nominalism raises the question: what did ancient Hellenistic scientists call themselves? There were well-established Greek words for practitioners of the individual sciences: ἰατρός "physician", ἀστρονόμος "astronomer, astrologer", μαθηματικός "mathematician, astrologer", μηχανικός "mechanician". There were also well-established words for the abstract investigation of nature: φυσικός "physicist", φιλόσοφος "philosopher". But there were no words for individuals who practiced multiple sciences, only characterizations of their learning: πολυμαθία "polymathy". Ancient Greek nominalism might suggest that no individuals worked across the sciences, on two or more different intellectual disciplines, or that there was even a recognized concept of the unified natural world which intellectual practice could pursue. These expectations are false. As the Introduction showed, there was a consistent notion of nature across intellectual disciplines in the Hellenistic period. Perhaps more interestingly, ancient Greek nominalism suggests that only philosophers (to which the φυσικοί "physicists" belonged) treated nature across different fields, or an abstract level. **No. 136** Eratosthenes for one acquired the title φιλόσοφος but he was also known by the disciplinary titles as well: μαθηματικός, ἱστορικός, γεώγραφος "mathematician, historian, geographer". These disciplinary distinctions imply that the ancient Greek cultural *persona* of the scientist remained closely affiliated to philosophy, as the parent discipline, and to the individual fields. So much for the *persona* of the Hellenistic scientist.

As for the ancient scientist's *persona* of courtier, my central contention is that the scientific *persona* at court was but a subspecies of court friendship. By "court friendship" I mean a cultural attitude of equality, esteem, reciprocity, and loyalty toward furthering the interest of the king. This description is a careful statement of social codes. Court friendship is both a way of speaking and a practice of exchange. The title to which courtiers aspired above all was *philos* "friend", the second highest rank a courtier could obtain at the most of the Hellenistic courts, including the Ptolemies'. (The highest rank was the language of blood-relations, which in the third century BCE extended only to members of the royal family or other monarchs.[13]) It is striking that a court society founded on hierarchy and dif-

---

13 See Introduction footnote 55.

ference aspired to equality. To be a "friend" of the king therefore denoted equality with royalty and the equitable exchange of interests both political and personal. To his friends the king gave positions of authority, combined with financial and social support. Konstan has argued ancient friendship is an emotional affective attitude approaching our own emotional attitudes toward friendship, a view probable as the core of ancient Greek notions of friendship.[14] Konstan suggests that the *philoi* of Hellenistic monarchs too conformed to an emotional attitude of friendship.[15] Rather, I suggest that Hellenistic political attitudes about the utility and pretense of equality are parasitic upon this emotive attitude at the core of ancient friendship.[16] Hellenistic court friendship was primarily an ideological function of monarchy, not a 'naturalized' friendship born of mutual affection. Still we should note that our evidence will not permit us to recover which friendships were genuine and which merely utilitarian.

Distinguishing genuine and feigned loyalty was as much a problem for ancient participants as us modern observers. The ambiguity was not a flaw but a staying power of the system. Court society was a hierarchy that demanded respect for the categories of power, even while the less powerful aspired to equality and reciprocation. Hellenistic court literature is filled with paradigms of genuine loyalty and feigned sociability: the true friend and flatterer emerge as typologies of courtiership. The surviving descriptive and prescriptive treatises for kingship or courtiership describe a distance between the real and idealized, between "true friend" and the "flatterer", that opens up a hermeneutic space of social action. A courtier might act within the discourses, sites, and social expectations of court friendship, while inner motives remained obscure.[17] What made the discursive formation of court friendship so widespread was the rich hermeneutic space of social action it enabled. In contradistinction with friendship, flattery does not hinge on truth-telling, does not presuppose moral reciprocity, and puts care of the self ahead of interest of the other. When king met courtier at the symposium, there was room for both the true friend and the flatterer. Both types played a role as well in a culture of intellectual exchange and patronage. Since court participants believed themselves to be engaged in an ethics of self-presentation and representation of friendship, my analysis is not itself explanatory: it is rather a thick description of the social practice of court science. Both friend and flatterer played a role in the culture of Hellenistic intellectual exchange and patronage. In recounting court friendship I

---

14 Konstan 1997: 1–23 is incisive and thoughtful.
15 Konstan 1997: 95–98.
16 Crook 2013: 69–73; Strootman 2014: 147–150 anticipate my analysis.
17 Feigning friendliness in the style of a flatterer at court is not the same mode of private revelation found in modern friendships. Konstan 1997: 14–18 has persuasively argued that modern friendship is marked by revelation and personal intimacy, as if there is a separate asocial and private self that discloses itself in friendship.

try to maintain a tension in my account between these poles of the spectrum of idealized friendship.

The *persona* of the court scientist in the guise of a courtier as friend pursued distinction, honor, and authority in common life, not by themselves. Ancient philosophers who propound the common life of friendship (such as the Peripatetics) were the ethical mouthpieces of this society; their philosophical opponents, such as the Cynics who propounded self-sufficiency (*autarkeia*) did not find adherents among Hellenistic scientists. Scholars of early modern science have noted the rhetoric of genteel respectability to which the new experimental scientists such as Robert Boyle gravitated among a society of fellows.[18] For these early modern practitioners of the experimental life, to be a scholar and a gentleman was akin to having obtained social credibility. For ancient scientists, the sought quality is not quite the same as social respectability among a society of fellows so much as the mark of social distinction in an agonistic society. The sought quality was called in Greek τιμή or δόξα, transliterated as *time* "honor" and *doxa* "renown". Both courtiers and patrons were said to be φιλότιμος or φιλόδοξος, "a friend-of-honor" and "a friend-of-renown" respectively.[19] (In light of the discourse of court friendship I have opted throughout for a stilted translation in order to emphasize the verbal association with being a friend, φίλος.) For a discourse of friendship runs throughout the products of court society, whereby the leveling of social distinction is directed outward from the king: the king is a "friend" of the truth (φιλαλήθης), the king is a "friend" to learning and knowledge (φιλομαθούντων), the king is a "friend" to doctors (φιλιατροῦντι), the king is a "friend" to the *technai* in general (φιλοτέχνων).[20] The discourse of friendship distributed the social status of the king and his political interests to crafts and learning.

Where in all this was the disposition toward practical action of the scientific self who claims new scientific knowledge? Court friendship offered a means of exchange between the court scientist and the socially superior patron of lay knowledge. It offered a discourse, a site, and practice of exchange. Friendship was a ready-made discursive formation for scientific attitudes such as a disposition for inquiry with truth-telling and for social relations with trust. To figure oneself as a friend of the king endowed the court scientist's work with the moral authority of truthfulness, a necessary virtue for any representation of the empirical world.

---

18 Shapin 1994.
19 LSJ *s. v.* φιλότιμος and φιλόδοξος offers "ambitious" and "loving fame" respectively. The passages about the king's *philoi* discussed in chapter 1 **nos. 11–18** testify to these usages among courtiers; passages discussing kings are referenced below. Weber 1993: 84.n2 had already seen the importance of φιλότιμος.
20 Ps.-Aristeas 206; Apollonius of Citium 16.11 K-K; Plutarch *Quomodo Adulator ab Amico Internoscatur* 58A; Philon *Belopoeica* 50.26 Th.

Friendship in the traditions of the Greek symposium provided a site for the possibility of intellectual and social exchange.

Since I have described the aspirational and ambivalent *personae* of friendship's practitioners, I now turn to describe other aspects of friendship: the ideology of kingship as a sponsor and object of friendship, the character types of friendship between truth and flattery, the communal meal as the cultural meeting place for court friendship, and modes of obligation and reciprocity between friends.

Kingship

Kings existed continuously throughout Greek history and Greek literature contains many portrayals.[21] The earliest Greek literature showed Thersites in the *Iliad* who "quarreled with kings" and Hesiod who denounced "the gift-devouring kings".[22] Yet criticism of the kings was tempered by the recognition that the king's right to rule was divine: "kings are from Zeus".[23] The idea of a divine right to rule blends quickly into claiming descent from divinity, as the Ptolemies did.[24] In such a way kingship was innate, natural, and marked on one's character: "a good character that has shared also in culture (*paideia*) is capable of ruling".[25] A cultured *ethos* or character formed by *paideia* was the prerequisite of the idealized king. At least in the Hellenistic world the image of Alexander the Great's kingship remained a cultural touchstone for the royal successors who came to power after his death. For example, a philosopher (Aristotle) had educated Alexander, so philosophers ought to educate royal heirs: hence we find philosophers called *tropheus* or educator, in charge the education of the royal heir. Providing a model for the ethics and status of Hellenistic court physicians, Alexander enrolled his physician Philip the Acarnian among his "most goodwilled friends" after Philip successfully cured him; more tales were told of Philip's loyalty by foreswearing to poison Alexander.[26] More

---

21 The literature on Hellenistic kingship is enormous. I have chosen to focus on primary sources rather than engage in controversies of the secondary literature. Readers may find collected here some of the main secondary scholarship, referred to intermittently below: Bikerman 1938, Gehrke 2013, Goodenough 1928 (mostly antiquated), Haake 2003, Haake 2013, Koenen 1993, Moyer 2011, Murray 2007, Strootman 2014. Since my focus is kings and courtiers, not inter-state relations, I take no account of how Hellenistic kings represented themselves before Greek city-states.
22 Homer *Ilias* 2.214 ἐριζέμεναι βασιλεῦσιν; Hesiod *Opera et Dies* 38–39 βασιλῆας δωροφάγους.
23 Hesiod *Theogonia* 96: ἐκ δὲ Διὸς βασιλῆες. Ptolemaic authors also attribute kingship to divine authorization: Callimachus *In Iovem* 79–90; Ps.-Aristeas 224.
24 Theocritus 17.13–50 (Philadelphus); *OGIS* 54.4–5 (Euergetes); 631 *BNJ* (Satyrus) F1 = Theophilus *Ad Autolycum* 2.7 (Philopator).
25 Ps.-Aristeas 290: ἦθος χρηστὸν καὶ παιδείας κεκοινωνηκὸς δυνατὸν ἄρχειν ἐστί.
26 Diodorus Siculus 17.31.4–7: κατέταξεν αὐτὸν [sc. Philip] εἰς τοὺς εὐνουστάτους τῶν φίλων; Quintus Curtius Rufus 3.6.

importantly, Alexander had urged some sort of cultural integration between his subjects the Greeks and *hoi barbaroi*, "barbarians" or non-Greek, even to the point of holding a mass wedding ceremony for his Macedonian generals and Persian noblewomen.[27] While the Macedonians to a man divorced their brides after Alexander's death, the cosmopolitan ideal of individual remit and virtue irrespective of ethnicity gained greater traction over the third century as Greek colonists lived alongside native populations. **No. 136** Eratosthenes, for example, praised Alexander for including non-Greeks among the "friends": "[Eratosthenes says] that Alexander, disregarding advice [sc. to separate Greeks and barbarians], benefited (εὐεργετεῖν) whomever among the noble men (εὐδοκίμων) he was able to receive at court".[28] Hellenistic kingship fused historic Greek and ancient Near Eastern traditions in an evolving presentation.

Whatever the particulars of Alexander's kingship and the practical circumstances of the individual Hellenistic kingdoms, there arose a Hellenistic literature with Greek predecessors about idealized kingship: that monarchy was the preferred form of government, what virtues the king ought to embody, how kings ought to act, receive friends, govern with justice, distribute wealth, and so on.[29] These treatises, generically called *peri basileias* "On Kingship", were written by philosophers.[30] At the court of Ptolemy **no. 15** Sosibius forged a treatise *On Kingship*, ascribing it to the Peripatetic Theophrastus; **no. 124** Sphaerus also wrote an *On Kingship*. In spite of the fragmentary state of the available evidence we can easily reconstruct the general content of this genre because these treatises were collections of *topoi* and rhetorical commonplaces about kingship. As Haake has pointed out, these were not theories of monarchy so much as idealizations of the good ruler, implying that the authors and addressees were more type-figures than individuals.[31]

The king's Zeus-given right to rule placed him hierarchically over his subjects, but the tradition of idealized kingship found it more tolerable to stress the rational basis for the king's rule: character made the man. Why were royal virtues so important? In fourth century Greek philosophical thought the king's virtues were necessary for the exercise of justice and governance.[32] It would seem as if a virtue-ethics theory of kingship was widespread in the Hellenistic world. Yet the Byzantine encyclopedia *Suda* records a different view of Hellenistic kingship:

---

[27] Plutarch *Alexander* 70.3.
[28] Strabo 1.4.9 = 66C.29–31 (166.29–31) Radt: διόπερ τὸν Ἀλέξανδρον ἀμελήσαντα τῶν παραινούντων, ὅσους οἷόν τ' ἦν ἀποδέχεσθαι τῶν εὐδοκίμων ἀνδρῶν καὶ εὐεργετεῖν.
[29] On these treatises, their history, and their possible contents see Murray 2007, Haake 2003, Haake 2013. My description of contents follows Murray 2007: 21–27.
[30] Diogenes Laertius 5.47, 5.49; see Haake 2003: 114.n25, Murray 2007.
[31] Haake 2003: 90. Haake 2013 suggests that the intended audience was the Greek *poleis*.
[32] Isocrates *Ad Nicoclem* 11 urged that the king's virtues ought to exceed his honors.

Kingship is rule without accountability, which is sustained in relationship to the wise alone. Kingship. Neither nature nor justice gives kingdoms to people but rather to people who can lead an army and handle affairs intelligently. Such was Philip and the Successors of Alexander. For kinship did not help the natural son at all because of his weakness of soul; but those of no relation became kings of nearly the entire known world.³³

The *Suda* presents a vision of kingship that owes nothing to biological descent nor legalism; it is a vision of δορίκτητος χώρα "spear-won land" or kingship by military claim, common to Alexander's immediate successors, his former generals, not his biological heir and half-brother Philip Arrhidaeus. Some modern scholarship, following the *Suda*, has portrayed Hellenistic kingship as a type of charismatic leadership.³⁴ But these two visions of kingship are perhaps not so opposed. A virtue-ethics typology of idealized kingship is more appropriate for the routine patterns of administration, as well as the rituals and formulas of royal presentation.³⁵ A charismatic kingship dependent on military success is more appropriate for the battlefield and moments of state crisis. Since our focus is the normal conditions of patronage of science at court, not the political fortunes of individual Hellenistic kingdoms, my account will focus on the virtue-ethics typology of idealized kingship.

Since the king's divine rule had no check but character, the king's competence and character had to be carefully formed. The later historian Diodorus Siculus wrote about the Egyptian pharaohs in an idealized portrait. This important text on Egyptian kingship was probably in imitation of a Hellenistic treatise on kingship, the genre of idealizing portraiture.³⁶ A central passage lays out the king's idealized courtiers, where character and competence were closely linked to social position.³⁷ The king's retainers were expected not to think of personal gain but royal interest; hence male offspring of the sanctified class in the prime of their physical prowess attended on the king. These men were πεπαιδευμένοι "cultured". They urged the king toward justice. Not only did the king's companions offer models of character

---

33 *Suda* β147: ἡ γὰρ βασιλεία ἀρχὴ ἀνυπεύθυνος, ἥτις περὶ μόνους ἂν τοὺς σοφοὺς συσταίη. Βασιλεία. οὔτε φύσις οὔτε τὸ δίκαιον ἀποδιδοῦσι τοῖς ἀνθρώποις τὰς βασιλείας, ἀλλὰ τοῖς δυναμένοις ἡγεῖσθαι στρατοπέδου καὶ χειρίζειν πράγματα νουνεχῶς· οἷος ἦν Φίλιππος καὶ οἱ διάδοχοι Ἀλεξάνδρου. τὸν γὰρ υἱὸν κατὰ φύσιν οὐδὲν ὠφέλησεν ἡ συγγένεια διὰ τὴν τῆς ψυχῆς ἀδυναμίαν. τοὺς δὲ μηδὲν προσήκοντας βασιλεῖς γενέσθαι σχεδὸν ἁπάσης τῆς οἰκουμένης.
34 Gehrke 2013.
35 Murray 2007: 23 "Thus justification of monarchy was in terms of the virtue of the king, and the checks on monarchy were checks of morality inherent in the character of the king himself. There was according to this view no need to worry overmuch about his relation to law; thus a problem that has afflicted all absolute monarchies was answered by defining it out of existence; for if a king ceased to be good, he became a tyrant, and the constitution was no longer one of the good constitutions." Haake 2013: 166.n4 points to Aristotle's remark that tyrants do not trust their friends.
36 Stephens 2003: 32–34 discusses these passages from Diodorus as if the author was Hecataeus of Abdera. I find the attribution plausible, although nothing in my argument depends on it.
37 Diodorus Siculus 1.70.2.

and competence but so did moral exemplars from books. When the king went to make sacrifices, a priest attended him. After the sacrifice "a scribe read aloud certain beneficial advices and actions from books of the most prominent men, so that he who held the rule of all things might, by contemplating the best interests in his own mind, be thus turned toward the governance ordained in each case."[38] While the text thematizes the Hellenistic contrast between learning through experience and learning through books, Diodorus also invokes the Greek notion that Egyptian kingship is a grand tradition of stability and constancy such that past exemplars remained notable paradigms of current conduct.[39]

Before the sacrifice Diodorus says that a priest ascended alongside the king to the altar and "anthologized his virtues in turn", an aretology of an idealized king.

> He said that the king was reverential toward the gods and most mild toward men. He was self-controlled, just, and great-souled, further without falsehood, a giver of goods, and was in general stronger than every desire. He set punishment for faults at lesser than their worth and returned thanks for good deeds at greater than their beneficence.[40]

Diodorus' passage provides an organizing structure for a virtue-based account of Hellenistic kingship. In the following I analyze the specific virtues of Hellenistic kingship by comparing Diodorus' account to the idealizing portraits of kingship found in Isocrates' *To Nicocles*, an exhortation to good kingship, and Ps.-Aristeas' *Letter to Philocrates*, whose central scene is a long dialogue about good kingship between the Ptolemaic king and Jewish scholars. I exclude documentary comparanda apart from Euergetes' Canopus Decree (*OGIS* 56) and Philopator's Memphis Decree (*Prose* 12–14), since these texts invest the monarchs with idealizing virtues. I pass over discussion of the king's self-control and piety, mentioned by Diodorus.[41]

---

38 Diodorus Siculus 1.70.9: ὁ μὲν ἱερογραμματεὺς παρανεγίνωσκέ τινας συμβουλίας συμφερούσας καὶ πράξεις ἐκ τῶν ἱερῶν βίβλων τῶν ἐπιφανεστάτων ἀνδρῶν, ὅπως ὁ τῶν ὅλων τὴν ἡγεμονίαν ἔχων τὰς καλλίστας προαιρέσεις τῇ διανοίᾳ θεωρήσας οὕτω πρὸς τὴν τεταγμένην τῶν κατὰ μέρος τρέπηται διοίκησιν.
39 Diodorus Siculus 1.1.1 introduces the widespread contrast between experience and book-learning at the beginning of his work. In order not to enumerate passages unduly I compare only authors of the second century BCE: Polybius 12.25e, Ps.-Aristeas 2, Hypsicles *Elements* 14 *praefatio*. On the traditions of Greek views of Egyptian kingship more broadly, see Stephens 2003: 20–64.
40 Diodorus Siculus 1.70.6: ἀνθομολογεῖσθαι δ' ἦν ἀναγκαῖον καὶ τὰς κατὰ μέρος ἀρετὰς αὐτοῦ, λέγοντα διότι πρός τε τοὺς θεοὺς εὐσεβῶς καὶ πρὸς τοὺς ἀνθρώπους ἡμερώτατα διάκειται· ἐγκρατής τε γάρ ἐστι καὶ δίκαιος καὶ μεγαλόψυχος, ἔτι δ' ἀψευδὴς καὶ μεταδοτικὸς τῶν ἀγαθῶν καὶ καθόλου πάσης ἐπιθυμίας κρείττων, καὶ τὰς μὲν τιμωρίας ἐλάττους τῆς ἀξίας ἐπιτιθεὶς τοῖς ἁμαρτήμασι, τὰς δὲ χάριτας μείζονας τῆς εὐεργεσίας ἀποδιδοὺς τοῖς εὐεργετήσασι.
41 These two virtues can be paralleled in the idealizing portraiture of kings. Self-control and abstention from desire: Isocrates *Ad Nicoclem* 29, 31; Ps.-Aristeas 209, 211. Piety: *Prose* nos. 12–14; Ps.-Aristeas 210, 215, 229. Euhemerus of Messene's *Holy Record* probably had a great influence on Hellenistic conceptions of piety in kingship. This work, which survives only in fragments and testimonia (*FGrH* 63), showed Zeus and other Greek gods as kings who through their piety and benefactions became deified as gods; cf. Stephens 2003: 36–39.

Diodorus' claim that the king was most mild (ἡμερώτατα) toward men was both an encouragement of the type of punishment as well as playing the part of a fair referee. An ideal of good kingship was the patience to listen to good advice. Sometimes this meant hearing unwanted advice. The Jewish scholars in Ps.-Aristeas offer the monarch advice about how to receive those of opposite opinion.

> "How ought [a king] be a friend-of-honor?" He replied: "All think that a king ought to be [so] toward those who are disposed to us in friendship. But I suppose that a king ought to have gracious friendship-in-honor toward those of the opposite opinion. In this way we might bring them to what is proper and advantageous."[42]

The term "friendship-in-honor" is the Greek term φιλοτιμία. The word is yet another discursive formation of the terminology of friendship at the court. Kings and courtiers were both said to be *philotimoi* when they desired honor, ambitious acquisition from its pejorative to its positive sense. Kings could also be *philotimoi* when they distributed honor, generous in the distribution of royal favor. Here in Ps.-Aristeas the scholar urges the king to the second, drawing upon the vocabulary of both distributing royal distinction (φιλοτιμίαν) and gift-giving (χαριστικὴν). Thus generosity toward people of the opposite opinion was an expression of the king's high-mindedness. Generosity towards people of the opposite opinion allowed the king to transcend smaller quarrels. Encouraged to mildness and generosity, the king was able to play the part of a listener capable of listening to multiple opinions.

To listen to multiple opinions without undue commitment to one side or another required not only self-control but also a discrimination of the truth. The discrimination of truth was reflected back on the judge, who was supposed to be above lying. A key ideology of kingship was the notion of truth-telling. Diodorus phrases this virtue negatively: the king must be untouched by falsehood (ἀψευδής). Isocrates advised that the king's preference for truth ought to render his talk more trustworthy than others' oaths.[43] A key passage in Ps.-Aristeas elaborates and expresses the idea positively:

> The king asked another: "How ought the king preserve the truth?" He replied: "By knowing that the lie brings great shame to all humans, but much more to kings: for having power to do what they want, why should they lie? You ought to understand, king, that god is a friend-of-truth."[44]

---

42 Ps.-Aristeas 227: πῶς τινα [mss.: πρὸς τίνα Hadas] δεῖ φιλότιμον εἶναι; ἐκεῖνος δὲ ἔφη· Πρὸς τοὺς φιλικῶς ἔχοντας ἡμῖν οἴονται πάντες ὅτι πρὸς τούτους δέον· ἐγὼ δ' ὑπολαμβάνω, πρὸς τοὺς ἀντιδοξοῦντας φιλοτιμίαν δεῖν χαριστικὴν ἔχειν, ἵνα τούτῳ τῷ τρόπῳ μετάγωμεν αὐτοὺς ἐπὶ τὸ καθῆκον καὶ συμφέρον ἑαυτοῖς.
43 Isocrates *Ad Nicoclem* 22.
44 Ps.-Aristeas 206: ὁ βασιλεὺς τοῦτον ἕτερον ἐπηρώτα· Πῶς ἂν τὴν ἀλήθειαν διατηροῖ; ὁ δὲ πρὸς τοῦτο ἀπεκρίθη· Γινώσκων ὅτι μεγάλην αἰσχύνην ἐπιφέρει τὸ ψεῦδος πᾶσιν ἀνθρώποις, πολλῷ δὲ μᾶλλον τοῖς βασιλεῦσιν· ἐξουσίαν γὰρ ἔχοντες ὃ βούλονται πράσσειν, τίνος ἕνεκεν ἂν ψεύσαιντο; προσλαμβάνειν δὲ δεῖ τοῦτό σε, βασιλεῦ, διότι φιλαλήθης ὁ θεός ἐστιν.

The more power a knower has, the less reason the knower has to misrepresent reality. A king is supposed to present reality. The knower's competence is indistinguishable from his character. The king must emulate the divinity, who is "a friend-of-truth" (φιλαλήθης). (As is typical in the symposia scene of Ps.-Aristeas, the Jewish scholar ends by offering the model of divinity to the king.) Aristotle had defined the *philalethes* "a friend-of-truth" as someone who spoke the truth even when it didn't matter.[45] To be a friend-of-truth was conduct one's self in a spirit of disinterestedness.

A disinterested king distributed fair and impartial justice. An ideology of Hellenistic kingship was that the distribution of justice was mercifully distributive, rather than proportional. Diodorus Siculus records that the kings were supposed to "set punishment for faults at lesser than their worth and returned thanks for good deeds at greater than their beneficence".[46] Likewise, Ps.-Aristeas has a Jewish priest tell the Ptolemaic king that "dealing with those who deserve punishment more mercifully than they deserve" will turn them from evil.[47] But justice was not just for malefactors. The king ought to have friends or *philoi*, who were those individuals especially privileged by royal beneficence. But how should he keep them εὔνοοι "loyal" toward him? Ps.-Aristeas records that the Jewish scholar tells the king that they should watch how he rules the masses, for "divinity benefits (εὐεργετεῖ) the whole species of humans".[48] The just king ought to supply the entire spectrum of humanity with his beneficence, even if he distributes greater quantities to his friends. Disinterested and merciful distributive justice fit the social heirarchy.

Thus the king ought to be kind or humane overall. The Greek word for the virtue of humaneness was φιλανθρωπία, literally "a friend-to-human". While not a new term in the Hellenistic period, the word was irresistible to Hellensitic court culture as another product of the discursive formation in the terminology of friendship. How ought I be a friend-to-human, asks the king in Ps.-Aristeas; the scholar replies that human fortune rises and falls, so the king should be piteous toward all.[49] The associated verb φιλανθρωπέω "to be kind" is common to Hellenistic royal language, applied to people, social institutions, or used intransitively.[50] Isocrates' *To Nicocles* contains a similar injunction that the king who is φιλάνθρωπος must

---

[45] Aristotle *Ethica Nicomachea* 1127b. I draw on the evidence of Aristotle's *Ethica Nicomachea* only to clarify the Hellenistic literature of idealizing kingship, although many more comparanda could be added.
[46] Diodorus Siculus 1.70.6.
[47] Ps.-Aristeas 188: κολάζων τοὺς αἰτίους ἐπιεικέστερον ⟨ἢ⟩ καθώς εἰσιν ἄξιοι, μετατιθεὶς ἐκ τῆς κακίας καὶ εἰς μετάνοιαν ἄξεις.
[48] Ps.-Aristeas 190: ὡς ὁ θεὸς εὐεργετεῖ τὸ τῶν ἀνθρώπων γένος.
[49] Ps.-Aristeas 208.
[50] Welles 1934: 373; *OGIS* 90.12.

look out for the welfare of the people.⁵¹ Kindness is a quality of mercy. So it must be distributed. The Ptolemaic monarchs so prized the act of kindly distribution, euergetism, that they named themselves after it. In the Canopus Decree of Ptolemy Euergetes "Ptolemy the Beneficent" the Greek term for the king's name and its associated cognates occur nineteen times.⁵² The king and queen "accomplish many deeds by benefiting (εὐεργετοῦντες) the temples throughout the land" and, after having provided grain to the populace in a year of failed rains, "left an immortal beneficence (εὐεργεσίαν) and the greatest memory of their excellence".⁵³ The survival of the people was the ever-living memory of the monarchs.⁵⁴

Diodorus also says that the king should be μεγαλόψυχος, literally "great-souled". This virtue does not have an easy English equivalent, although "magnanimous" is the Latinate translation. In Greek the word refers to the holding of social distinction or honor (τιμή). Aristotle had defined the μεγαλόψυχος in his table of virtues near the beginning of the *Nicomachean Ethics*.⁵⁵ As is typical of Aristotle's virtue ethics, he defined greatness of soul as a quality of personal worth positioned as the mean between two extremes. Greatness of soul is holding honor in neither too high nor limited a fashion.⁵⁶ Aristotle encouraged the great-souled man to error on the side of excess in his acquisition of honor, rather than less. The great-souled takes contests for large honors, not petty trifles. The ambitious man, Aristotle points out, is called friend-of-honor. Both Aristotle's and Ps.-Aristeas' account of friendship-in-honor praise the seeking of honors, or marked ambition, befitting the ideal of a noble aristocrat born into positions of high social influence. Diodorus and the Hellenistic tradition of idealized kingship follow Aristotle in joining of greatness of soul with friendship-in-honor.

Finally, Diodorus says that the ideal king should be a giver of goods. Diodorus' point goes beyond the euergetism of the Ptolemies discussed above. For the king distributed not only moral qualities, such as mercy and justice, but also financial

---

51 Isocrates *Ad Nicoclem* 15.
52 OGIS 56.5, 7, 8 (bis), 18, 21, 22, 23, 25, 31, 32, 33, 36, 44, 46, 47, 54, 58, 75.
53 OGIS 56.8: διατελοῦσιν πολλὰ καὶ μεγάλα εὐεργετοῦντες τὰ κατὰ τὴν χώραν ἱερά, 56.18 ἀθάνατον εὐεργεσίαν καὶ τῆς αὐτῶν ἀρετῆς μέγιστον ὑπόμνημα καταλείποντες.
54 Cf. Isocrates *Ad Nicoclem* 36 "Prefer to leave behind as a memorial the images of your virtue rather than of your body."
55 Aristotle *Ethica Nicomachea* 1107b.21–30 "Regarding honor and dishonor the mean is μεγαλοψυχία, an excess being called empty vanity and a deficit being called μικροψυχία. Just as we said that generosity has a relation to magnificence (that is, differing by being concerned with small amounts of money), so there also exists a virtue in relation to μεγαλοψυχία (which is concerned with great honor), while this virtue is concerned with small honor. For it is possible that one must to strive for honor [in the right way] more than less. The man who exceeds in his striving is called φιλότιμος, the one who is deficient is called not-φιλότιμος, and the mean is unnamed."
56 Cf. Isocrates *Ad Nicoclem* 25: "Consider great-minded those not attempting more than they can accomplish but those aiming for the good who can accomplish what they attempt."

rewards and gifts. Ps.-Aristeas' scholars warned that the king must exert self-control over gift-acquisition, one of the vices of kingship.[57] Rather, the king should give and receive in a more even fashion. Isocrates urges monarchs to use wealth to secure their friends and look toward the future.

> Display your magnificence not in one of those rich expenditures that disappear at once, but in the previously mentioned ways, namely the beauty of your possessions and in the benefits for your friends. Such expenditures will remain good for you during your lifetime, and for your descendants you will leave behind a more worthy expenditure than what you have spent.[58]

Magnificence (μεγαλοπρέπειαν), Aristotle had said in his discussion of the great-souled man, is the outlay of a large amount of money, compared to generosity. Isocrates' advice to the monarch aims to establish his legacy. The king should purchase artifacts that are beautiful; the king should spend for the benefit of his friends. Isocrates' language of friends and beneficence (ταῖς τῶν φίλων εὐεργεσίαις) is the discourse of Hellenistic court society.

Ps.-Aristeas ranks the recipients of the king's largesse in a descending order from parents, to friends, then all people who exist in a state of friendship toward the king.[59] Is this an ancient injunction to love one's parents? While the Ptolemaic kings did prominently feature their genealogy in their decrees and Philopator made a family shrine, terms of friendship and kinship were used in honorific ways in Hellenistic court society.[60] Ideologically kinship was a useful tool. "What use is kinship?" asks the king in Ps.-Aristeas. The scholars reply that, if suffering and evil come, one will still retain "reputation and success" among one's kin.[61] The biological image retained its force in describing at least the fruits of royal virtue. "For all to whom the king has sowed graces will branch forth goodwill."[62]

---

57 Ps.-Aristeas 209.
58 Isocrates *Ad Nicoclem* 19: τὴν μεγαλοπρέπειαν ἐπιδείκνυσο μηδ' ἐν μιᾷ τῶν πολυτελειῶν τῶν εὐθὺς ἀφανιζομένων, ἀλλ' ἔν τε τοῖς προειρημένοις καὶ τῷ κάλλει τῶν κτημάτων καὶ ταῖς τῶν φίλων εὐεργεσίαις· τὰ γὰρ τοιαῦτα τῶν ἀναλωμάτων αὐτῷ τε σοὶ παραμενεῖ καὶ τοῖς ἐπιγιγνομένοις πλείονας ἄξια τῶν δεδαπανημένων καταλείψεις.
59 Ps.-Aristeas 228.
60 Genealogies in Theocritus 17.13–38, *OGIS* 54.1–5, *OGIS* 56.1, *Prose* nos. 12–13; the Ptolemaic family shrine was called either *Soma* "Body" or *Sema* "Tomb" according to Zenobius 3.94. Ps.-Aristeas' injunction to honor family contradicts the *Suda*'s vision of a military kingship, in which natural family ties (*physis*) counted for less than charismatic military leadership.
61 Ps.-Aristeas 242: δόξα καὶ προκοπή.
62 Ps.-Aristeas 230: πᾶσι γὰρ χάριτας ἔσπαρκας, αἳ βλαστάνουσιν εὔνοιαν.

Courtiership

The evidence for a normative account of ancient Hellenistic courtiership comes again from treatises about idealized kingship.[63] In this idealized account courtiers who attended the king too had their expected virtues. But who should surround the king? In Diodorus' story of idealized Egyptian kingship the sons of the leading aristocracy formed the companions of the monarch to urge him to virtue. Alexander remained a moral exemplar here too, as in Eratosthenes's praise of Alexander for selecting cultured, noble men at his court without regard to ethnic background. These courtiers became the king's friends. The king could pick his courtiers, unlike his family. Isocrates advises monarchs to pick their friends carefully.

> Acquire as friends not all those who want to be [your friends] but only those worthy of your nature; nor acquire as friends those with whom you pass the time best but those with whom you will govern the city best.[64]

Isocrates reduces the monarch's friendship to its political function: ruling. Hellenistic royal friendship was an ideological function of monarchy, not a 'naturalized' friendship born of mutual affection.

The king's long association with his chosen subordinates allowed him opportunity to observe them. The Ptolemaic king in Ps.-Aristeas asks his Jewish scholar-guests how to recognize "those acting in plot against him".

> If the king should observe their behavior when free, their good order in receptions, councils, and other exchanges, the king would observe them not exceeding propriety in their congratulations and the other parts of their behavior.[65]

The social meetings of court society became a stage for observing behavior of courtiers. The receptions (ἀσπασμοί) might be the general morning levée or some other regular reception, the councils being irregular business of the morning or midday, with symposia and drinking parties starting in the late afternoon.[66] These social occasions allowed the courtiers to share the king's fortunes and sorrows in the regular administration of the kingdom.

---

[63] Polybius has many scattered remarks on courtiers. Those relevant to the courts of Euergetes and Philopator are recounted in chapter 1 under **nos. 15, 16, 121.**

[64] Isocrates *Ad Nicoclem* 27: φίλους κτῶ μὴ πάντας τοὺς βουλομένους, ἀλλὰ τοὺς τῆς σῆς φύσεως ἀξίους ὄντας, μηδὲ μεθ' ὧν ἥδιστα συνδιατρίψεις, ἀλλὰ μεθ' ὧν ἄριστα τὴν πόλιν διοικήσεις.

[65] Ps.-Aristeas 246: εἰ παρατηροῖτο τὴν ἀγωγὴν ἐλευθέριον οὖσαν, καὶ τὴν εὐταξίαν διαμένουσαν ἐν τοῖς ἀσπασμοῖς καὶ συμβουλίαις καὶ τῇ λοιπῇ συναναστροφῇ τῶν σὺν αὐτῷ, καὶ μηθὲν ὑπερτείνοντας τοῦ δέοντος ἐν ταῖς φιλοφρονήσεσι καὶ τοῖς λοιποῖς τοῖς κατὰ ἀγωγήν.

[66] Morning levée for the reception of courtiers: Ps.-Aristeas 304, Polybius 5.56.10, Diodorus Siculus 1.70.4. See Bikerman 1938: 34–35 for further documentation of the court's daily routine.

In addition to culture and good noble breeding, courtiers bound to the king were expected to show loyalty or goodwill toward the king. After all, the king was supposed to trust himself to courtiers who "conversed with the king on account of goodwill (εὔνοιαν), not those who out of fear nor from assiduousness brought all business to profit."[67] The courtier's interests reflected the king's, not his own. Correspondingly, the king was urged to honor courtiers who were loyal to his interests. Isocrates urges the king to "honor with offices those of your friends most familiar but those in truth those most loyal."[68] Family lineage was no substitute for personal loyalty among courtiers. The idealized vision of courtiership reflected a charismatic monarchy, in which the courtiers were bound to the king of out personal affection. An official ideology of personal affection to the charismatic king often translated in practice to family loyalty to dynasties. For example, **no. 15** Sosibius, most powerful *philos* of Philopator had sons and a daughter participate in service to the court in official positions such as bodyguards and priesthoods. Sosibius is the subject of an especially perverse example of a courtier's loyalty toward the king. Since Sosibius had Philopator's mother and brother murdered on the Philopator's orders, the Alexandrians remembered Sosibius in the grisly proverb "The murderer is good-willed (εὔνους)."[69]

Another essential self-representation for any courtier was honesty, assuring the king of his friendship and loyalty. Honesty was the virtue separating cultural *personae* of real and feigned friendship. If courtiers were idealized as 'friends', court literature was filled with exemplars of true friendship and friendship perverted: the figures of *kolax* or flatterer and *parrhēsiast* or truth-teller represent the dual stock figures of lying and truth-telling before individuals of power.[70] Hellenistic flattery arose in the context of presumed personal intimacy and independence from obsequiousness, as the flatterer feigned moral virtue, loyalty, culture, and honesty expected in true friendship. Plutarch wrote in a picturesque analogy: "They say that the gadfly settles against the ear with bulls and the tick with dogs; the flatterer

---

[67] Ps.-Aristeas 270: τοῖς διὰ τὴν εὔνοιαν, εἶπε, συνοῦσί σοι, καὶ μὴ διὰ τὸν φόβον μηδὲ διὰ πολυωρίαν, ἐπανάγουσι πάντα πρὸς κερδαίνειν.
[68] Isocrates *Ad Nicoclem* 20: τίμα ταῖς μὲν ἀρχαῖς τῶν φίλων τοὺς οἰκειοτάτους, ταῖς δ' ἀληθείαις αὐταῖς τοὺς εὐνουστάτους.
[69] Zenobius 3.94: εὔνους ὁ σφάκτης.
[70] Helpful literature includes Engberg-Pedersen 1996, Konstan 1996, Konstan et al. 1998, Tsouna 2007: 91–142, Strootman 2014: 172–175. I assume throughout that the flatterer is representing himself as a friend, rather than straightforwardly as a flatterer. Hence the problematic for the social superior was to distinguish the true intentions of his 'friend'. I will use the transliteration *kolax* for the main characterization of the flatterer, since this is the dominant word in sources analyzed here. Ancient Greek words for flatterers include κόλαξ, παράσιτος, and ἄρεσκος, translated respectively as "flatterer", "parasite", and "obsequious". Ancient sources distinguish more clearly the obsequious man from the other two, as someone who flatters everyone indiscriminately without an aim for himself. See *RE* 18.4: 1381–1397 *s.v.* 'parasitos' and footnote 82 below for fuller discussion.

takes hold of the ears of friends-of-honor with praise and, once fixed, is hard to drive off."[71] Flatterers as gadflies seem annoying but hardly a real threat. Yet the *Gergithius*, a treatise about flattery named after one of Alexander the Great's flatterers written by the Peripatetic Clearchus of Soli, explains why the flattering behavior of courtiers themselves threatened idealized Hellenistic kingship. Clearchus' *Gergithius* connects the consequences of flattery with Aristotle's portrayal of the great-souled man.

> Flattery makes the character of flatterers lowly, because they are contemptuous of those around them. Evidence is the fact that they endure everything while knowing what others dare do. By making the character of the people flattered who are puffed up by flattery vain and empty, flattery makes them suppose that all things are in excess for them.[72]

The characters of people who are flattered become vain and empty (χαῦνα καὶ κενά), the very terms Aristotle uses for the individual who has an excess of self-conception in regard to honor, the great-souled man. Flattery suggested to the patron that he was great-souled and so embodied the qualities of the ideal Hellenistic friend. Flatterers, with their disavowal of truth-telling, did not offer a moral check on the king's character. Flatterers failed to act as friends of the king in the way the tradition of idealized kingship demanded.

Distinguishing true friends and flatterers thus became a *topos* of the literature of idealized kingship. Isocrates juxtaposes two types of courtiers, the flatterer and the truth-teller.

> Think trustworthy those who do not praise whatever you say or do but censure your failings. Give *parrhēsia* to those are well-minded, in order that you may have them to judge with you when you are in doubt. Separate those who flatter you skillfully and those who serve you with goodwill, in order that the wicked do not do better than the good.[73]

Good courtiers act with *parrhēsia* (within limits) in adjudging both good and bad for the king. These individuals serve with the king's best interest: they loyally act with goodwill toward the king. Bad courtiers indiscriminately flatter the king. Iso-

---

71 Plutarch *Quomodo Adulator ab Amico Internoscatur* 55E: τοῖς μὲν οὖν ταύροις τὸν οἶστρον ἐνδύεσθαι παρὰ τὸ οὖς λέγουσι καὶ τοῖς κυσὶ τὸν κρότωνα· τῶν δὲ φιλοτίμων ὁ κόλαξ τὰ ὦτα κατέχων τοῖς ἐπαίνοις καὶ προσπεφυκὼς δυσαπότριπτός ἐστιν.
72 Clearchus of Soli fr. 19.2–6 Wehrli = Athenaeus 6.255d: τὴν κολακείαν ταπεινὰ ποιεῖν τὰ ἤθη τῶν κολάκων καταφρονητικῶν ὄντων τῶν περὶ αὐτούς. σημεῖον δὲ τὸ πᾶν ὑπομένειν εἰδότας οἷα τολμῶσι. τὰ δὲ τῶν κολακευομένων ἐμφυσωμένων τῇ κολακείᾳ χαῦνα καὶ κενὰ ποιοῦσαν ⟨αὐτήν⟩, πάντα ἐν ὑπεροχῇ παρ' αὐτοῖς ὑπολαμβάνεσθαι κατασκευάζεσθαι. I follow Wehrli's text.
73 Isocrates *Ad Nicoclem* 28: πιστοὺς ἡγοῦ μὴ τοὺς ἅπαν ὅ τι ἂν λέγῃς ἢ ποιῇς ἐπαινοῦντας, ἀλλὰ τοὺς τοῖς ἁμαρτανομένοις ἐπιτιμῶντας. δίδου παρρησίαν τοῖς εὖ φρονοῦσιν, ἵνα περὶ ὧν ἂν ἀμφιγνοῇς ἔχῃς τοὺς συνδοκιμάσοντας. διόρα καὶ τοὺς τέχνῃ κολακεύοντας καὶ τοὺς μετ' εὐνοίας θεραπεύοντας, ἵνα μὴ πλέον οἱ πονηροὶ τῶν χρηστῶν ἔχωσιν.

crates recognizes the ability of flattery to be artful, but bold flattery does not help the king when he needs advice. The monarch needs honest advisors particularly whenever he is of two minds about an issue. "A good counselor is the best and most kingly of all possessions: think that these make your kingdom great, who can best aid your own mind."[74] Ps.-Aristeas confirms that Hellenistic friends continued to act as ideal courtiers with *parrhēsia*: "friends advise with *parrhēsia* for what is good".[75]

The paradigmatic ancient virtue of truth-telling separating flattery from truth-telling was *parrhēsia*, "frank criticism" or "freedom of speech". How did this happen? The term *parrhēsia* has come to be associated in modern thought with "speaking truth to power", the characteristic of an activist criticizing a powerful figure. But for ancient peoples the noun had wider valence. To use *parrhēsia* encompassed free speech, honest speech, or critical speech, depending on its public or private context and application. Originally forged in the agonism of public debate, *parrhēsia* was the distinctive Athenian political virtue of democratic citizenship. In democratic Athens *parrhēsia* juxtaposed the freedom of speech given to the citizen against the limitations of speech imposed on the non-citizen classes, such as metics and slaves.[76] In the Hellenistic period *parrhēsia* underwent a well-documented transition from public to private virtue.[77] It became a personal virtue between friends, as in discussions of courtiership. *Parrhēsia* was a chief virtue among friends in the symposium.[78] Among friends *parrhēsia* indicated honest unconstrained speech: it underlined intimacy between friends rather than undermining social hierarchy.[79] The man who spoke freely also acted freely; so according to Aristotle the great-souled man was also a *parrhēsiast*, or a free-speaking man.[80] *Parrhēsia* was a mark of a free man who spoke his mind and didn't depend on others to support him, marking a courtier's noble status.

Clearchus' *Gergithius* is an important piece of evidence not only because it typologizes the court actors of the paragon of Hellenistic kingship, Alexander, with Aristotelian virtues.[81] The *Gergithius* also suggests that Peripatetic treatises articu-

---

[74] Isocrates *Ad Nicoclem* 53: ὅτι σύμβουλος ἀγαθὸς χρησιμώτατον καὶ τυραννικώτατον ἁπάντων τῶν κτημάτων ἐστίν. ἡγοῦ δὲ τούτους μεγίστην σοι ποιεῖν τὴν βασιλείαν, οἵτινες ἂν τὴν διάνοιαν τὴν σὴν πλεῖστ' ὠφελῆσαι δυνηθῶσιν.
[75] Ps.-Aristeas 125: συμβουλευόντων παρρησίᾳ πρὸς τὸ συμφέρον τῶν φίλων.
[76] Raaflaub 2004: 223 "free speech is almost synonymous with citizenship".
[77] Schlier 1942.
[78] Peterson 1929; Konstan et al. 1998: 3–5.
[79] Isocrates *Ad Nicoclem* 3. Teles *De Fuga* 23.4–12 equates **no. 30** Hippomedon's function of counselor and advisor (πάρεδρος καὶ σύμβουλος) to the king with *parrhēsia*.
[80] Aristotle *Ethica Nicomachea* 1124b.
[81] In Athenaeus' long excerpt (Clearchus fr. 19 Wehrli = Athenaeus 6.255c–257d) Clearchus discusses the social origins of flattery, as its practitioners migrated from Cypriot courts to Persian monarchs to Macedonian courts. Gergithius, Alexander's flatterer, does not appear in the surviving fragments (Athenaeus only mentions him as an introduction to the treatise) but Gergithius' name

lated both descriptive and prescriptive accounts of flattery and friendship in relation to court contexts.[82] Although fragmentary, these Peripatetic treatises must have represented early Hellenistic predecessors to our surviving Roman-period treatises describing truth-telling in friendship, Plutarch's *How to Tell a Flatterer from a Friend* and Philodemus of Gadara's *On Frank Criticism*. I therefore treat Plutarch's and Philodemus' anxieties over friendly Greek approaches to Roman power as manifestations originating with Hellenistic concerns about *parrhēsia*.

Philodemus, an Epicurean philosopher in residence at the bay of Naples, treats *parrhēsia* in the form of truth-telling as an essential technique for philosophical progress within the Epicurean community; Philodemus considers it the essence of friendship. Acccording to Philodemus *parrhēsia* is, in turn, a method of admonition from teacher to pupil to correct mistakes or a method of self-reflection to bring to consciousness one's own mistakes. Thus, while most of Philodemus' account focuses on the particular Epicurean psychological use of truth-telling, his treatise does confirm that *parrhēsia* was used in symposia and offers insight into truth-telling between different social classes.[83] One section of the treatise is entitled: "Why is is that, *ceteris paribus*, those illustrious in wealth and honors endure *parrhēsia* less?"[84] Philodemus replies to his own question:

> They [sc. the illustrious] do not pleasantly welcome others refuting them, because they think that the many find fault due to envy, and because they are accustomed to being associated with by everyone for the purpose of charm. ... They assume those associating with them forthrightly are becoming friends-of-renown (φιλοδοξεῖν) in order that they may be called *parrhēsiasts*. They consider such association as near insolence and to their own discredit. Kings, on account of their power in general, are not pleasantly inclined toward the people spoken of.[85]

---

is clearly derived from one of the flattering groups, the Gergitha descended from the Gerginoi. Presumably Clearchus' treatise *Gergithius* described the cultural life of Greece, ending in the *personae* of contemporary prominent individuals associated with Alexander's court.

82 Clearchus for example also wrote a treatise called *On Friendship*. Although surviving fragments are few, Clearchus fr. 17 Wehrli discusses a sympotic context for friendship. The earliest treatises specifically on flattery seem to date from the late 4th century, Clearchus' *Gergithius* and Theophrastus' *On Flattery*. Aristotle discusses flattery in *Ethica Nicomachea* 1108a26; Theophrastus *Characteres* 2, 5 includes two flattering types. Modern scholarship has recognized the importance of the Peripatetics in discussions of flattery, although extant evidence is more dominated by literary representations from Middle and New Comedy primarily preserved in Athenaeus *Deipnosophistae* book 6. Discussions of the Peripatetic analysis include Ribbeck 1883, RE 18.4: 1381–1397 s. v. 'parasitos', Fortenbaugh 2011: 204–206. I follow Ribbeck 1883 in treating the ancient terms *kolax* and *parasitos* as equivalents.

83 Philodemus *De libertate dicendi*, P.Herc. 1471 fr. 48, col.XVIb Konstan.

84 Philodemus *De libertate dicendi*, P.Herc. 1471 col.XXIIb.10–13 Konstan: διὰ τι τῶν ἄλλων ἐπ' ἴσης ἐχόν|των ἧττον φοροῦσ[ι]ν ⟨οἱ κ⟩αὶ ταῖς | περιουσίαις κα[ὶ] ταῖς δόξαις | λαμπ[ρ]οί;

85 Philodemus *De libertate dicendi*, P.Herc. 1471 col.XXIIIa.1–6, col.XXIIIb6–15 Konstan: ἐ]ξελέγχοντας [ο]ὐχ ἡδέω[ς | προσδέχονται, [ὅτι] διὰ φθό|νον πολλοὺς ἐπιτ[ι]μᾶν ἑαυ|τοῖς νομίζουσι, [κ]αὶ συ|νειθισ|μένοι ε[ἰ]σί πως [ὑ]πὸ πάντων | πρὸς χάριν ὁμιλεῖσθαι ... ὑπολαμβάνου|σιν μᾶλλον καὶ φιλο|δοξεῖν | τοὺς ἀν[υ]ποστόλως ὁμιλοῦν|τας ὑπονοοῦσιν, ἵνα καλῶν|ται παρρησιάσται, καὶ πα[ρ']

Philodemus provides insight into the mentality of wealthy patrons. They think that social criticism at their expense rebounds to the credit of the truth-teller. Truth-telling might be too strong a term, since the illustrious thought the criticism motivated by envy rather than a regard for truth. The word "truth" appears little in Philodemus' treatise; rather, charm and reputation, elements of socially superior standing, mean more to the establishment. For Philodemus *parrhēsia* indicates moral criticism and rebuke.[86]

Plutarch's treatise, *How to Tell a Flatterer from a Friend*, more directly elucidates the court context of truth-telling. While Plutarch's treatise (like Philodemus') argues that truth-telling in the form of *parrhēsia* is the essence of friendship, the treatise is dedicated to a normative account of true *parrhēsia*.[87] Plutarch constructs a detailed analogy between the application of moral criticism in *parrhēsia* and a doctor's application of medication to a patient: medication needs strength, the right time for application, and the right mixture for the individual constitution.[88] So too the implicit moral criticism of truth-telling in *parrhēsia* needs context to work appropriately. Flatterers, Plutarch says, abuse words of moral quality and with *parrhēsia* drive their patrons toward vice.[89] But true *parrhēsia* corrects mistakes.[90] Highlighting the democratic origins of truthful speech, Plutarch compares the flatterer to the misuser of *isēgoria* or equal speech, another democratic term for *parrhēsia*: the flatterer is "just like the freedman in a comedy who thinks that abuse is the enjoyment of *isēgoria*."[91] *Parrhēsia* was not a neutral term and truth-telling in one context might be perceived as malicious intent, surliness, or agonism in another. "You can find people who think that they are employing *parrhēsia* when they make abuse and find fault."[92] Plutarch's essay contains a description of a flatterer who, in simulating *parrhēsia*, comes closer to general social hostility about minutiae.[93] In a more serious example **no. 17** Ptolemaeus, *philos* of Philopa-

---

ὕ|[β]ριν ἡγο[ῦ]ντα[ι] τὸ τοιοῦτο | καὶ ἀτιμ[ί]αν ἑαυτῶν. οἱ δὲ | βασιλεῖ[ς διὰ τὸ] καθόλου δύν[α|σθ]αι π[ρὸ]ς το[ὺ]ς [ε]ἰρημένο[υς | οὐχ ἡδέως τρέψονται].

86 It is worth noting that Philodemus also wrote a treatise on flattery, which he considered an antagonist of friendship; see Konstan et al. 1998: 6–7. Unfortunately the papyrus containing Philodemus' treatise on flattery is in much worse condition than his treatise on *parrhēsia*. For a discussion of what remains see Tsouna 2007: 126–142.

87 Plutarch *Quomodo Adulator ab Amico Internoscatur* 51C–D.

88 Plutarch *Quomodo Adulator ab Amico Internoscatur* 74D, lengthy analysis of normative *parrhēsia* at 66E-74E.

89 Nature of flatterer at Plutarch *Quomodo Adulator ab Amico Internoscatur* 56C, 62B, 64F.

90 Plutarch *Quomodo Adulator ab Amico Internoscatur* 59E.

91 Plutarch *Quomodo Adulator ab Amico Internoscatur* 66D: ὥσπερ ἀπελεύθερον ἐν κωμῳδίᾳ τὴν κατηγορίαν ἰσηγορίας ἀπόλαυσιν ἡγούμενον. See Raaflaub 2004: 221–225 on the near equivalence of *isēgoria* and *parrhēsia* and their common democratic origin.

92 Plutarch *Quomodo Adulator ab Amico Internoscatur* 66A: εὕροις ... οἰομένους, ἂν λοιδορῶσι καὶ ψέγωσι, παρρησίᾳ χρῆσθαι.

93 Plutarch *Quomodo Adulator ab Amico Internoscatur* 59E–F.

tor, feigned *parrhēsia* to worm his way into an exiled king's inner circle, only to advise Philopator's soldiers to keep close watch on the exile.[94] The acceptance and subsequent betrayal is difficult to understand unless there was near equality in social relations, with moral encouragement and criticism shared between parties. *Parrhēsia* alone was not a guarantee of true friendship.

To sum up, the discourse of friendship at court represented an idealizing picture of its two cultural *personae*, king and courtier. The most important virtues for the king's patronage in court friendship were that the king was cultured, honest, a fair judge above factionalism, humane, and beneficent. Courtiers were expected to be noble, cultured, and loyal, while negotiating the line between honest criticism and too conspicuous flattery. In the idealizing picture of court friendship, both king and courtier were great-souled and friends-of-honor whose mutual intimacy fostered honor and glory.

Symposia

We pass now from the discourse of court friendship to its social sites. Diodorus Siculus' anthology of the virtues of idealized kingship portrayed a ritualized encounter, when the ideal courtiers praised the king after a sacrifice in front of the people. Leading ritual sacrifices was one of the formal functions of Hellenistic kingship. The king's virtues were also displayed in other formal functions of everyday kingship: taking petitions, making correspondence with governors of provinces, hosting guests, hunting, sympotic dining. The king's courtiers too displayed their characteristic virtues during these functions. Not every courtier was present for all these functions: the king's "chief-huntsman" did not sit in on governing councils nor did the *hypomnematographos* go hunting regularly. Yet there was a space where the general court was gathered together on a regular basis: the symposium.

We have much evidence of sympotic gatherings at Hellenistic courts. Hellenistic literature is full of its representations and Athenaeus' *Deipnosophistae* contains extensive anecdotes and reminisces about Hellenistic royal symposia.[95] Although a *symposion* was technically the drinking portion following the *deipnon* or banquet, I will use the term 'symposium' in reference to the specifically Hellenistic version of a communal meal including both eating and drinking. Here king met courtier; scientists were not absent from these gatherings. To situate scientists within the royal symposia, I will describe sympotic practices, the idealized role of courtiers

---

94 Plutarch *Agis et Cleomenes* 57.2 "there was some intimacy and *parrhēsia* between them".
95 We have already met some of these descriptions in **nos. 9, 12, 103, 107–108**. A very descriptive example is Athenaeus 4.128c–130d, the summarized and quoted contents of a letter by Hippolochus describing an early third-century Macedonian symposium for a wedding feast.

as advisors, the virtues exercised by both kings and courtiers at symposia, and the discourse of the contribution for the symposium.

Court actors participated in two kinds of symposia.[96] Symposia were once smaller affairs for male friendship and erotics within Greek city-states; in the Hellenistic age of kingdoms and colonies, symposia were fit for the dignity of the king. In a private symposium, the king dined in the Greek style as a host among friends. The couches were arranged within a particular private room of the palace in the traditional rectangular shape against the wall of the room; fewer than ten people took part on the couches. These smaller meals emphasized equality and reciprocity, although they could still be elaborate. The other symposium was more public, yet differed from the open performances for public consumption such as the extravagant parade put on by Philadelphus.[97] Larger meals with subordinates reinforced hierarchy, while allowing for opportunities of display and interchange between smaller parties. In this second symposium, effectively a kind of state banquet, the king represented himself as the chief power, often sitting on a hemicircular special couch with room for several elite peers in front of his guests on other couches laid out in rows. In Ps.-Aristeas's idealized representation of the royal reception of the Jewish scholars, the scholars sat on two rows of couches with the king in the middle so that half were to his left and the other half to his right, as if in the midst of friends; yet there are seventy or seventy-two scholars in a formal reception room especially prepared with moveable furniture.[98] Ps.-Aristeas' description thus preserves the majesty and status of the king's resources, while emphasizing the equality and exchange between the king and his advisors. At these closed banquets, the king feasted with his political friends and generals, sometimes hosting distinguished visitors, while dining and entertainment staff circulated.

Cultural practices at the symposium (beyond hedonism) included performances, such as singing and dancing, and the recitation of written work. Wit was prized and rewarded with social and financial prizes. Court friends beyond jesters might participate in these games. An example details the wit of the doctor Philotas at the table of Marc Antony's son in Alexandria.[99]

> Philotas often dined with him [sc. the oldest of Antony's sons] among his other companions, whenever he didn't dine with his father. Once there was a doctor dining with them, bold and offering trouble, but Philotas stopped him from talking with a witticism (σοφίσματι): "To the

---

96 On the entire subject see Nielsen 1988, Murray 1996, Strootman 2014: 61–65, 155–156, 188–191.
97 Callixeinus' account of Philadelphus' symposium tent, separate from the parade: *BNJ* 627 (Kallixeinus of Rhodes) F2 = Athenaeus 5.196b–197c.
98 Ps.-Aristeas 183.
99 Philotas of Amphissa was a minor physician who assumes a historical importance thanks to his role as an oral source in Plutarch's *Antonius*. At the time of the narrative frame, Antony was living in Alexandria and enjoying his storied romance with Cleopatra. On Philotas as a physician, see Scarborough 2012: 10–14.

patient who has something of a fever cold water must be given; every fever-victim has something of a fever; therefore every fever-victim must be given cold water." The man was struck silent. Antony's son was pleased and laughed: "Philotas I grant (χαρίζομαι) all these things to you" pointing out a table full of many large cups.¹⁰⁰

Whatever picturesque details have been embellished in the remembering and retelling, Philotas' story describes the sympotic place of a physician. He was a member of the companions, part of the retinue of Antony's son Marcus Antonius Antyllus. The physician regularly dined with his social superior. Medical language was welcomed at the symposium especially in the form of cleverness (σοφίσματι). What is the heft of Philotas' repartee to the other doctor? On the level of medical advice it elides the usual medical niceties of qualifications by degree about severity of illness and application of therapy for the universal. On the level of the symposium, however, it is advice to temper one's (hot) wine with (cold) water and avoid drinking wine neat, a form of barbarism commonly opposed to civilized conversation in the symposium.¹⁰¹ Philotas verbally completes his opponent's remark or caps him by seizing the category of civilized sympotic behavior. Apart from medical language then the physician was indistinguishable from other sympotic guests. He aspired to grants and favors (χαρίζομαι) of material goods from his superior in recompense for his cultural knowledge or *paideia*. The symposium offered the court scientist an opportunity to display knowledge about the natural world.¹⁰²

The sympotic practices of wit, agonism, performance, capping, and conversation might be used in the service of a hedonistic, drunken party or a civilized, philosophical exchange.¹⁰³ An example of this tension comes from the Peripatetic Clearchus of Soli's *On Riddles:*

> Inquiry into riddles is not foreign to philosophy: the ancients used to make a display of their learning (παιδείας) in these matters. For they proposed inquiries to their fellow drinkers not just as people now ask questions, what sex position or which and what kind of fish is sweetest and most seasonable, or what is particularly good to eat after Arcturus or the Pleiades or after the Dog-star. Prizes for winners are kisses, worthy of disgust from people with decent sensibilities; the penalty for failing is drinking unmixed wine, which they drink in preference

---

100 Plutarch *Antonius* 28.7–9: συνδειπνεῖν παρ' αὐτῷ μετὰ τῶν ἄλλων ἑταίρων ἐπιεικῶς, ὁπότε μὴ δειπνοίη μετὰ τοῦ πατρός. ἰατρὸν οὖν ποτε θρασυνόμενον καὶ πράγματα πολλὰ παρέχοντα δειπνοῦσιν αὐτοῖς ἐπιστομίσαι τοιούτῳ σοφίσματι· "τῷ πως πυρέττονι δοτέον ψυχρόν." πᾶς δ' ὁ πυρέττων πως πυρέττει· παντὶ ἄρα πυρέττοντι δοτέον ψυχρόν. πληγέντος δὲ τοῦ ἀνθρώπου καὶ σιωπήσαντος, ἡσθέντα τὸν παῖδα γελάσαι καὶ εἰπεῖν "ταῦτα ὦ Φιλῶτα χαρίζομαι πάντα σοι", δείξαντα πολλῶν τινων καὶ μεγάλων ἐκπωμάτων μεστὴν τράπεζαν.
101 As for example in Callimachus *Aetia* fr. 178; see Harder 2012: 2.969.
102 Chapters 3 and 4 return to the place of the natural world in sympotic discussions and performance.
103 On some of these themes see Kwapisz 2014.

to health. Obviously such matters are characteristic of readers of Philaenis and Archestratus, or any fan of the so-called gastrologies.[104]

Clearchus' moralizing testifies to the new habit of parading one's knowledge of fashionable cookbooks and sex manuals. From a traditional perspective it was bad taste to be so conversant with prose texts of banausic content. After the quoted passage Clearchus contrasts cookbooks and sex manuals with the older sympotic topics: reciting iambic poems, reciting poems with the same number of syllables, alphabetical geography, picking out Homeric commanders. In the contrast between topics of the new symposium and old, Clearchus clearly sided with traditional, elitist themes that showcased a symposiast's intimate knowledge of Greek poetry. While Kwapisz has suggested that Clearchus means to demonstrate that the Hellenistic symposium was undermined as an important venue for *paideia*, I suggest that Hellenistic *paideia* rather encompassed both old and new aesthetics.[105] Demonstrations of new *paideia* were an awareness of presentist themes and prosaic, technical knowledge. Elite poetry's supremacy lost ground at the expense of banausic prose.

We find other confirmation that technical prose was extensively discussed with pleasure at Hellenistic symposia, still involving the practices of wit and agonism characteristic of older symposia.

> Epicurus reproved [the thesis] that 'the wise man is a friend-of-spectacle and alongside a host rejoices in lectures and festival sights, not giving space at drinking parties to learned queries and *topoi* of literary scholars but exhorting even kings who are friends-of-the-Muses to soldiers' tales and common tomfoolery rather than to remain talking in symposia about learned and literary problems.'[106]

---

104 Clearchus of Soli fr. 63 Wehrli = Athenaeus 10.457c–e: τῶν γρίφων ἡ ζήτησις οὐκ ἀλλοτρία φιλοσοφίας ἐστί, καὶ οἱ παλαιοὶ τὴν τῆς παιδείας ἀπόδειξιν ἐν τούτοις ἐποιοῦντο. προέβαλλον γὰρ παρὰ τοὺς πότους οὐχ ὥσπερ οἱ νῦν ἐρωτῶντες ἀλλήλους, τίς τῶν ἀφροδισιαστικῶν συνδυασμῶν ἢ τίς ἢ ποῖος ἰχθὺς ἥδιστος ἢ τίς ἀκμαιότατος, ἔτι δὲ τίς μετ' Ἀρκτοῦρον ἢ μετὰ Πλειάδα ἢ τίς μετὰ Κύνα μάλιστα βρωτός· καὶ ἐπὶ τούτοις ἆθλα μὲν τοῖς νικῶσι φιλήματα μίσους ἄξια τοῖς ἐλευθέραν αἴσθησιν ἔχουσι, ζημίαν δὲ τοῖς ἡττηθεῖσιν τάττουσιν ἄκρατον πιεῖν, ὃν ἥδιον τῆς ὑγιείας πίνουσι· κομιδῇ γάρ ἐστι ταῦτά γέ τινος τοῖς Φιλαινίδος καὶ τοῖς Ἀρχεστράτου συγγράμμασιν ἐνῳκηκότος, ἔτι δὲ περὶ τὰς καλουμένας γαστρολογίας ἐσπουδακότος.
105 Kwapisz 2014: 211. That Clearchus prefers poetry is clear from his *On Riddles*, frr. 84–95 Wehrli, that include numerous examples of playful, sometimes puzzling poems. At the same time Clearchus' own definition of a riddle (fr. 86 Wehrli = Athenaeus 10.488c) does not exclude sympotic discussions of prose: "A riddle is an amusing problem, commanding one to find in his mind through inquiry that which is proposed, said for the sake of profit or penalty".
106 Epicurus fr. 5, 20 Usener = Plutarch *Non Posse Suaviter Vivi secundum Epicurum* 1095C: ἀτοπίαν ὧν Ἐπίκουρος λέγει φιλοθέωρον μὲν ἀποφαίνων τὸν σοφὸν ἐν ταῖς διαπορίαις καὶ χαίροντα παρ' ὁντινοῦν ἕτερον ἀκροάμασι καὶ θεάμασι Διονυσιακοῖς, προβλήμασι δὲ μουσικοῖς καὶ κριτικῶν φιλολόγοις ζητήμασιν οὐδὲ παρὰ πότον διδοὺς χώραν, ἀλλὰ καὶ τοῖς φιλομούσοις τῶν βασιλέων παραινῶν στρατηγικὰ διηγήματα καὶ φορτικὰς βωμολοχίας ὑπομένειν μᾶλλον ἐν τοῖς συμποσίοις ἢ λόγους περὶ μουσικῶν καὶ ποιητικῶν προβλημάτων περαινομένους. Plutarch immediately adds that

The philosopher Epicurus offers a moralizing picture of a normative symposium. Bawdy soldiers' tales ought to give way to questions of intellectual worth. Epicurus evidently distinguished between high and low subjects for conversation. Whether Epicurus' preferred sympotic topics follow Clearchus' intended archaism remains unclear.

But it is clear that these two philosophers and spokesmen for a communal life, a Peripatetic and an Epicurean, valued a certain sort of *paideia* at symposia. They were opposed to the sympotic hedonism of unmixed wine, performance sights, books about sex and food and tales of martial bravado. Instead, they spoke for learned, philosophical discussion, involving questions of high literature and logic. Their sympotic practices idealized the role of *paideia* in shaping group identity. Nor was this surprising, when the tradition of idealized kingship presented the king's *paideia* as the moral prerogative for kingship.[107] An anecdote from Plutarch links Philopator's pretense to *paideia* with his lack of moral character: "When Ptolemy seemed to be a friend-of-learning, his flatterers argued with him about rare words, single verses, and histories half the night but when he employed cruelty, hubris, beat the drum, and participated in mysteries no one rose against him."[108] Plutarch underscores the role of flatterers in advancing sympotic *paideia*, following Clearchus' distinction between old and new symposia that, in Plutarch's view, did not properly form the king's character. The symposium as a social site for court friendship allowed a hermeneutic space between debased and idealized modes of friendship.

Court intellectuals had interests in presenting idealized symposia as moral exemplars. The exemplar of a symposium in Greek culture remained the so-called golden verses of Homer *Odyssey* 9, when Odysseus is entertained at the court of the Phaecians. **No. 136** Eratosthenes, for one, corrected the text of the Homeric symposium to include the phrase "when wickedness is absent".[109] Normative symposia had a moral purpose. The court literature of idealized kingship and friendship prized certain virtues on behalf of kings and courtiers alike. Since the king's character needed guidance, there was a role for advisors in idealized Hellenistic

---

Epicurus wrote these lines in a book called περὶ βασιλείας *On Kingship*, the genre of idealizing portraiture discussed above. Epicurus also wrote a treatise *Symposium*, whose fragments are collected in Usener 1887: 115–119. Only philosophers appeared in *Symposium*; Epicurus *Symposium* fr.56 Usener = Athenaeus 5.177b.
107 Ps.-Aristeas 290.
108 Plutarch *Quomodo Adulator ab Amico Internoscatur* 60A: οὕτω δὲ καὶ Πτολεμαίῳ φιλομαθεῖν δοκοῦντι περὶ γλώττης καὶ στιχιδίου μαχόμενοι καὶ ἱστορίας μέχρι μέσων νυκτῶν ἀπέτεινον· ὠμότητι δὲ χρωμένου καὶ ὕβρει καὶ τυμπανίζοντος καὶ τελοῦντος οὐδεὶς ἐνέστη τῶν τοσούτων. I understand γλώττης, στιχιδίου, and ἱστορίας to be generalizing singulars.
109 Athenaeus 1.16d–e. Eratosthenes replaced κατὰ δῆμον ἅπαντα "throughout all the people" with κακότητος ἀπούσης "when wickedness is absent", corresponding to the criticism of Plato *Respublica* 3.390a–b that the Homeric lines need to be understood with reference to self-control. Eratosthenes thereby reinforced the elitism of the Homeric exemplar.

kingship. The literature of idealized kingship envisioned moral and political advisors from among the king's friends, his chief courtiers, and the age-mate sons of the leading nobles of the kingdom; this literature also envisioned a role for expert knowledge in ruling the kingdom. The experts were philosophers, scientists, wisemen, and priests. These experts advised the monarch about human behavior, the nature of the gods, and the nature of the natural world. Ps.-Aristeas captures the king's need for expert advisors.

> "[The king] asked the next scholar: 'How ought [a king] be a friend-to-lectures (φιλήκοος)?' He replied: "By understanding that it is useful to know everything."[110]

Yet the utility of knowledge over all domains is not the same as the aspiration to know everything. A contemporary Peripatetic philosopher, Aristo of Ceos, parodied the character of the παντειδήμων "know-it-all" in a treatise on vices, the sort of person who attempts all branches of knowledge (τῶν μαθημάτων ἀντιποιούμενος πάντων).[111] The so-called know-it-all represented the non-modern man who did not recognize the differences between branches of knowledge and who, continues Aristo, like the sophist Hippias of Elis pretended to achieve independently the accomplishments of communal life. Expert knowledge was specific and not general. Even the ambition of the Alexandrian library was to collect specialized learned knowledge of all sorts, not incorporate it into a single individual.

The monarch's pose of friendship was intended to harness the disciplinary knowledge of individuals for utility in governing. The king needed to exercise his virtues of cultured disinterestedness, without regard to factional opinions, and honesty when welcoming expert knowledge to sympotic discussion.

> "How ought a king to pass his time in symposia?" The scholar replied: "By taking up conversation alongside friends-of-learning and men capable of suggesting what is useful to the kingdom and to the lives of its subjects – you would find nothing more harmonious or more Muse-like. For these men, friends-of-gods, have educated their thoughts toward the fairest things."[112]

The king entertained a variety of experts, since expert knowledge might prove useful for all possible facets of human life in the kingdom. Court scientists thus envisioned knowledge of the natural world serving the political order. The utility of expert knowledge corresponded to the interests of the political elite, a conclusion

---

110 Ps.-Aristeas 239: Πῶς ἂν φιλήκοος εἴη; ἐκεῖνος δὲ εἶπε· Διαλαμβάνων ὅτι πάντα συμφέρει γινώσκειν.
111 Aristo of Ceos fr. 21i.11–38 SFOD = *P.Herc.* 1008.18.
112 Ps.-Aristeas 286–287: Πῶς δεῖ διὰ τῶν συμποσίων διεξάγειν; ὁ δὲ ἔφησε· Παραλαμβάνοντα τοὺς φιλομαθεῖς καὶ δυναμένους ὑπομιμνήσκειν τὰ χρήσιμα τῇ βασιλείᾳ καὶ τοῖς τῶν ἀρχομένων βίοις – ἐμμελέστερον ἢ μουσικώτερον οὐκ ἂν εὕροις τι τούτων· οὗτοι γὰρ θεοφιλεῖς εἰσι πρὸς τὰ κάλλιστα πεπαιδευκότες τὰς διανοίας.

I will reinforce below. The king's promotion of expert knowledge for use in his kingdom rendered him a friend-to-humanity.

Symposia offered a site for courtiers, like the king, to exercise their virtues of court friendship – honesty, loyalty, cultured honor – in social exchanges. The host encouraged his friends to speak out honestly with *parrhēsia* and to discuss topics of importance to society and high culture. As Plutarch's story of Philotas showed, gift-exchange was the characteristic sympotic practice by which court friends pledged loyalty to each other's interests and sustained equal standing. Guests pledged a contribution for the communal enjoyment of the symposium, called a *symbolē* or *eranos*, while the symposiarch host would offer friends and participants some gift as a token of appreciation.[113] The gift-exchanges that happened at the symposia were ambiguous: one on the one hand, a gift followed true friendship and concretizes the prior expectation of reciprocity, but gifts given freely also allowed the power differential to be preserved because they could be given without expectation of moral reciprocity.

Strikingly textual works written or performed for court society followed the practices of the symposium. Courtiers represented themselves as friends and framed their books as their contributions to sympotic exchange. The Hellenistic historian Demetrius of Skepsis recorded an anecdote about recitation at a royal symposium that became a sympotic contribution.

> Demetrius of Skepsis in the fifteenth book of *On the Trojan Order* said: "At symposium next to Antiochus the king, surnamed The Great, there danced to armor not only the king's *philoi* but also the king himself. But when the order of dancing came to Hegesianax of Alexandria Troas who wrote the histories, he rose and said: 'Do you wish, king, to see me dance badly or do you want to listen to me pronounce my own works well?' Upon being asked to speak he sat, so that the king was recompensed for a contribution, and Hegesianax became one of the *philoi*."[114]

Hegesianax of Alexandria Troas was both a poet and historian as well as a friend of Antiochus the Great, opponent of Philopator at the battle of Raphia.[115] Here it is

---

113 Pun on *symbolai*, banquet contributions at Athenaeus 1.4b. Further examples of *symbolai* given in Athenaeus 8.364f–365e. Is Ps.-Aristeas 246 συμβουλίαις a pun, indicating both councils of state as well as private exchanges in symposia? See Strootman 2014: 190.n18 for comment on the distribution of sympotic gifts.

114 Demetrius of Scepsis fr. 7 Gaede = Athenaeus 4.155b: Δημήτριος δ' ὁ Σκήψιος ἐν τῷ πέμπτῳ καὶ δεκάτῳ τοῦ Τρωικοῦ Διακόσμου, παρὰ Ἀντιόχῳ, φησί, τῷ βασιλεῖ τῷ μεγάλῳ προσαγορευθέντι ἐν τῷ δείπνῳ πρὸς ὅπλα ὠρχοῦντο οὐ μόνον οἱ βασιλέως φίλοι, ἀλλὰ καὶ αὐτὸς ὁ βασιλεύς. ἐπεὶ δὲ καὶ εἰς Ἡγησιάνακτα τὸν Ἀλεξανδρέα ἀπὸ Τρῳάδος τὸν τὰς ἱστορίας γράψαντα ἡ τῆς ὀρχήσεως τάξις ἐγένετο, ἀναστὰς εἶπε· Πρότερον, ὦ βασιλεῦ, κακῶς ὀρχούμενόν με θεάσασθαι βούλει ἢ καλῶς ἀπαγγέλλοντός μου ἴδια ποιήματα θέλεις ἀκροάσασθαι; κελευσθεὶς οὖν λέγειν οὕτως ᾖσε τὸν βασιλέα ὥστ' ἐράνου τε ἀξιωθῆναι καὶ τῶν φίλων εἷς γενέσθαι. I follow Olson's 2007–2012 text.

115 Antiochus used Hegesianax in diplomatic roles for negotiation with the Romans, confirming that he fulfilled one of the political functions of court friendship; *BNJ* 45 (Hegesianax of Alexandria Troas) T4a.

unclear whether Hegesianax is offering to recite his prose works or poems, although the latter seems more likely. The anecdote recorded testifies to the expectation that guests would provide an *eranos* or contribution to the banquet. While normally a financial contribution, the *eranos* could take other forms. In this case, Antigonus expects his *philoi* to dance as part of their contribution. Hegesianax offers an alternate *eranos*, which he will accomplish much better than the dancing: he will read his own works.[116] The book as content or material object becomes the sympotic contribution expected of the guest.

A saying attributed to the Peripatetic advisor of Ptolemy I Soter, Demetrius of Phalerum, reversed the *topos* between oral friendship and written advice: "What friends dare not urge to kings is written in books."[117] The usual reading of this proverb is that courtiers cannot be honest with kings and so write out what they fear to say in person. Our post-Enlightenment world is inclined to believe the power of the written word in such Hellenistic *exemplaria* as the poet Sotades' repost to Philadelphus on his incestual marriage ("You're shoving your dick into an unholy hole"), but such outright confrontation with or criticism of monarchs is rare in extant Hellenistic literature.[118] Plutarch himself frames Demetrius' proverb within the social context of building the Alexandrian library: friends are no replacement for the specialized knowledge found in individual books. Rather, Demetrius' advice is itself a literary *topos* about the written word as a substitute for a personal advisor.

The book became the contribution of an otherwise absent advisor.[119] The absent advisor is a theme running through court literature, as much as the extensive network of correspondence necessary to sustain the bureaucratic apparatus of royal governance. (Selecus I is said to have remarked that he would not have taken up the crown, if he had known how much reading and writing was involved in kingship.[120]) Isocrates' advice to King Nicocles opens with this very *topos*:

> Some are accustomed, Nicocles, to bring you kings clothes, bronze, worked gold, or some other such possession, which they have need of but you have plenty of; they seem to me clearly not to be making a gift but commerce and sell it with more professionalism than people who engage in trade. But I would think that this gift would be most beautiful, most useful,

---

[116] Cameron 1995: 71–103 collected many examples of performance and recitation of texts at Hellenistic symposia. Not all examples are certain evidence but the general point seems secure.
[117] Demetrius of Phalerum fr. 63 Wehrli = Plutarch *Regum et Imperatorum Apophthegmata* 189D: ἃ γὰρ οἱ φίλοι τοῖς βασιλεῦσιν οὐ θαρροῦσι παραινεῖν, ταῦτα ἐν τοῖς βιβλίοις γέγραπται.
[118] Sotades fr. 1 CA = Athenaeus 14.621a: εἰς οὐχ ὁσίην τρυμαλιὴν τὸ κέντρον ὠθεῖς. Athenaeus calls this an "inopportune *parrhēsia*". Other court poets celebrated the marriage: Callimachus fr.392 Pfeiffer, Theocritus 17.128–135.
[119] Chapter 3 resumes this discussion with reference to extant treatises of Hellenistic court science.
[120] Plutarch *An Seni Respublica Gerenda Sit* 790A.

and especially fitting – for me to give and you to receive – if I should able to define by partaking and abstaining from which needs you would best govern both your city and your kingdom.[121]

Isocrates' opening coopts the discourse of gift-exchange for his advice letter. The behavior that courtiers should imitate is that of a good friend; the ideal courtier was the true friend that the king himself would choose. Isocrates' court friendship was a product of the intellectual needs of state, not a naturalized friendship born of mutual affection.

Scientists in Gift-Exchange

Who was the scientific self at the Ptolemaic courts? A self who aspired to the virtues of the courtier combined with the authority of an expert in a specialized knowledge. Scientific knowledge is a form of truth-telling. Observational accuracy is secured through the experimenter's able-bodiedness and socially trustworthy witnesses. Authority does not necessary follows from epistemic originality. I suggest that the scientific practitioner at court adopted a sympotic *persona* with epistemic virtues good for empirical knowledge-making: the friendship type of the *parrhēsiast* or the flatterer. The scientific self as a friend-of-learning became a species of court friendship. The *persona* of the ancient Ptolemaic courtier aspired to virtues of truth-telling, *parrhēsia*, *paideia*, and loyal friendship within the culture of the symposium. As a true friend the scientific self combined these epistemic virtues with specialized expertise useful for the lives of the kingdom's inhabitants. As a flatterer the scientific self pretended to truth-telling and offered expert knowledge as a complement to royal entertainment such as dancing and heavy drinking. The scientific self of courtly culture initiated a reciprocal exchange of goods: the scientific practitioner gave intellectual knowledge as a sympotic *eranos*, while the king in his capacity as symposiarch in recompense gave finances and distinction. A social dynamics of patronage likely underlay the empirical knowledge of Ptolemaic court scientists. Hence we finally come to an explicit discussion of gift-exchange.

We have already seen evidence in earlier chapters how scientists approached royal power with knowledge about the natural world. In order to gain Euergetes'

---

121 Isocrates *Ad Nicoclem* 1–2: οἱ μὲν εἰωθότες, ὦ Νικόκλεις, τοῖς βασιλεῦσιν ὑμῖν ἐσθῆτας ἄγειν ἢ χαλκὸν ἢ χρυσὸν εἰργασμένον ἢ τῶν ἄλλων τι τῶν τοιούτων κτημάτων, ὧν αὐτοὶ μὲν ἐνδεεῖς εἰσιν, ὑμεῖς δὲ πλουτεῖτε, λίαν ἔδοξαν εἶναί μοι καταφανεῖς οὐ δόσιν, ἀλλ' ἐμπορίαν ποιούμενοι καὶ πολὺ τεχνικώτερον αὐτὰ πωλοῦντες τῶν ὁμολογούντων καπηλεύειν. ἡγησάμην δ' ἂν γενέσθαι ταύτην καλλίστην δωρεὰν καὶ χρησιμωτάτην καὶ μάλιστα πρέπουσαν ἐμοί τε δοῦναι καὶ σοὶ λαβεῖν, εἰ δυνηθείην ὁρίσαι ποίων ἐπιτηδευμάτων ὀρεγόμενος καὶ τίνων [ἔργων] ἀπεχόμενος ἄριστ' ἂν καὶ τὴν πόλιν καὶ τὴν βασιλείαν διοικοίης.

favor **no. 137** Conon, the court astronomer, recognized Queen Berenice's stolen lock as a catasterized constellation in the night sky, part of a cluster of Egyptian mythological figures keeping cosmic order. **No. 132** Caphisophon, possible royal physician, interceded with Euergetes to free his friend from prison. These two court scientists employed different means to achieve their political ends. Conon used his knowledge of the natural world in pursuit of patronage, in what I have termed court science. Caphisophon, on the other hand, played the part of an intimate friend of the king, not a scientist knowledgeable about the natural order, to secure his friend's release. Court scientists played the part of scientist or courtier to achieve some political end.

The career of the court physician Apollophanes of Seleucia illustrates the discursive formation of court friendship. Apollophanes was the court physician to Antiochus III the Great, the Seleucid king who was Philopator's opponent at the battle of Raphia. Now Antiochus was a young king and so a particular courtier named Hermeias had been appointed to an important post, but Hermeias had gradually eliminated his rivals and gathered more power to himself with the guardianship of the king's infant son as his object. Fearing for himself and the king, Apollophanes intervened with the king, who also feared Hermeias.

> Antiochus said that he had great thanks for having dared to speak to him about these things with care ... Together Apollophanes and the king agreed and made a plan: as if spells of fainting had afflicted the king, they would release the court retinue for a few days but retain for consulting individually the *philoi* on an excuse of medical investigation.[122]

So the trap was laid. The doctors ordered a morning walk in the cold for the benefit of the king, when court literature records that the king had a morning audience.[123] So Hermeias and the *philoi* privy to the plot came at the appointed time, but the rest of the retinue was late because of the change to the king's routine. The group left the camp and in a remote area the king gave the signal to kill Hermeias. Apollophanes' influence with the king continued; at a council of the king's friends he gave a major speech prompting the campaign against Philopator that ended in the battle of Raphia.[124] Whether Apollophanes was among the royal *philoi*, he was certainly prominent at court and beyond: he was honored by the king in surviving inscriptions and honored as a patron of a local shrine of Zeus.[125] In addition to

---

122 Polybius 5.56.4, 7–8: ἐκείνῳ δὲ μεγάλην χάριν ἔχειν φήσαντος [sc. Antiochus] ἐπὶ τῷ κηδεμονικῶς τετολμηκέναι περὶ τούτων εἰπεῖν πρὸς αὐτόν ... συμφρονήσαντες μετὰ ταῦτα καὶ προβαλόμενοι σκῆψιν ὡς σκοτωμάτων τινῶν ἐπιπεπτωκότων τῷ βασιλεῖ, τὴν μὲν θεραπείαν ἀπέλυσαν ἐπί τινας ἡμέρας καὶ τοὺς φίλους ἔλαβον ἐξουσίαν οἷς βούλοιντο κατ' ἰδίαν χρηματίζειν διὰ τὴν τῆς ἐπισκέψεως πρόφασιν.
123 Ps.-Aristeas 304.
124 Polybius 5.58.7–8.
125 Samama nos. 133, 233.

his political prowess Apollophanes' pharmaceutical recipes circulated in the Greek medical tradition into the Byzantine era, with compounds for snake-bites, burns, and hemorrhoids.[126]

Neither Apollophanes' scientific nor political interests were novel; for this reason they are all the more illustrative about the ethical self of the court scientist. Apollophanes represented the categories of good friendship presented in this chapter: like a good courtier he represents the king's interests at every turn with loyalty, he plays the *parrhēsiast* when speaking of Hermeias' designs on the king's life, his speech in favor of war emphasizes the territory to be gained as a hearth, central to kingship's conception of familial power. In setting a trap for Hermeias Apollophanes placed his scientific knowledge within the king's political interests. In turn for his good friendship Antiochus granted Apollophanes social and financial honors, as the inscriptions record. The relationship between Apollophanes and his king testifies to the primacy of friendship as a discursive construction of gift-exchange in court life.

The named intervention and court friendship of Apollophanes stand in contrast to an anonymous example of scientists' activation of gift-exchange. I refer to most important astronomical advance during Euergetes' reign, seemingly a calendar reform, which is explained in the Canopus Decree promulgated under Euergetes in 238 BCE.[127] The Canopus Decree addresses a mismatch between the religious calendar governed by the stars and the seasonal calendar governed by the sun; the reform promulgated prevents a future mismatch. The ancient Egyptian year began in the month of Thoth, roughly our September; in the Canopus Decree it is celebrated in Pauni, roughly our June, in conjunction with the fruit harvest and the annual Nile flood. The star of Isis, better known as Sirius or the dog star, opens the religious year according to the hieroglyphs. The regular solar Egyptian year was 360 days distributed throughout twelve months with five additional days annually added at the end of the cycle, for a total of 365 annual. But due to the physical fact that the Earth's axis of rotation itself shifts with gravity, the length of the solar year (the sun's rising and setting on the horizon) is not exactly equivalent to the length of a sidereal year (a star's rising and setting on the horizon). The star Egyptians used to mark their new religious year, Sirius, adds exactly one day to its rising over four solar years. The seasonal year, governed by the sun, became disjoined from the religious year, governed by the rising of Sirius. The Canopus Decree established that one intercalated day be inserted into the seasonal calendar every four years in order to keep the seasonal calendar in alignment with the sidereal calendar.

Euergetes' court had three Greek astronomers capable of advising him on the calendar change promulgated by the Canopus Decree: **nos. 136–138** Eratosthenes,

---

[126] *EANS* 117–118.
[127] *OGIS* 56.33–46.

Conon, and Dositheus. Geus has argued that Eratosthenes of Cyrene was responsible for the calendar reform of the Canopus Decree. He points in particular to Eratosthenes' treatise *On the Eight-Year Cycle*, mentioned by the later astronomer Geminus.

> For this reason the Isis-festival was celebrated at the winter-solstice and earlier still at the summer solstice, as also Eratosthenes mentions in his *On the Eight-Year Cycle*. It will be celebrated again in the future in fall, summer solstice, spring, and winter solstice. For in 1460 years it is necessary that every festival pass through all seasons of the year and again be established at the same time of the year.[128]

Geus argues that the problem addressed by Eratosthenes in *On the Eight-Year Cycle* is that treated in the Canopus Decree: the festivals of Isis mentioned by Eratosthenes are synonymous with the festivals that inaugurate the Egyptian religious year; the religious year, governed by a sidereal calendar, was out of alignment with the solar calendar.[129] Yet it is not clear how far Geminus is quoting Eratosthenes, since Geminus states that the religious festival will continue to shift seasons until the necessary 1460 years (a so-called Sothic cycle) have elapsed to synchronize sidereal and solar calendars once more. If Eratosthenes was responsible for the calendar reform of the Canopus Decree, his *On the Eight-Year Cycle* was written before the decree was promulgated since it indicates that a mismatch will continue between solar and sidereal calendars.

But Eratosthenes was not the only court astronomer to write a treatise called *Eight-Year Cycle*; Dositheus is reported to have done so as well.[130] The problem handled in such treatises concerned the conjunction of the solar calendar and lunar calendar: the moon's phases reoccur on the same day of the solar year every eight years.[131] Establishing a conjunction of lunar phases that governed the civic and religious calendars of the individual Greek cities with the solar year was a preoccupation of fifth and fourth century BCE Greek astronomers. Prominent earlier astronomers were credited with titles concerning this cycle; Eratosthenes apparently disputed the authenticity of one such treatise.[132] Yet to construe Eratosthenes' or Dositheus' treatise as the impetus for the calendar reform of the Canopus

---

[128] Geminus *Isagoge* 8.24: ὅθεν τὰ Ἴσια πρότερον μὲν ἤγετο κατὰ τὰς χειμερινὰς τροπάς, καὶ πρότερον δ' ἔτι κατὰ τὰς θερινὰς τροπάς, ὡς καὶ Ἐρατοσθένης ἐν τῷ περὶ τῆς ὀκταετηρίδος ὑπομνήματι μνημονεύει. καὶ ἀχθήσεται πάλιν κατὰ φθινόπωρον καὶ κατὰ τὰς θερινὰς τροπὰς καὶ κατὰ τὸ ἔαρ καὶ κατὰ τὰς χειμερινὰς τροπάς. ἐν ἔτεσι γὰρ ͵α υξ' ἅπασαν ἑορτὴν διελθεῖν δεῖ διὰ πασῶν τῶν τοῦ ἐνιαυτοῦ ὡρῶν καὶ πάλιν ἀποκατασταθῆναι ἐπὶ τὸν αὐτὸν καιρὸν τοῦ ἔτους.
[129] Geus 2002: 207–210.
[130] Censorinus *De Die Natali* 18.5.
[131] In physical reality the conjunction is not consistent: there is roughly a day to day-and-a-half variance.
[132] Achilles Tatius *In Arati Phaenomena* 19 (47.22–24 Maass).

Decree means imputing a technical content at odds with the title: the eight-year cycle concerned the conjunction of solar and lunar calendars; the Canopus Decree concerns the conjunction of solar and sidereal calendars. While it is hard to believe from technical grounds that any treatise written by Eratosthenes or Dositheus continued to dispute the two hundred year-old problem of lunar and solar conjunction, our sources are too thin to infer that Eratosthenes or Dositheus treated the title *Eight-Year Cycle* as a generalized topic concerning the alignment of a sidereal year with a solar year.

Against a possible Greek background to the Canopus Decree Pfeiffer has argued that the calendar reform was initiated by the Egyptian priests.[133] His primary objection is that the so-called calendar reform is, in fact, not a calendar reform at all. At this time in the Ptolemaic state there were three calendars: the Egyptian religious calendar, the Macedonian calendar, and the fiscal calendar. The Canopus Decree adds an additional day into the Egyptian calendar, which is not reflected in the other calendars seen in surviving papyri. Furthermore, while according to the hieroglyphs the new year began with the rise of Sirius, in the year 238 BCE the seasonal calendar with the fruit harvest and the rise of the Nile was three months behind the rise of Sirius. To have brought the Egyptian calendar into cosmic alignment would have meant subtracting an entire three months in 238 BCE.[134] The leap year day did not last beyond the reign of Euergetes; later Ptolemaic dates make no reference to it. Pfeiffer argues that the intercalated day, to be celebrated as an additional festival day of the *theoi Euergetai*, is therefore best understood not as a kingdom-wide calendar reform but as a particular religious honor to the *theoi Euergetai* within their lifetimes.

Does recognizing that the Ptolemic court was a multicultural society force us to choose between Greek or Egyptian agency? Rather than delimit the agency to one ethnic group, I would prefer to emphasize the social dynamic driving the process. The Canopus Decree shows some scientific knowledge being put it at the service of royal power in the form of court science. Whether it was Egyptian observation, Greek theory, or some combination, the agents responsible for the intercalated day were just as effective courtiers and petitioners of royal power as they were observers of and theorists about the risings of Sirius. It is unfortunate that we do not know the names of these scientists who used their knowledge of the natural world to offer a royal patron an honor.

And what an honor to give to royal patrons – a whole new day, named and celebrated in their honor! Gold and silver are fine things worthy of kings but a new

---

133 Pfeiffer 2004: 249–257.
134 Pfeiffer 2004: 255: "Um das [sc. the synchronization] zu erreichen, hätte man den gesamten Kalender um drei Monate nach vorne verschieben müssen. Das Dekret betont hingegen mit Nachdruck, daß die Jahreszeiten 'das Gehörige machen gemäß der *jetzigen* Beschaffenheit des Kosmos.'"

day is very rare indeed and worthy of gods. The Canopus Decree twice specifies exactly the movement of Sirius: "And whenever the rising of the star happens to change into another day in four years", "since the star changes one day in four years".[135] Sirius' movements were well observed and charted over extended time; a fact about the natural world was established. One intercalated day set to bring seasonal and sidereal calendars into alignment is a simple and effective solution to the social problems that arise from this natural fact. Once every 1460 days a new solar free day is added to the calendar to keep the Sun and Sirius in alignment. The sun's rising and setting on that intercalated day is a cause for religious celebration in honor of political power.

> [It is decreed] that from the present moment a one day festival of the *theoi Euergetai* be intercalated in four years after the five added days before the new year, so that everyone may know that the previous defect in the arrangement of the seasons, the year, and the tradition about the entire revolution of the pole has happened to be corrected and filled by the *theoi Euergetai*.[136]

A natural fact has become transformed into a statement of political will: everyone will recognize the achievement of King Ptolemy Euergetes and Queen Berenice, *theoi Euergetai*, for being gods able to maintain the correct alignment of other cosmic powers. The Canopus Decree does not credit court scientists for the intercalated day because the Decree is a celebration of the power of the *theoi Euergetai*.[137]

Kings in Gift-Exchange

Since gift-exchange involves two individuals, I conclude by describing royal motive to instigate scientific research. The ideology of kingship was thick with gift-exchange.[138] Diodorus Siculus' idealizing picture of Ptolemaic kingship listed being

---

[135] *OGIS* 56.38: ἐὰν δὲ καὶ συμβαίνηι τὴν ἐπιτολὴν τοῦ ἄστρου μεταβαίνειν εἰς ἑτέραν ἡμέραν διὰ τεσσάρων ἐτῶν; *OGIS* 56.41–42: τοῦ ἄστρου | μεταβαίνοντος μίαν ἡμέραν διὰ τεσσάρων ἐτῶν.

[136] *OGIS* 56.44–46: ἀπὸ τοῦ νῦν μίαν ἡμέραν ἑορτὴν τῶν Εὐεργετῶν θεῶν ἐπάγεσθαι διὰ τεσσάρων ἐτῶν ἐπὶ ταῖς πέντε ταῖς | ἐπαγομέναις πρὸ τοῦ νέου ἔτους, ὅπως ἅπαντες εἰδῶσιν διότι τὸ ἐλλεῖπον πρότερον περὶ τὴν σύνταξιν τῶν ὡρῶν καὶ τοῦ ἐνιαυτοῦ καὶ τῶν νομιζο|μένων περὶ τὴν ὅλην διακόσμησιν τοῦ πόλου διωρθῶσθαι καὶ ἀναπεπληρῶσθαι συμβέβηκεν διὰ τῶν Εὐεργετῶν θεῶν.

[137] Pfeiffer's 2004 objections to the *opinio communis* also concern differences between the Greek and hieroglyphic texts at this very point: the Greek διὰ τῶν Εὐεργετῶν θεῶν suggests that Euergetes took some agency, whereas the demotic and hieroglyphic point toward an honor for or because of the *theoi euergetai*; see Pfeiffer 2004: 142–144.

[138] Plutarch's moralizing picture of the good Spartan king Cleomenes in exile (*Agis et Cleomenes* 53–54) shows a king who dispenses wisdom without elaborate court ceremony or social intrigues. Compare also the picture of Cleomenes' symposia in Athenaeus 4.142c–f.

a giver of goods among the virtues of a good king. The king as a patron of elite culture maintained a judicious disinterestedness and a gracious ability to listen to all sorts of specialized knowledge. The king's patronage of expert knowledge allowed him to exercise his ideal virtues of humanity and beneficence. The king's patronage alighted on agonistic scientists who preserved goodwill toward the king while negotiating the line between honest criticism and too conspicuous flattery. But what projects did kings support?

Of all scientific endeavors, Hellenistic kings were interested in war and building projects; so they supported the necessary technology. A ready example is Archimedes, whom ancient sources implicate in royal initiatives about the natural world. From proving forgeries in metal crowns to building war-machines Archimedes' ancient reputation was built on engineering exploits initiated by royal interest.[139] Archimedes also built an enormous ship for king Hiero of Syracuse whose decoration featured all the aspects of a city – columns, horse-stable, fish-tank, library, gymnasia, turrets, and a symposium room with an inlaid floor depicting the *Iliad*.[140] Few harbors except Syracuse or Alexandria could accommodate the size of the ship, so Hiero sent it to Ptolemy as a gift, complete with a shipment of grain because Egypt's harvest was short.[141] The financial value of the ship was clearly less important than its symbolic value of Hiero's royal power and political friendship: the ship concretizes Hiero's royal virtues of magnificence and justice while displaying his worth as a friend. **No. 139** Philon identified the royal initiative for supporting technology: "The craftsmen in Alexandria accomplished this [a particular war-machine] first, having gotten large financial support on account of the fact that they secured the patronage of kings who were friends-of-renown and friends-of-the-*technai*."[142] In the language of court friendship, kings were φιλόδοξοι. It was characteristic of the great-souled man that he sought glory.

The language of honor and value shows that gift-giving in ancient science was an exchange between two private individuals, typically already members of the elite. So the patronage of ancient science was not euergetism, the conspicuous giving for the common polity from the rich to the poor.[143] Rather, it shared the language of euergetism in the exchange of favors granted and the motivations of generosity of spirit characteristic of the great-souled man; but, unlike a euergetism that offered something to the people, it adhered to a vision of technological utility

---

139 Vitruvius 9.pr.9–10; Plutarch *Marcellus* 14.7–15.
140 Athenaeus 5.206e–209b. Hiero completes his technological wonder in the Greek style by commissioning an epigram from a poet.
141 Perhaps the intended recipient was Euergetes, since the Canopus Decree (*OGIS* 56.15–18) documents a grain shortage in the early years of his reign.
142 Philon *Belopoeica* 50.24–26 Th, cited above in **no. 139**.
143 See Veyne 1990: 70–200 on Greek euergetism. While focusing primarily on donations to the cities, Veyne's description of the ideology of the *euergetai* applies equally to kings.

and aesthetic appreciation of the natural world intended for the elite of court society. Gift-giving in ancient science was an activity of an elite class, for the pleasure of that class. We have no evidence that the inventions or discoveries produced through the patronage of ancient science were widely adopted, commercialized, or instrumentalized to provide what was economically "useful for the kingdom".[144] We know of some agricultural innovations adopted in the Ptolemaic Fayum but otherwise the notion of "utility", widespread in description of Hellenistic scientific material, must refer to its value as an object of cultural prestige.[145] Court patronage promoted elite knowledge and novelty, not democratic utilization. If the people happened to profit from court science, that was an additional consequence of the gift's intended effect – to the greater glory of the kingdom and its rulers. For example, the Canopus Decree's proclamation of a new day created a festival day for the people, but the alignment of the solar and sidereal calendar is presented as a triumph honoring the *theoi euergetai*. The patronage of ancient science was a product of the king's desire for glory, his quest to be the ideal friend.

A central theme in contemporary scholarship of the sociology of science is that financial incentives and institutional structures alter the focus and legitimation procedures of the production of scientific knowledge. The application of that conclusion to the material under discussion here should be patent. It is wrong to say that the ideology of kingship and the practice of gift-exchange corrupted science, as if the contemporary ideology of democracy and practice of peer-review were normative matrices for the production of scientific knowledge. Rather, the conclusions of this chapter point in two directions. First, we should give more credit to the ideologies and cultural practices of historic non-democratic political structures in achieving successful production of scientific knowledge useful within its historical situation and valid in transhistorical review. It is a simple truism, still easy to forget, that uncovering truth about the natural world happens under multiple po-

---

144 Ps.-Aristeas 286.
145 On agriculture in the Fayum see Hölbl 2001: 62–63. Meißner 1999: 161 has pointed to *SB* 6.9302 as indicating the general relationship of utility between scientist and king in the third century BCE. This papyrus is a petition (ἔντευξις) from a land-holder in a regional Egyptian city to the king. The petitioner wishes to announce a μηχανή or machine capable of bringing an end to the drought (ἀβροχία) that has afflicted the land and prevented the Nile from rising much for three years. It would seem as if the petitioner's machine was some sort of irrigation device to divert river water to fields along the riverbanks that had not received their customary inundation. Since the Greek term μηχανή covers simple as well as complicated devices, without further evidence we cannot determine the complexity, sophistication, or even feasibility of the announced device. Insofar as maintaining the Nile's flood and accompanying agricultural fertility was part of the good kingship expected of the pharaoh (cf. Stephens 2003: 96–100), this petition appeals to the king's expected beneficence. Therefore I cannot share Meißner's view that *SB* 6.9302 shows the general valuation of "utility" between scientist and king; rather it shows the specific example of a scientific applicant's recognition of the king's desire for glory.

litical orders and publication regimes. Second, the *longue durée* of scientific patronage might provide useful analytic tools for considering the contemporary knowledge production emerging from non-institutional sites and for reviewing the changes occurring to financial support in a digital, sharing economy. Perhaps the strongest point of departure between Hellenistic court science and patronage in its many contemporary forms is that in antiquity scientific inventions and discoveries were never monetized or commoditized. The comparison reminds us that money was a derivative outcome, not driving function within ancient gift-exchange. Such a conclusion is no more than the broader point, common to scholarship on premodern gift-exchange, that the gift-economy is not a precursor to a market economy. Although gift-exchange is a transactional matter, other things were on offer in antiquity. Both king and courtier sought honor or renown (τιμή, δόξα) in friendship with each other. The historiographical point enunciated at the beginning of this chapter stands: the mechanism of ancient gift-exchange was embedded within a complex negotiation of social roles and cultural resources. To search for ancient financial transactions as a precursor to understanding gift-exchange neglects the important contextual discourse of friendship that dominates all aspects of Hellenistic court.

3 An Entertaining Genre

> I see that you are disposed in a way friendly-to-doctors (φιλιάτρως), King Ptolemy ... In order that the details be very easy to follow for you, I will set out first the passages of Hippocrates, afterwards arranging the more ready ways of resettings as a sort of contribution (ἔρανον) from the events themselves, which happen through the labor of assistants.[1]

Apollonius of Citium's introduction to his treatise on Hippocrates' works on bone-setting plunges us at once into the literary remains of ancient science in the gift-exchange market. We see the parameters of court friendship described in chapter 2. Apollonius addresses his patron with the language of court friendship (φιλιάτρως). Ptolemy's figuration as the ideal friend distributes his social standing to the *technai*. Apollonius frames his work on Hippocrates as an *eranos*, as if the medical resetting of joints was a fit spectacle and useful knowledge for sympotic participants. The treatise substitutes for Apollonius as an absent advisor, providing the king and his circle useful specialized expertise from the historical Greek paragon of medical knowledge. Apollonius' treatise will act as an explanation of Hippocrates' text to provide a less technical account to the king as a lay reader.

Apollonius' treatise is what I will call a "court science treatise", any literary text written by a practicing scientist to a court official with political power. The court science treatises derive from a well-defined social situation and share several aesthetic features. We may therefore call such literature a genre, although the grouping is a modern analytic category. This chapter treats the literary features of this genre.[2] I provide a close literary reading of the four extant prose treatises written by Hellenistic scientists at court: Archimedes of Syracuse's *Sand-Reckoner* written for King Gelon of Syracuse, **no. 136** Eratosthenes of Cyrene's *Letter to Ptolemy* written to King Ptolemy (probably **no. 1** Ptolemy III Euergetes of Egypt), Biton's *Construction of War-Machines and Artillery* for King Attalus (probably Attalus III of Pergamum) and Apollonius of Citium's *Treatise on Hippocrates' On Joints* written for King Ptolemy (perhaps Ptolemy XII Auletes of Egypt).[3] Archimedes and Eratosthenes probably wrote in 240s through 220s BCE, while Biton wrote c.150–130 BCE and Apollonius wrote 90–70 BCE.[4]

---

1 Apollonius of Citium 10.3, 10.15–12.1 K-K: θεωρῶ φιλιάτρως διακείμενόν σε, βασιλεῦ Πτολεμαῖε ... ἵνα δὲ πάνυ εὐπαρακολούθητά σοι τὰ [παρὰ τἀνέρος] κατὰ μέρος γένηται, πρότερον τὰς τοῦ Ἱπποκράτους λέξεις ἐκθήσομαι ἑτοιμοτέρους τοὺς τρόπους τῶν ἐμβολέων ὑποταξάμεν(ος) οἷόν τινα ἔραν(ον) ἀπ' αὐτῶν τῶν ἔργων, ⟨αἳ⟩ διὰ τῆς συμπαραλαμβανομένων ἀνδρῶν ὑπηρεσίας γίνονται.
2 My analysis is indebted to Langslow 2007, Netz 2009a, and Roby 2016.
3 Ps.-Scymnus' *Periegesis ad Nicomedem* is a treatise on geography dedicated to a minor Hellenistic king. I have excluded it here because it is written in iambic verse rather than prose and thus carries additional literary aims beyond those of the prose treatises, but believe that it too supports my analysis of the rhetoric of patronage in Hellenistic science.
4 Insofar as I am concerned with authorial *personae* and the literary techniques of these treatises, my argument is not substantially affected by chronology. Yet none of the attributions and dates

Origins

Before turning to the extant treatises I entertain a speculation about the social and generic origin of these treatises. Apollonius' treatise quoted above does not open with a formal epistolary greeting, such as Ἀπολλώνιος χαίρειν βασιλεῖ Πτολεμαίῳ "Apollonius sends greetings to king Ptolemy". Fraser has nonetheless called Apollonius' opening a dedicatory epistle, correctly in my view.[5] Importantly Massar has suggested that the origins of the court science treatise derive from a particular kind of letter.[6] I believe her hypothesis is correct and offer a defense of this position.[7]

There is a large amount of pseudepigraphic material from Greco-Roman antiquity purporting to be letters or other works written by court scientists to kings.[8] The historical implausibility of these letters casts doubt on other, plausibly genuine scientific letters of the Hellenistic period, such as those fragmentarily attested from physicians to political authorities: Erasistratus' letter to King Ptolemy, Aristogenes' letter to Antigonas Gonatas, **no. 141** Andreas' letter to Philopator's *philos* **no. 15** Sosibius, Zopyrus' letter to king Mithridates VI of Pontus.[9] This chapter's narrow focus on extant and genuine material offers a solid basis for future historical investigation of the pseudepigraphic corpora. But how did the letter come to be the generic origin of our court science treatises?

A particular piece of evidence from among the fragments of Hellenistic sympotic writers suggests a plausible answer for the origins of the court science treatise.

---

advanced here are entirely secure. Knorr 1978 contains the most extensive discussion of the chronology of the Archimedean corpus, placing the *Arenarius* at an early place in Archimedes' career and relatively soon after Gelon's co-regency (mid 240s?). The most secure dating of Eratosthenes' treatise remains Wilamowitz's inference that Eratosthenes' epigram refers to his status as *tropheus* of Philopator; cf. discussion in **no. 136**. Biton is very difficult to date, with possibilities from c.250 through 130 BCE; see discussion in Marsden 1971: 78 and Roby 2016: 19.n1. I defend a view below that Biton means to demonstrate his *paideia* through his discussion of technology rather than its immediately deployment by Attalus. Consequently I prefer a later dating under Attalus III. Apollonius' dedicatee has usually been thought to be either Ptolemy XII Auletes or his brother Ptolemy of Cyprus; see Berrey 2014a: 164.n76 for discussion. Since I am unable to advance an argument in favor of either Auletes or his brother, I refer without confidence to his addressee as Auletes. These treatises have infrequently been discussed in depth in scholarship. For Archimedes' *Arenarius* see Dijksterhuis 1987: 360–373, Netz 2003, Netz 2009a, Mendell 2016. I discuss Eratosthenes' *Letter to Ptolemy* more extensively in chapter 4 where I list bibliography. For Biton see Marsden 1971, Meißner 1999, Roby 2016. For Apollonius, see Roselli 1998, Berrey 2015, Roselli 2015.

5 Fraser 1972: 1.363.
6 Massar 2005: 187–189.
7 The genre of the scientific letter has received only cursory treatment in scholarship: Langslow 2007, Taub 2013.
8 Good discussions in von Staden 1989: 579–581, van der Eijk 2001: 352–358, Massar 2005: 187–188.
9 Erasistratus fr. 268 Garofalo = Caelius Aurelianus *Tardum Passionum* 5.50 = 884.4–6; *Suda* α3910 mentions Aristogenes' letter; Andreas fr. 9 vS = Soranus *Gyneacia* 4.1 was discussed under **no. 141**; Zopyrus fr. 267 Deichgräber = Galen *De Antidotis* 14.150K.

Hegesander of Delphi, a Greek historian writing in the middle of the second century BCE, records an anecdote about the doctor Menecrates, who wrote unsolicited to King Philip, father of Alexander the Great. This Menecrates called himself Zeus and took to referring to his cured patient by the names of various gods.

> In a letter to King Philip Menecrates wrote: "Menecrates as Zeus sends greetings to Philip. You are king of Macedon, I am king of medicine. You can destroy the healthy whenever you wish; but I can save the sick and keep the healthy living disease-free until old age, if they obey my instructions. Macedonians serve you as bodyguards, but those who will be blessed serve me. For as Zeus I offer them life." Philip wrote back to him as if he were mad: "Philip sends good health to Menecrates."[10]

The anecdote concludes with Menecrates receiving his comeuppance at a symposium hosted by Philip. Menecrates was seated among the statues of the gods and correspondingly offered the divine gifts of libations and incense, while the human guests ate normal food. Whether this anecdote is real or not, Hegesander reflects a belief that Menecrates' fault lay in his claim to divinity, not his medical abilities nor his initiating correspondence with Philip.[11] The evidence of Hegesander is more secure evidence than the pseudepigrapha for establishing the generic history of the scientific advice letter since the author deprecates his source and the king, not the doctor, triumphs in a contest of wits. For all its triviality, the anecdote offer up the virtues from the *personae* of court friendship documented in chapter 2. The doctor Menecrates presents an ethical self grounded in equality and truth-telling (speaking as a *parrhēsiast* king to king); the king in turn receives the letter with mildness and self-control and acts piously toward the gods. Hegesander offers evidence that Hellenistic scientists wrote uninitiated to kings in order to offer them scientific advice; and further that both scientists and kings had models of self-presentations, the one to figure himself as an advisor for the kingdom, the other to respond with dignity while entertaining or dismissing the offer. I therefore suggest that the letter offering service to the king stands as the social and generic origin of the court science treatise.

---

10 Hegesander fr. 5 *FGH* = Athenaeus 7.289c–d: καὶ ἐπιστέλλων Φιλίππῳ τῷ βασιλεῖ οὕτως ἔγραψεν· Μενεκράτης Ζεὺς Φιλίππῳ χαίρειν. σὺ μὲν Μακεδονίας βασιλεύεις, ἐγὼ δὲ ἰατρικῆς, καὶ σὺ μὲν ὑγιαίνοντας δύνασαι ὅταν βουληθῇς ἀπολλύναι, ἐγὼ δὲ τοὺς νοσοῦντας σῴζειν καὶ τοὺς εὐρώστους ἀνόσους οἳ ἂν ἐμοὶ πείθωνται παρέχειν μέχρι γήρως ζῶντας. τοιγαροῦν σὲ μὲν Μακεδόνες δορυφοροῦσιν, ἐμὲ δὲ καὶ οἱ μέλλοντες ἔσεσθαι· Ζεὺς γὰρ ἐγὼ αὐτοῖς βίον παρέχω. πρὸς ὃν ὡς μελαγχολῶντα ἐπέστελλεν ὁ Φίλιππος· (Φίλιππος) Μενεκράτει ὑγιαίνειν. I follow Olson's 2007–2012 text.

11 I cautiously accept the historicity of the anecdote. While Hegesander has not enjoyed much modern scholarly credibility, Dalby 2000 makes the case that he is repeating earlier genuine sources. If the anecdote is historically accurate, this passage would be the earliest surviving Greek letter to a king offering medical advice, contemporary with Mnesitheus of Athens frr. 20, 45 Bertier (cited by van der Eijk 2001: 352). These are dietetic fragments written as letters, thus suggesting dating the medical advice letter to the mid-fourth century. The anecdote of Hegesander is preserved precisely because it damages the doctor's reputation and embellishes that of the king.

An Entertaining Genre

The insight that the court science treatise derives from the letter offering service to the king opens up a series of social questions. What sort of service? In presence or by distance? Is the author known or unknown to the king? How does the author position himself as an advisor to the king? Extant textual evidence alone can provide possible answers to these social questions. One way of treating the extant texts would be a philological analysis. Langslow has helpfully distinguished between letters that consistently use a second-person grammatical address throughout and those treatises that are prefaced with a second-person grammatical greeting.[12] In first case we have a genuine letter, in the second a treatise prefaced with a dedicatory letter. Yet the extant treatises only partially map onto this distinction. Apollonius' and Biton's works contain second person forms throughout yet contain no formal epistolary address or conclusion. Eratosthenes' text contains both second person grammatical forms throughout and also contains a formal epistolary introduction. Archimedes' treatise contains a formal opening and conclusion, but only rarely second-person forms in the main body of the text. Instead of a philological analysis I suggest that the literary style of the writer provides a better means of organizing the generic features of the court science treatise. Archimedes and Eratosthenes wrote in a self-aware, literary style, while Biton's and Apollonius' treatises depended on its accompanying illustrations to enliven a pedestrian exegesis. I suggest that Archimedes and Eratosthenes wrote to kings to whom they were already known, but that Biton and Apollonius had no personal relationship to their addressed kings before writing.

I emphasize the epistolographic origins of the court science treatise because it grounds certain specific features of the genre from its social context. I am advocating that we deliberately adopt a "naive" reading of the court science treatise and take seriously their named royal addressees as representing their intended audience. Contemporary scholarship has balked at naive readings of these treatises, as if the named reader should determine the text's intended effects.[13] First, the named addressees probably did not always understand the technical content written in front of them. Second, court science treatises do sometimes acknowledge a readership beyond the named royal addressee.[14] Yet the royal addressee was no veneer

---

12 Langslow 2007.
13 I accept the position of Roby 2016: 34 with qualifications: "A patron who is explicitly signaled in the preface, and perhaps alluded to more subtly later in the text, is at best the centerpiece of a wide-ranging possible readership with equally wide-ranging expectations." But in my view scholarship has not taken on board how the asymmetry in expertise between scientific authors and their lay readers governs the literary themes of court science treatises discussed below.
14 Archimedes *Arenarius* 236.17–22 Heiberg: "I suppose it useful that the naming of the numbers be said, so that even those readers who do not come upon my writing *To Zeuxippus* do not imagine that nothing has been said earlier about the naming in this book."

of a label or calculated misdirection of readership for these authors. I suggest that the addressee refers to *personae* rather than persons: courtiers as *personae* seek to appeal to the *personae* of kings. The conclusion that the court science treatise derives from a letter offering service to the king hits at the fundamental asymmetry in court relations. Chapter 2 observed that asymmetry governed ethical *personae* in the market of ancient gift-exchange. Since court friendship was an ideological function of the need to rule over a hierarchical arrangement of members of the royal house, bureaucrats, ministers of state, military advisors, and entertainers, gifts given and exchanged between these individuals and the king did not follow symmetrical social relations and reciprocal obligations. Gifts given freely allowed the power differential to be preserved because they could be given without expectation of moral reciprocity. If asymmetry held for social relations at court, it also held for domains of expertise. The king was a lay individual without specialized knowledge in the natural world. The courtier-scientist, on the other hand, was an expert in a particular domain, offering his specialized knowledge as advice to a ruling monarch.

How might the court scientist communicate his specialized knowledge to a lay reader? A well-known ancient anecdote has the mathematician Euclid remarking to King Ptolemy that there is no royal road to geometry.[15] This mythos – for it is nothing but a historiographical mythos[16] – captures well modern sensibilities about the value of intellectual labor and meritocratic reward promoted in contemporary education in the natural sciences. The anecdote has meaning with an ancient context too. After all, the study of the natural world can be intellectually difficult. Success is due to intellectual capability, not social status. In studying the natural world, all students (ancient or modern) start from an equivalent starting point. Here lies the rub for the ancient court scientist. The dictum "there is no royal road to scientific knowledge" poses the problematic of providing expert instruction when the audience may not understand it.

The social context of the origins of the court science treatise provides the literary theme observed in the chapter 2, namely the *topos* of the absent advisor. Court science treatises are always written by experts to non-experts. The fundamental asymmetry of expert knowledge between writer and recipient produces both a literary strategy of didacticism and entertainment for the genre. Court science treatises

---

[15] Proclus *In Primum Euclidis Elementorum Librum* 68.15–17.
[16] Euclid is the most reticent ancient mathematical author: he tells us nothing about himself and we know nothing about him. Proclus *In Primum Euclidis Elementorum Librum* 68–70 has trouble in attributing anything to Euclid except describing the content of his works. Proclus can offer no more evidence of Euclid's Platonism than the construction of the Platonic solids in *Elementa* 13. The anecdote tells us more about Proclus' Platonist view of mathematics, with its notions of intellectual ascent and anti-utilitarianism, than about the historical Euclid or what relationship he might have had to Ptolemy I Soter.

presume a role for expert knowledge and that the recipient will find this knowledge helpful. Hence these are not protreptic treatises in the sense that they urge study of scientific material. Rather they seek to communicate expert results to a lay reader. They do not attempt to justify the study of some facet of natural knowledge. Rather they assert their conclusions are useful. Since the written word is the substitute for the absent advisor, entertaining the reader is an important literary goal of the court science treatise. Court scientists adapted various rhetorical strategies for maintaining the royal reader's interest. Strabo recognized as much when he accuses Eratosthenes of writing for pastime and fun, as if scientific knowledge were an entertaining game.[17]

The literary features of the court science treatise derive from the generic format of the letter offering service, the cultural value of the results of the study of the natural world, and historically-specific features of Greek literary production. The court science treatise participates in the same dynamic of the social life of texts as that older generic format of scientific texts that emphasizes egoism. The author assures the reader and potential employer that he aspires to similar culture (*paideia*); the author makes a display of his expertise, including positioning his own work within its larger traditional context. The court science treatise also highlights the cultural values of utility associated with an understanding of the natural world to the glory of the king. In addition the court science treatises under discussion here are marked by specific aesthetics, such as belatedness, that mark literary production in the Hellenistic period. The court science treatise *had* to be an entertaining genre, for, as written substitutes for the absent advisor, the treatise needed to hold the attention of the royal reader. Authors of court science treatise engage in several thematic strategies to deliver written entertainment.

The close reading of extant court science treatises pursued in this chapter therefore concerns the authorial strategies court scientists used in writing to potential patrons. I am less concerned with analyzing the scientific claims made in these treatises. By focusing on the means of persuasion I intend to view the wide variety of rhetorical techniques as a product of the need to master the asymmetry between expert and lay person. As didactic texts the court science treatises implicate the reader to see, act, or understand in a certain way. Yet my interest here lies in the meaning authors intended to convey to rather than the way readers received it. I treat the following literary themes: belatedness and praise, text and image, expertise and then courtiership in authorial *personae*. The first two literary themes con-

---

[17] Strabo 1.2.2 = 15C.19–22 (34.19–22) Radt makes the comment in the midst of a longer passage on Eratosthenes' character, association with philosophical figures, and trustworthiness as a writer: "Therefore Eratosthenes was in between wanting to be a philosopher and not yet daring to apply himself into this discipline, but succeeded only to seem so, or as someone making even some digression from general knowledge for amusement or even for fun."

cern the context of the treatises themselves; the second two concern the author's voice and ethical relationship to the reader.

Belatedness and Praise

A reading of belatedness and praise seemingly conflates two different aesthetics. Belatedness is the specific Hellenistic aesthetic of referring to a prior literary work as a model. Praise in court science treatises, on the other hand, refers to the ideology of kingship articulated in chapter 2. The sophisticated, erudite openings of Archimedes' *Sand-Reckoner* and Eratosthenes' *Letter to Ptolemy* stylishly combine belatedness and praise. Each text opens with a literary reminiscence of some sort. The literary allusion or quotation serves to direct the topic at once to an elite lay audience, whose knowledge of *belles-lettres* was appreciably greater than their command of technical literature. Literary references also assure the addressee that the subject matter has been treated previously and on that basis is not obscure or impractical. For the writer the allusion to a previous writer or *topos* offers the opportunity to define his own contribution to the subject matter. Praise of royal power is more implicit than explicit in these openings. The differences among extant treatises are revealed in their closings. Since Archimedes and Eratosthenes write in a self-aware literary style, their treatises close in a ring composition by returning to themes in praise of human reasoning and kingship. Apollonius' and Biton's treatises contain little to no explicit praise of royal power but engage in belatedness by positioning their works as researches into previous technical literature (*historiai*).

Archimedes' *Sand-Reckoner* frames its discussion of large numbers as a challenge and praise to human reasoning. The *Sand-Reckoner* opens with a proverb about the grains of sand:

> Some think, King Gelon, that the number of grains of sand (ψάμμος) is unlimited in amount. I mean not only that around Syracuse and the rest of Sicily, but even all the land, both inhabited and uninhabited. And there are others who think not that it is unlimited, but that no such number can be named whose amount it exceeds.[18]

The Greek word for sand, ψάμμος, refers both to the material ("sand" in English) and individual members of that material ("grains of sand" in English). Here in

---

18 Archimedes *Arenarius* 216.1–9 Heiberg: οἴονταί τινες, βασιλεῦ Γέλων, τοῦ ψάμμου τὸν ἀριθμὸν ἄπειρον εἶμεν τῷ πλήθει· λέγω δὲ οὐ μόνον τοῦ περὶ Συρακούσας τε καὶ τὰν ἄλλαν Σικελίαν ὑπάρχοντος, ἀλλὰ καὶ τοῦ κατὰ πᾶσαν χώραν τάν τε οἰκημέναν καὶ τὰν ἀοίκητον. ἐντὶ τινες δέ, οἳ αὐτὸν ἄπειρον μὲν εἶμεν οὐχ ὑπολαμβάνοντι, μηδένα μέντοι ταλικοῦτον κατωνομασμένον ὑπάρχειν ἀριθμόν, ὅστις ὑπερβάλλει τὸ πλῆθος αὐτοῦ.

Archimedes' opening ψάμμος clearly refers to individual members. The measurement of sand provided Greek literature, like many others, with a metaphor of human limitation. The poet Pindar, alluding to various impossible tasks for a human being, states: "Sand has fled number".[19] In a still more revealing passage, the historian Herodotus recounts the oracle given to the king of Lydia when he set a puzzle to determine which oracular responses came from the true god: "I know the number of sand and the measures of the sea, I understand the dumb man and I hear the man who does not speak."[20] The oracle's claim to know the number of grains of sand indicates that the prophecy comes from the true god, Apollo. To return to Archimedes' opening, the laconic "some" is the collective voice of Greek tradition: to claim to know the number of grains of sand is the task of the god, not the human scientist. The claim of Archimedes' second group is more mathematically accurate and within the ken of human ability: "no such number can be named whose amount it exceeds". Sand has, as Pindar put it, fled number. Archimedes' task in this treatise is to transcend the previous limits of human knowledge and name a number bigger than the number of grains of sand in the world and the number that would fill the cosmos.[21]

Archimedes' *Sand-Reckoner* returns to the theme of human reasoning in its ending. Throughout the *Sand-Reckoner* presents a spectacle of ever-increasing numbers. Netz has memorably called this spectacle of magnitudes "the carnival of calculation".[22] After a blistering series of calculations Archimedes ends his piece with the definite answer to the question he had initially posed, a no-longer impossibly-large number. He concludes by addressing Gelon:

> I suppose that these matters, King Gelon, will seem not well-believable to the many and those who have not shared in learning, but to those who have understood and given thought about the distances and the sizes of the earth, the sun, the moon, and the entire universe they will be believable on account of the proof. Therefore I think that to you too it is not unfitting to have considered these matters.[23]

Archimedes recapitulates the themes of his court science treatise: the social divide between a lay audience and those with definite knowledge, the trustworthiness

---

[19] Pindar *Olympia* 2.98: ψάμμος ἀριθμὸν περιπέφευγεν.
[20] Herodotus 1.47: οἶδα δ' ἐγὼ ψάμμου τ' ἀριθμὸν καὶ μέτρα θαλάσσης, / καὶ κωφοῦ συνίημι καὶ οὐ φωνεῦντος ἀκούω.
[21] Archimedes *Arenarius* 216.14–218.1 Heiberg.
[22] Netz 2009a: 30–32.
[23] Archimedes *Arenarius* 258.5–12 Heiberg: ταῦτα δέ, βασιλεῦ Γέλων, τοῖς μὲν πολλοῖς καὶ μὴ κεκοινωνηκότεσσι τῶν μαθημάτων οὐκ εὔπιστα φανήσειν ὑπολαμβάνω, τοῖς δὲ μεταλελαβηκότεσσιν καὶ περὶ τῶν ἀποστημάτων καὶ τῶν μεγεθέων τᾶς τε γᾶς καὶ τοῦ ἁλίου καὶ τᾶς σελήνας καὶ τοῦ ὅλου κόσμου πεφροντικότεσσιν πιστὰ διὰ τὰν ἀπόδειξιν ἐσσεῖσθαι· διόπερ ᾠήθην καὶ τὶν οὐκ ἀνάρμοστον εἶμεν [ἔτι] ἐπιθεωρῆσαι ταῦτα.

and value of Archimedes' contributions, and the addressee Gelon's now explicit inclusion in the select group of individuals with insight into the natural world.[24] The language becomes generalized. Archimedes no longer speaks of definite numbers but knowledge in general. Trust in the logic of mathematics has given credibility to counter-intuitive conclusions. Whereas traditional Greek thought supposed that sand escapes number, Archimedes' arithmetical calculations have arrived near a conclusion known only to the Pythia, oracular voice of the gods.[25] Deep knowledge of the natural world is within the grasp of reasoning humans. Thanks to Archimedes' treatise Gelon is now included in the group of wise humans who understand the deep structure of the natural world.

Eratosthenes' letter to Ptolemy Euergetes thematizes royal power. It opens with direct quotation from an unknown tragedian.[26]

> They say that one of the old tragedians brought on Minos, who was preparing a tomb for Glaucus, and upon learning that it would be altogether one hundred feet said: "You have named a small enclosure for a royal tomb: let it be double, and, that it not fail in its beauty, speedily double each side of the tomb." He seems to have erred. For, if the sides are doubled, a plane-figure becomes four-times greater and a solid eight-times greater.[27]

What appears at first sight as Eratosthenes' polemic toward the innumeracy of tragedians is a charming deflection from a potentially more pointed object of criticism, the king. After all, who is the subject of the understated jab "he seems to have erred": the tragedian or Minos? The building of the tomb for Glaucus is a minor episode in a larger mythological story of which ends with Glaucus being restored to life by a seer.[28] The mythographers Apollodorus and Hyginus tell the story.[29] When Glaucus was boy he fell into a jar of honey while playing and died; his parents searched for him in vain but received word from an oracle that whoever

---

24 See Netz 2009a: 164–166 on this passage.
25 Netz 2009a: 30–33 points out that Archimedes *fails* to measure the number of grains of sand. Rather, Archimedes exchanges increasingly larger sets of inequalities in an attempt to bound the number of grains of sand. Pindar's dictum "Sand has fled number" remains true, even at the end of the treatise.
26 Fragmenta adespota 166 TGrF. In spite of the verses' contextual reference to Polyidus, Kannicht and Snell 1981: 62 deny on the authority of Wilamowitz that these lines are from the three chief Attic tragedians. Netz 2009a: 161 suggests that the quotation might be an invention of Eratosthenes.
27 Eutocius *In Archimedis de Sphaera et Cylindro Libros II* 88.5–13 Heiberg: τῶν ἀρχαίων τινὰ τραγῳδοποιῶν φασιν εἰσαγαγεῖν τὸν Μίνω τῷ Γλαύκῳ κατασκευάζοντα τάφον, πυθόμενον δέ, ὅτι πανταχοῦ ἑκατόμπεδος εἴη, εἰπεῖν· Μικρόν γ' ἔλεξας βασιλικοῦ σηκὸν τάφου· / διπλάσιος ἔστω, τοῦ καλοῦ δὲ μὴ σφαλείς / δίπλαζ' ἕκαστον κῶλον ἐν τάχει τάφου. ἐδόκει δὲ διημαρτηκέναι· τῶν γὰρ πλευρῶν διπλασιασθεισῶν τὸ μὲν ἐπίπεδον γίνεται τετραπλάσιον, τὸ δὲ στερεὸν ὀκταπλάσιον.
28 Hyginus *De Astronomia* 2.14 = 578–584 Viré states that Eratosthenes credited Asclepius with the revitalization of Hippolytus whereas others credit Asclepius with the restoration of Glaucus.
29 Apollodorus *Bibliotheca* 3.17–20; Hyginus *Fabulae* 136.

could solve a riddle would find the boy. The seer Polyidus solved the riddle and subsequently found Glaucus. But, said Minos, since Polyidus had found Glaucus dead, he would shut him up in a tomb with the boy's dead body. A snake entered the tomb with Polyidus, who killed it; another snake entered and with a certain herb restored the dead snake to life. Polyidus used the herb on Glaucus, who was brought back to life and restored to his parents. In this story Minos was a paradigm of bad kingship: the child was left unattended, the oracle misused, the seer shut away. That the king (or the tragedian) should be confused about the proportional means to double the volume of the tomb is among his lesser sins. Eratosthenes' mythological opening alludes to the improper royal exercise of power over the revelatory authority – both the oracle and the seer. The literary allusion thematizes justice, proportionality, building, kingship and fatherhood. Eratosthenes offers a literary counterexample to the mathematical advice he will give Euergetes.

The treatise ends in an epigram summarizing the usefulness of Eratosthenes' invention to the problem enunciated in the beginning. The instrument is useful for enlarging the proportions of existing structures, whether cuboid buildings, cattle-pens, well, or grain stores.[30] The epigram returns to the theme of correctly increasing building proportions initially introduced at the start of the letter by the quotation from the unknown tragedian. The final lines of the epigram in particular tout Eratosthenes' own contributions in the treatise and praise Ptolemy with pharaonic ideology.

> O Ptolemy, happy (εὐαίων) that as father coming into your youth (συνηβῶν) with your son you yourself gave all that is friendly to the Muses and Kings! Later, o Heavenly Zeus, may he grasp the scepters from your hand. May these come to pass and someone might say looking at this dedication that it was the work of Eratosthenes of Cyrene.[31]

The conclusion thematizes fatherhood and good kingship; once again the reader is returned to the themes of the introductory quotation. The epigram has philological similarities to epigrams for royal recipients in its final verses.[32] Furthermore, the address to "heavenly Zeus" is positioned in the same metrical *sedes* as other epi-

---

30 Eutocius *In Archimedis de Sphaera et Cylindro Libros II* 96.10–13 Heiberg. I discuss these aspects further in chapter 4.

31 Eutocius *In Archimedis de Sphaera et Cylindro Libros II* 96.22–27 Heiberg: εὐαίων, Πτολεμαῖε, πατὴρ ὅτι παιδὶ συνηβῶν / πάνθ', ὅσα καὶ Μούσαις καὶ βασιλεῦσι φίλα, / αὐτὸς ἐδωρήσω· τὸ δ' ἐς ὕστερον, οὐράνιε Ζεῦ, / καὶ σκήπτρων ἐκ σῆς ἀντιάσειε χερός. / καὶ τὰ μὲν ὣς τελέοιτο, λέγοι δέ τις ἄνθεμα λεύσσων / τοῦ Κυρηναίου τοῦτ' Ἐρατοσθένεος.

32 The phrase εὐαίων, Πτολεμαῖε in line 13 recalls the same conjunction of metrical *sedes* and royal name in Callimachus epigram 51.3 *HE*: εὐαίων ἐν πᾶσιν ἀρίζηλος Βερενίκα "happy among all women enviable Berenice". This Berenice is probably **no. 2** Berenice II. A fragmentary epigram quoted at **no. 8** dedicated to Philopator contains the formula in the same sedes, εὐαίων Πτολεμ[αῖος, at the start of the line; see *SH* 979.2, where Lloyd-Jones and Parson supplement the nominative.

grams of **no. 110** Callimachus. The appeal to "heavenly Zeus" is surprisingly uncommon and occurs nowhere else in Hellenistic literature outside of Eratosthenes' epigram and the citations of Callimachus.³³

Moreover, Eratosthenes' reference to good kingship in this case appears in an Egyptian form. On the Rosetta Stone Ptolemy V Epiphanes is titled in Greek "the living image of Zeus, son of the Sun, Ptolemy the ever-living, beloved by Ptah",³⁴ while the corresponding hieroglyphic section titles him "living image of Amun, son of Re, Ptolemy, living-forever, beloved by Ptah."³⁵ Koenen has explained that the Rosetta Stone portrays Ptolemy both as the son of Amun-Re and the incarnate form of Amun-Re.³⁶ Similarly I suggest that Eratosthenes evokes the perpetual youthfulness (εὐαίων, συνηβῶν) of the pharaoh as both the son and living image of Amun-Re, identified as heavenly Zeus, in the wish of everlasting kingship by reference to the coronation ceremony. The coronation ceremony of the Egyptian pharaoh ended with a hymn setting out the king as distant son of the sun god Amun as he took up the crook, one of the scepters of Egyptian kingship.³⁷

> You stand up on it, this land that came from Atum, the spittle that come from the Beetle, and evolve on it and become high on it, and your father sees you, the Sun see you. He has come to you, his father: he has come to you, O Sun ... [seven other epithets are given] that you may make this [pharaoh] grasp the Cool Waters and receive the Akhet; that may make this [pharaoh] rule the Nine and provide the Ennead; that may give in this [pharaoh's] arm the crook that lowers the head of the Delta and the Nile Valley.³⁸

The pharaoh holding the scepters of rule over Egypt is an iconic image of Egyptian kingship. Eratosthenes' concluding lines in his epigram, describing how the next Ptolemy takes up scepters of his rule over Egypt from the Sun god himself, reference the future coronation of the new pharaoh. Eratosthenes' praise of Ptolemy at the conclusion of his treatise is thus a wish for proper kingship expressed in the language and ideology of the Ptolemaic pharaohs. Eratosthenes wishes Euergetes to be a proper king, unlike Minos introduced by the ancient tragedian, who prospers his kingdom with mathematically proportionate building.

While Archimedes and Eratosthenes write with a literary self-awareness, Apollonius' and Biton's more prosaic treatises do not lack belatedness. These treatises

---

33 Callimachus epigram 52.3 HE ναίχι πρὸς εὐχαίτεω Γανυμήδεος, οὐράνιε Ζεῦ "yes by long-haired Ganymede, heavenly Zeus" and Callimachus *In Iovem* 55 καλὰ μὲν ἤέξευ, καλὰ δ' ἔτραφες, οὐράνιε Ζεῦ "beautifully you matured, beautifully you grew, heavenly Zeus."
34 OGIS 90.3–4: εἰκόνος ζώσης τοῦ Διός, υἱοῦ τοῦ Ἡλίου, Πτολεμαίου αἰωνοβίου, ἠγαπημένου ὑπὸ τοῦ Φθᾶ.
35 I cite the translation from Koenen 1993: 50.
36 Koenen 1993: 60–61.
37 On the pharaonic coronation ceremony see Frankfort 1948: 105–109.
38 I quote *Pyramid Text* Utterance 222 in the translation of Allen 2015: 42–43.

thematize respectively the technical history of bone-setting and war-machines. Apollonius and Biton examine particular historical texts as a type of *historia*, a Greek word meaning both "investigation" as well as the dominant genre of that investigation, "history".

Apollonius introduces his work as a composition on an older composition. I quoted a portion of this introduction above, where I highlighted its language of court friendship. Now I focus on Apollonius' generic expectation of belatedness.

> I observe that you are disposed in a way friendly-toward-doctors, King Ptolemy, and, since you see that I am eagerly accomplishing your orders, (of those things devised for the aid of men by the most divine (θειοτάτου) Hippocrates in his writing on instruments) I thought it good to interpret those things written by him about dislocation, necessarily starting with the [things written] about the setting of the shoulder, which you ordered me to present you at present.³⁹

The Greek king ordered a physician to write a commentary on a treatise of injuries commonly occurring in wrestling written by the legendary founder of Greek medicine: here is Greek cultural traditionalism, writ scientific.⁴⁰ Apollonius' text as a commentary on an older physician's work is the generic framework of the Empiricist medical sect (to which Apollonius belonged) called *historia*. Apollonius was heir to a one hundred and twenty year tradition of commentary and interpretation within the Empiricist medical sect. The Empiricist medical sect had come to trust Hippocrates' observations and explanations through an intensive reading culture of internal consistency and explaining Hippocratic usage through his own words.⁴¹ Since ancient reading practices rarely treated the *persona* of the author differently from the author's historical person, Apollonius' composition on Hippocrates' works properly begins with Hippocrates' divine person (θειοτάτου). Hippocrates' lineage had been an object of discussion since he rose to prominence, treated also by the earlier Ptolemaic court scientists **no. 136** Eratosthenes and **no. 141** Andreas.⁴² Andreas, a Herophilean and a sectarian opponent of the Empiricists, propagated a story of Hippocrates' dishonesty, claiming that he burned down the medical archive at Cnidus after acquiring his knowledge there. By contrast, Apollonius' epithet reconfirms Hippocrates' glorified status in the Empiricist medical community, linking his ancestry to the demi-god hero Asclepius, who was deified for his medi-

---

39 Apollonius 10.3–8 K-K: θεωρῶ φιλιάτρως διακείμενόν σε, βασιλεῦ Πτολεμαῖε, καὶ ἡμᾶς δὲ σὺ ὁρῶν προθύμως τὰ ὑπό σου προσταχθέντα διαπρασσομένους, τῶν ὑπὸ Ἱπποκράτους τοῦ θειοτάτου συγγράψαντος περὶ ὀργάνων εἰς ἀνθρώπων βοήθειαν ἐπινενοημένων μεταλαμβάνειν καλῶς ἔχειν ἐνόμιζον τὰ περὶ ἐξαρθρήσεων αὐτῷ συγγραφέντα, δεόντως ἐπιλαβὼν καὶ τὰ περὶ ὤμου καταρτισμοῦ, ὃ κατὰ τὸ παρὸν ἐπέταξας μεταδοῦναί σοι.
40 Apollonius 10.13–14 recognizes the specifically Greek cultural context of the injury: "although the resetting of the joint happens publicly at the palaistra".
41 Berrey 2015.
42 See discussion under **no. 141**.

cal knowledge. Apollonius therefore situates himself and his subject within an agonistic intellectual tradition of the Hellenistic medical sects. Apollonius' treatise is three books long, whereas each other court science treatises is only the length of a single papyrus scroll. Apollonius has an introduction at the beginning of each book, although he quotes so much text from Hippocrates that his own contribution occupies relatively little space. The multiple books might be a strategy of luxurious production.[43]

Like Apollonius' text, Biton's treatise invokes belatedness because he recounts previous texts in his listing of war-machines. Biton describes six war-machines, listing the author of each. His description of the machines outlines their basic design in the fashion of Greek mechanicians, specifying the general parts and labeling them with alphabetic icons. Biton describes the war-machines in terms of their parts, from the inside out, rather than a finished, gross object. Biton's presentation is thus a *historia* of artifacts.[44] The identification of Biton's work as a *historia* raises the possibility of antiquarianism, as if Biton were describing dated technology for literary purposes. Did Biton mean to advance these machines for King Attalus' employment in war? Or did Biton discuss these works as a means of demonstrating his *paideia* in technical knowledge? There is no evidence either internal or external to the text that Biton's artillery artifacts were the leading edge of technology. Since the issue revolves around how we understand "utility", I understand Biton's *historia* according to the notion of utility set out for court science in chapter 2. I argue that Biton's *historia* is a demonstration of his familarity with technical knowledge as a presentation of *paideia*.

Text and Image

What is the relationship between text and image?[45] This question is not the replication of a familiar tension between reality and representation in scientific literature. For ancient science we possess only ancient scientists' representations of reality.

---

[43] Apollonius neatly divides his subject matter between books: book 1 covers dislocations of the shoulders and upper arms; book 2 covers dislocations of the forearms and spine; book 3 covers dislocations of the leg.

[44] Cf. Roby 2016: 19–21, 39, 216–221 on Biton's *historia*: (39) "Biton's text resembles, more closely than any other surviving text in which mechanical artifacts are described, what Asper calls the "Rezeptstil", a form characterized by discrete sections that are internally structured and marked off from one another with formulaic language." (221) "Instead of presenting himself as a technological innovator, Biton casts himself in the role of technologically savvy anthologist, collecting and reshaping technologies devised by others and presenting them in a text that increases their accessibility and utility for his own audience."

[45] Netz 1999: 12–88, Roby 2016 discuss the question at length.

Even then our access to ancient scientists' representation of reality is limited: while all four ancient science court treatises originally contained images in addition to their text, the surviving manuscript tradition of Archimedes' *Sand-Reckoner* has lost all original diagrams.[46] Instead of a tension between reality and representation I pose a literary question about the relationship between written text and images embedded in the text. The material object of the book presented to the king contained a seamless presentation of text and image together on the page. In what way does the author invite his reader to understand the text-image complex: are they united? Is image subordinate to text? The relationship between the verbal and the pictorial is another aspect of ancient scientists' authorial strategies in writing to social superiors. Didactism emerges as an important theme. What scientific authors find difficult to express in non-technical language can be more easily represented in pictorial form for a lay audience.

There is an uneven distribution of images surviving from the different ancient sciences. While historians of science have always stressed the importance of images and graphic representation to the scientific enterprise, historians of ancient mathematics have led the way in investigating the ancient scientific properties of the graphic representations of mathematical objects that survive in manuscript.[47] Ancient people's pictorial representation has yielded important results about ancient cultural understandings of mathematical space, continuity, realism, and impossibility. Medical diagrams, by contrast, have been lesser studied as sources for evidence about the intellectual claims of ancient medical authors. I will try to make some headway against the scholarly imbalance between branches of knowledge by foregrounding the medical work of Apollonius.

Apollonius' text is filled with images. Long passages of the work simply quote Hippocrates' words, to which Apollonius appends a picture explicating the technical procedure previously described.[48] Illustration is thus a key pedagogical re-

---

[46] It is not easy to find the manuscript illustrations for several of the court science treatises. Marsden 1971 includes no original diagrams in his critical text of Biton; J. L. Heiberg 1913 drew his own diagrams for Archimedes *Arenarius* since no manuscript diagrams survive. The manuscript diagrams for Eratosthenes' text are described in Netz 2004: 296. Manuscript illustrations for Biton's work are included in black-and-white in Rehm and Schramm 1929, although the origins contain color. Color images from the manuscript *Plut.* 74.7 for Apollonius of Citium's work are included in color in Bernabo 2010: plates 9–38. Since I have not personally inspected any manuscripts, this section will have to serve as a protreptic to a future study of the manuscript images *in situ*.

[47] Important scholarship includes Netz 1999, Saito and Sidoli 2012, Roby 2016.

[48] Apollonius 20.1–5 K-K provides his justification to avoid a lengthy retelling of every Hippocratic passage: "There is no need in such cases to explain excessively the details done in treatment. For in these cases the doctor [i.e. Hippocrates] rather clearly has given orders [ἕκαστα ⟨ἐφέστα⟩κε K-K *sequuntur* ἕκαστα ⟨διέσταλ⟩ται Dietz *apud* Schöne: εκαστα και *mss.*] in what way one must understand every aspect. Therefore no one would suppose [ὑπολάβοι K-K: ὑπολάβῃ Schöne] that we acted idly to circumvent the prose narration after the passages, but rather we think that saying it twice is troublesome."

source for Apollonius. His authorial strategy emphasizes his tools and attempts at conceptual clarity, while simultaneously recognizing the difficult and technical nature of the subject matter. For instance, Apollonius opens his second book laying stress on the combination of text and image.

> I will lay out [Hippocrates'] passages first then afterwards I will lay out the manner of the reductions through the diagrams themselves to the point that, just as was done in the previous ones, they be easily understandable for you and the theory about the joints, which is not unuseful, be transmitted to you as a friend-of-doctors not incomplete. I know that what is accomplished in surgery can be difficult to understand in words: therefore, if something is not clear to you, I hold the nature of the subject responsible, not me. There is a way of presentation with a diagram that leads to utility and activates a vivid (ἐναργῆ) understanding from the circumstances themselves.[49]

Apollonius apologizes to Ptolemy for gap between the visible demonstration of the surgical intervention and its textual description. Text and image together lead toward *enargeia*, realism or vividness, itself an important Hellenistic aesthetic principle.[50] Illustration is pedagogically useful to explicate a complicated, technical text. At the very end of his treatise Apollonius reiterates the usefulness of illustrations: "We have clearly established what has been said by Hippocrates regarding the resetting of the joints in this and the other two books by illustrations."[51] So utility is a key goal throughout the treatise. In the opening of his second book, quoted above, Apollonius announces that the goal of reading his explication of Hippocrates is utility. The goal of "utility" is both in accord with Apollonius' Empiricist practice of reading and the ideal of court knowledge to contribute to the kingdom. Since one is hard pressed to imagine King Ptolemy personally resetting joint reductions in the Hippocratic manner, the utility of Apollonius' Hippocratic exegesis lies in the *paideia* it transmits.

Apollonius' work is preserved in a manuscript that includes images depicting healers and patients. The manuscript illustrations are not schematic or outlines, as

---

49 Apollonius 38.7–15 K-K: πρότερον δὲ τὰς λέξεις αὐτοῦ καταχωριῶ, εἶτ' εἰρομένως τὸν τῶν ἐμβολῶν τρόπον δι' αὐτῶν τῶν ὑποδειγμάτων ὑποτάξω πρὸς τό, καθάπερ καὶ ἐν τοις πρότερον, καὶ τούτοις εὐπαρακολούθητά σοι γενέσθαι τήν τε περὶ ἄρθρων θεωρίαν, οὐκ ἀνωφελῆ καθεστῶσαν μὴ ἀτελείωτον φιλιατροῦντί σοι παραδοθῆναι. οὐκ ἀγνοῶ δέ, διότι τὰ διὰ χειρουργίας ἐνεργούμενα δυσκόλως διὰ λόγου καταλαμβάνεσθαι δύναται ὅθεν, ἐάν τινα μὴ σαφῆ σοι γίνηται, μὴ ἡμᾶς, ἀλλὰ τὴν τοῦ πράγματος αἰτιῶ φύσιν. τὸ[υ] μὲν γὰρ ὑποδείγματος ἔχει [*mss.*: ⟨οὐκ⟩ ἔχει Kollesch] τρόπον ἐπὶ τὴν χρείαν μεταγόμενον, τὸ δ' ἀπ' αὐτῶν τῶν συμβαινόντων ἐναργῆ τὴν κατάληψιν ἐγχειρί(ζει). Kollesch and Kudlien 1965 leave Kollesch's insertion of the negation in the final sentence in the *apparatus* but include it in the German translation. In my view it is unnecessary.
50 On *enargeia* in ancient science see Roby 2016: 73–78, 90–91.
51 Apollonius 112.7–9 K-K: τὰ μὲν οὖν παρ' Ἱπποκράτει περὶ ἄρθρων καταρτισμοῦ διασεσαφημένα καὶ διὰ τούτου μὲν καὶ διὰ τῶν πρὸ αὐτοῦ δὲ δύο βιβλίων διὰ τῶν ὑποδειγμάτων ἐμφανῆ κατεστήσαμεν.

if they should have been drawn like geometrical figures to indicate solely where force was to be applied or by what means. Instead individuals and devices are drawn in a realistic manner, with fleshed out bodies whose shading indicates three dimensions positioned over natural chairs, doors, slings, and other recognizable human artifacts. The realistic perspective looks out from a human viewer toward a vertical plane, framed by columns and an architectural cornice, where participants stand with their sides diagonal to the viewer's perspective. The images are presented in a manner easily accessible to a lay viewer, who needs no imagination or technical clarification to understand the image on the page.[52]

Despite this realism, one of Apollonius' preserved illustrations does not yield information about an important piece of Hippocrates' technology. The main manuscript from which the other extant manuscripts were copied, *Plut.* 74.7, contains two different images of a platform Hippocrates describes for the reductions of the legs. The platform was later called the Hippocratic bench. Hippocrates had described the construction and workings of the bench, a passage quoted by Apollonius: "It was said before that it is worthwhile for one who practices in a populous city to have acquired a quadrangular (τετράγωνον) plank, six cubits long or rather more, and about two cubits broad; while for thickness a [single] span will suffice."[53] Since Hippocrates' description has long been opaque, some modern scholars have turned to Apollonius' treatise for clarification. *Plut.* 74.7 on folio 211v depicts an image of the Hippocratic bench quoted immediately after the above description of the bench's construction. In contrast to the same manuscript's other realistic images of patients and healers, the image on f.211v is quite difficult to understand. The bench is not depicted from the viewer's horizontal perspective but rather from the vertical, as if one is looking down onto a horizontal plane.[54] Since the frame blocks the viewer's perspective from planes underneath the bench, the vertical supports and legs of the bench jut away awkwardly from the frame. It appears that there are either two or four supports that jut out above the board.[55] Grooves mentioned in Hippocrates' text are not visible in this image. But the shape

---

[52] Roby 2016: 162, 191 remarks that Apollonius' illustrations highlight the artifact's use and show the social context in which mechanical technology is embedded. She suggests that Apollonius' illustrations, while differing from the geometric tradition, may (162) "draw on a visual tradition common to mechanical illustrations".

[53] Hippocrates *De Articulis* 72 = 4.296L *apud* Apollonius 74.12–15 K-K: εἴρηκα δὲ καὶ πρόσθεν ἤδη ὅτι ἐπάξιον, ὅστις ἐν πόλει (πολυ)ανθρώπῳ ἰητρεύει, ξύλον κεκτῆσθαι τετράγωνον ὡς ἑξάπηχυ ἢ ὀλίγῳ [ἢ] μεῖζον, εὖρος δὲ δίπηχυ, πάχος δὲ ἀρκέσει σπιθαμιαῖον.

[54] The perspective may explain why the bench is the only Apollonian image in manuscript *Plut.* 74.7 lacking the framing device of two columns and a cornice.

[55] Those supports appearing on the left of the image are meant to represent the bench's legs underneath the board. If all left-pointing staves are meant to be understood underneath the board, then the two left leaning-staves on the right are also under the board. In this interpretation only the two upright staves in the middle would be above the board.

of the board is clear: it is a square, not a rectangle. In other images of patients being reduced on this bench in this same manuscript, f.217r or f.219r, the bench appears as a rectangle. Consequently there are two different images of the bench portrayed in *Plut.* 74.7.

I argue that the two different illustrations of the bench derive from different sources. The rectangular bench on f.217r or f.219r were copied by the scribe from an image existing in the exemplar manuscript. But the square bench on f.211v was likely drawn by the scribe from his reading of Hippocrates' text. For Hippocrates' text uses the word τετράγωνον, properly meaning "square", but Hippocrates' subsequent dimensions – 6 by 2 by 1 cubits – indicate that the bench is a rectangle, ὀρθογώνιον in Greek. So perhaps the famous image of the Hippocratic bench resulted from an illustrator reading the description given by Hippocrates and drawing by habit what he read, rather than him copying an earlier illustration. That is, the illustrator simply read the Greek τετράγωνον and drew a square rather than understanding the text in full, since Hippocrates goes on to describe the dimensions of a rectangular rather than square bench.[56] The speculation implies that the manuscript from which *Plut.* 74.7 was copied either likely did not have an image of the Hippocratic bench or it had been irreparably damaged by the time it was copied into *Plut.* 74.7. I suggest that the famous image of the Hippocratic bench on f.211v does not derive from an ancient exemplar and, consequently, we must look to the other images of the bench for some depiction which goes back to antiquity.

The conclusion raises the question whether the images transmitted in *Plut.* 74.7 are originally Byzantine or are accurate copies of Apollonius' autograph. Schöne has given reason to think that these images go back to an ancient exemplar, although whether they are accurate copies of Apollonius' origins remains unclear.[57] Apollonius himself describes the images this way: "I will present to you the subsequent ways of resetting not with commentaries but, in way visible to the eye, with a shadow-painting illustration the sight of the resettings in detail and the reduction

---

56 In the technical language of historians of mathematics, the geometric shape of the board in the manuscript image is overspecified. The illustrated figures impose more stringent conditions on a mathematical object than the textual description of the object warrants. Cf. Saito and Sidoli 2012: 140–141: "One of the most pervasive features of the manuscript figures is the tendency to represent more regularity among geometric objects than is demanded by the argument. For example, we find rectangles representing parallelograms, isosceles triangles representing arbitrary triangles, squares representing rectangles [*note:* this is the case of *Plut.* 74.4 f.211v], and symmetry in the figure where none is required by the text." Still, the parallel is not exact to mathematical usage, since mathematical texts would write the name of the figure specifically: they would write ὀρθογώνιον not τετράγωνον. Hippocrates' usage of τετράγωνον here may instead mean the moral sense of "perfect, without blemish" as in LSJ *s.v.* τετράγωνον II. Thus an alternate translation of Hippocrates *De Articulis* 4.296L is: "It was said before that it is worthwhile for one who practices in a populous city to have acquired a plank without blemish (τετράγωνον)".
57 Schöne 1896: xxvi–xxix.

of the joints."⁵⁸ "Shadow-painting" is an ancient term for depicting shading and depth to an image. It remains unclear what other means of pictorial representation Apollonius means to exclude by this term.⁵⁹ In *Plut.* 74.7 all objects within the vertical plane of the images seem to exist at the same depth and do not exhibit perspectivism. Only by placing objects in front of others and removing the depiction of the participant half-hidden by, for instance, a chair does the manuscript illustration show depth.⁶⁰

Apollonius is aware of the possibilities of graphically representing medical procedures. For example, after a passage from Hippocrates describing dislocations of the elbow forward or backwards and its accompanying resettings, Apollonius claims that it is only possible to depict one direction of the resetting.

> It is not possible to draw with diagrams the resetting of the elbow backwards – for when an unnatural bend happens [the situation] offers only extension of the dislocation and treatment of the backwards ⟨dislocations⟩ by the hands – but I will record the resetting that happens to the front for the sake of being able to visualize the cause of benefit.⁶¹

The image Apollonius places immediately following is entitled in the manuscript ἑτέρα ἐμβολὴ ἀγκῶνος "another reduction of the elbow".⁶² The image f.195v shows the patient standing at right, diagonal to the viewer, with an extended right arm whose hand is tucked under the physician's left armpit. The physician stands at left holding the patient's arm, with the physician's right hand on the muscles of the patient's forearm and the physician's left hand on the muscles of the patient's upper arm. The physician's left leg is slightly raised, in order to stabilize himself on the right leg before using his own weight as a lever to pull back or push forward on the patient's elbow joint, as the case might be. That is, the image shows possi-

---

58 Apollonius 14.8–11 K-K: τοὺς δὲ ἑξῆς τρόπους τῶν ἐμβολέων ⟨οὐ⟩ [ἐμβολῶν ⟨οὐ⟩ Schöne: 'μβολέων K-K, a misprint] δι' ὑπομνημάτων, ζωγραφικῆς δὲ σκιαγραφίας τῶν κατὰ μέρος ἐξαρθρήσεων παραγωγῆς τε τῶν ἄρθρων ὀφθαλμοφανῶς τὴν θέαν αὐτῶν παρασχησόμεθά σοι. Schöne adds the negation on the basis of 20.1–5 K-K, cited above in footnote 48.
59 For other possibilities see Roby 2016: 162–166, who considers Vitruvius' typology of architectural illustrations.
60 The manuscript consistently uses this means of exhibiting depth, even in the image of the Hippocratic bench when viewed from the vertical plane. Apollonius does not verbally describe any materials missing from the diagrams, unlike mathematicians' and mechanicians' use of νοεῖν to mark objects inadequately drawn in diagrams; cf. Roby 2016: 169.
61 Apollonius 42.29–44.5 K-K: τὴν μὲν οὖν εἰς τοὐπίσω τοῦ ἀγκῶνος ἐξάρθρησιν οὐκ ἀναγκαῖον διὰ τῶν ὑποδειγμάτων ὑπογράφειν – ἐκ γὰρ τῆς παρὰ φύσιν γεγενημένης κάμψεως μεταστάσεως ἔκτασιν περιέχει μόνον καὶ τῶν ὄπισθεν ταῖς χερσὶ καταρτισμόν – τὴν δὲ εἰς τὸ ἔμπροσθεν συμβαίνουσαν καταγράψω χάριν τοῦ δυναμένην συνθεωρη⟨θῆ⟩ναι παραιτίαν ὠφελείας γενέσθαι.
62 Bernabo 2010: plate 19, f.195v. It is called "another" because the previous passage and image picture the resetting of the elbow that has dislocated to the inside or outside in the horizontal plane.

bilities for both a forward and backward resetting. But to match the verbal description, the illustration must be read as showing the physician's weight only pulling backward in extending the dislocation forwards then squeezing the patient's arm muscles in order to flex the arm, for together the motions of extension and flexing match the forward resetting Hippocrates prescribes.

Yet Apollonius' text denies that the backward resetting can be shown, a claim I cannot explain. After all, Hippocrates says that "the treatment is the same" for the forward and backward resetting.[63] Apollonius' objection is apparently not one of representing a multiplicity of therapeutic interventions. The previous passage and image described by Apollonius represent a resetting for the elbow dislocated to the inside or outside: that is, the previous image represented resetting in two directions. I noted above that the physician's raised leg and levered weight in f.195v might be read as if the physician were ready to push forward in counterextension, a necessary treatment for an elbow dislocated backwards. But Apollonius' text accompanying the image illustrating "another reduction of the elbow" precludes reading the physician's weighted stance as a treatment for such a backwards dislocation. Crucially text and image match for intended meaning only in one reading of both. The text then instructs the viewer how to "read" the accompanying image. The images are subordinated to the text; they illustrate the text in a didactic way. The text, with its explicit praise of the king and formal declarations of court friendship, contains the intended meaning of the text-image complex.

Studying Apollonius' book dedicated to King Ptolemy suggests broader conclusions about the relationship between the text and image in their joint complex, when we apply those conclusions to the genre of the court science treatise. The images downplay technical details; they complement the meaning of the text. The text instructs lay readers how to "read" the image. Together the text-image complex compromises a cultural object that offers didacticism and *paideia*. A reading of the three other court science treatises bears out the authorial strategies adopted by Apollonius to explain the relationship between text and image.

Biton's text also emphasizes the relationship between word and image. Biton's treatise describes the named war-machines of five engineers: two catapults without torsion, a movable siege tower, a moveable siege ladder, and two belly-bows. After each description Biton closes with some variation of the following: "The figure as it occurs lies below".[64] At the end of the treatise Biton reminds the king "The figures and the measurements have been written above".[65] It appears then that Biton followed the description of each individual machine with a drawing of the machine. The text labels parts with Greek letters in the style of a Greek mathematical

---

[63] Hippocrates *De Articulis* 19 = 4.132L: ἴησις ἡ αὐτή. Apollonius does not quote this remark.
[64] Biton 48.1–2 W: τὸ δὲ σχῆμα οἷον τυγχάνει ὑπόκειται, 51.4, 56.6–7, 61.1, 64.2–3, 67.3–4.
[65] Biton 68.1 W: τὰ δὲ σχήματα καὶ τὰ μέτρα προγέγραπται.

work and gives dimensions for parts. The manuscript diagrams transmitted follow this labeling, although occasionally also label parts by their mechanical name (ἀγκών, τράφηξ).⁶⁶ Since the text itself does not sufficiently explain how some parts are to be connected, Biton's ancient readers must have taken this information from the manuscript illustrations. Yet Roby points out that Biton adopts a casual attitude toward the mathematical labeling his diagrams:

> In just one passage of his *Kataskeuai* Biton varies his conventions for labeling beams several times: the catapult he is describing has a pair of beams referred to collectively as AB (in the diagram A corresponds to one end of each beam, and B to the other), two more beams referred to as ΓΓ, and yet another pair of beams which are first referred to as ΔΕ, and then changed to Δ when Biton focuses closely on the upper side.⁶⁷

Biton's labels are indexical features of the drawing accompanying the text. They orientate the viewer toward certain relevant aspects of the machine. While adopting the appearance of Greek mathematical texts, the labels themselves are not part of geometric proof. The image is subordinate to the text. Still, the text and illustrations complement each other by presenting the machine directly to the reader as an assemblage of notable parts. The technical account therefore has a didactic character for a general reader. Biton clearly envisages a lay rather than professional reader for his casual technical exposition.

Eratosthenes' letter preserves a better record of the relationship between text and image than Biton, chiefly since the letter itself describes the mathematical figures used as well as their placement on the inscribed *stele*. At the close of the dedicatory letter Eratosthenes records the contents of the *stele*.

> The bronze instrument is in the dedication and is joined to the crown of the stele by lead; under it the proof is enunciated somewhat concisely, then the figure, and after it an epigram. Let these things have been subscribed below for you, so that you have them as in the dedication. Of the two figures the second is drawn on the stele.⁶⁸

---

66 The manuscript illustrations provided by Rehm and Schramm 1929: 8 from three different manuscripts appear to derive from a common ancestor. See Marsden 1971: 9–14 on the manuscript tradition. The machines are not drawn in projection, with the exception of the city-sacker. It may not be a coincidence that the city-sacker is also the only diagram pictured from the outside of the machine so that its interior workings are not visible to the reader. Marsden 1971: 62 argues that the autograph diagrams included several perspectives of the machine, perhaps drawn in in *scaenographia*; on ancient perspectivism of mechanical diagrams see Roby 2016: 157–165. Roby 2016: 184–190 has commented on the double reading necessary to interpret mechanical diagram labeled both in alphabetical notion and with part-names in what she calls the "layered diagram".
67 Roby 2016: 175. Roby refers to Biton's first described machine, 45–47 W, the catapult designed by Charon of Magnesia.
68 Eutocius *In Archimedis de Sphaera et Cylindro Libros II* 94.8–14 Heiberg: ἐν δὲ τῷ ἀναθήματι τὸ μὲν ὀργανικὸν χαλκοῦν ἐστιν καὶ καθήρμοσται ὑπ' αὐτὴν τὴν στεφάνην τῆς στήλης προσμεμολυβδοχοημένον, ὑπ' αὐτοῦ δὲ ἡ ἀπόδειξις συντομώτερον φραζομένη καὶ τὸ σχῆμα, μετ' αὐτὸ δὲ ἐπίγραμμα. ὑπογεγράφθω οὖν σοι καὶ ταῦτα, ἵνα ἔχῃς καὶ ὡς ἐν τῷ ἀναθήματι. τῶν δὲ δύο σχημάτων τὸ δεύτερον γέγραπται ἐν τῇ στήλῃ.

The contents of the *stele* follow, transcribed in the order laid out by the text. In replicating the image and text from the *stele* Eratosthenes thus offers to Ptolemy verisimilitude. Like Apollonius, Eratosthenes expects images to enliven and explicate a technical subject. The text presents not so much an ecphrasis of an image as a transcription of it. The format of the presentation – on a *stele*, surmounted by a bronze object, prose engraved with a drawing, and finished with an epigram – hybridizes generic categories. The bronze artwork reminds the viewer of a statue; a prose text monumentalizes its subject; the epigram suggests egotism and memorialization of its author, as Eratosthenes' own name closes the final line.[69]

Eratosthenes records that the letter to Ptolemy contained two mathematical figures, just as our manuscripts transmit. The second figure, contained on the *stele*, draws attention to the instrument's product of the mathematical means: the viewer's eye alights on the triangle called ΑΕΚ, whose cross-struts ΒΖ and ΓΗ are the required mean proportionals that the problem seeks. Placing this figure on the *stele* directly after the proof (analogous to the position of ancient mathematical diagram on papyrus after its proof) highlights the mathematical results of the machine. But there is another object on the *stele*, Eratosthenes' instrument. If we assume that the machine was soldered onto the stone crown of the *stele* as three bronze squares on a sliding rail, then the machine would have looked very much like the first illustration preserved in our manuscripts. The machine would look like a piece of decorative art atop the *stele*. It is therefore tempting to envision Eratosthenes' dedicatory *stele* as a vertical tapestry, showing two aesthetic pieces – the machine and the poem – surrounding two mathematical pieces – the proof and the figure. The mathematical account was subordinated to a grander and more culturally specific purpose of monumentalizing art.

Archimedes also considers the relationship between text and image in the themes of didacticism, *paideia*, and the subordination of pictorial content to textual expression. The thematic aim of the *Sand-Reckoner* is to express an idea verbally. Archimedes *names* a number of the grains of sand in the universe in his own complex formulation.[70] Since Archimedes' system is opaque to the modern reader, scholarship on the *Sand-Reckoner* usually introduces the subject-matter of the treatise by describing the contemporary Greek numbering system.[71] The traditional system was a means of expressing in graphical representation the naming of numbers. The largest named unit was the myriad or 10,000. One could write the graphic representation of a myriad, the majuscule M, and over the top of the myriad write the

---

[69] Eutocius *In Archimedis de Sphaera et Cylindro Libros II* 96.27 Heiberg.
[70] Archimedes *Arenarius* 256.25–30 Heiberg: "And since the thousandth unit of the seventh numbers is proportionally two-and-fifty from the unit and the myriad-times myriad myriads is thirteenth from the unit of the same proportion, it is clear the product will be four and sixty from the unit by the same proportion."
[71] Dijksterhuis 1987: 361.

number of myriads. The traditional system was capable of expressing in graphical representation numbers up to a myriad-myriad, or ten-thousand ten-thousands, named one hundred million in our counting. Archimedes' *Sand-Reckoner* develops a system for naming numbers in excess of a myriad-myriad, not representing them in graphical form. Archimedes supplements the meaning of his text in two different ways with graphic representation: while he employs the technical graphic representation of his contemporary mathematics at two points, he otherwise describes verbally two experiments he has conducted to obtain boundary values for his calculations.

Archimedes uses graphical representation for both numbers and geometrical demonstrations at two points. One of the *Sand-Reckoner*'s four astronomical assumptions states that the sun's diameter exceeds a certain value.[72] The resulting argument depends first on an experimental measurement, to which I will return, and a geometrical proof about similar triangles. In making his calculations Archimedes uses contemporary Greek graphical representation for the numbers in ratios as well as a geometrical diagram for the comparison of similar triangles.[73] These are complicated passages, written in the technical formal language of Greek mathematics.[74] Although the proof depends on no higher mathematics than what is found in Euclid's chapters on circles, angles, and triangles (that is, no mathematics beyond the *Elements*), it is difficult to believe that Archimedes' expressed recipient, Gelon, completely understood the mathematical text in front of him. Later in the text Archimedes again uses contemporary graphic representation for "a fact useful also to know", a geometrical progression in arithmetic.[75] Archimedes proves that the product of two numbers in geometric continuous progression can be known from their distance from each other in relation to the first term of the series. He expresses the series in the alphabetic notation of Greek geometry that he uses for arithmetic series in his other works, such as *Quadrature of the Parabola*.

Yet apart from those graphic representations, Archimedes employs only verbal description in the *Sand-Reckoner*. His writing, while written in the wordy expressions characteristic of precise Greek mathematical language, conveys technical content to the lay reader effectively. I suggest that Archimedes' verbal descriptions are a means of conveying technical information without pictures. In particular, Archimedes gives two experiments he has done in order to provide dimensions for the sun's diameter and the size of a grain of sand.

Experimentation serves as a rhetorical technique for establishing trust in Archimedes' authority. The values Archimedes needed to establish were difficult to ob-

---

[72] Archimedes *Arenarius* 222.3–6 Heiberg.
[73] Archimedes *Arenarius* 226.12–232.26 Heiberg.
[74] Even the use of νοείσθω at Archimedes *Arenarius* 226.24 Heiberg is a formal feature of Greek mathematics, as Netz 2009b has shown.
[75] Archimedes *Arenarius* 240.19–20 Heiberg: χρήσιμον δέ ἐστι καὶ τόδε γιγνωσκόμενον.

tain: "Since it is not easy to take precisely the angle on account of the fact that eyesight, or hand technique, or the instrument by which the angle is taken are not trustworthy at declaring precision".[76] Archimedes first described an experiment obtaining boundary values for the angle at the apex of a triangle between the edges of the sun and his observer's eye. By sighting along a ruler with a bored cylinder Archimedes was able to establish a value for an angle that necessarily implies a value for the diameter of the sun. But Archimedes emphasized his own authority in devising, building, and conducting the experiment: "for I had to take the angle of the matter at hand [sc. the sun's incident ray] for the proof".[77] With its authority in the combination of reproducibility and manual dexterity, Archimedes' *handling* of the experiment becomes both a didactic theme as well as an entertaining one. The moving, arranging, sighting, and measuring of the sun's angle becomes both a model for a lay reader to follow Archimedes' exemplar as well as a way of enlivening a complicated set of mathematical relationships for the same lay reader.

Consider Archimedes' second experiment about measuring the volume of sand.[78] I give a brief summary. Archimedes assumed that a volume of sand not larger than a poppy seed would contain a myriad grains of sand. He then set out to measure the diameter of a poppy seed in order to provide a comparative magnitude for the sand. He found that twenty-five poppy seeds placed end-to-end were the length of one dactyl, a Greek unit of measurement. To obtain a boundary value he deliberately overestimated the number of seeds at forty in one dactyl. By inference the diameter of a myriad of grains of sand is therefore one fortieth of a dactyl in length, a value useful in estimating the number of grains of sands when the diameter of the universe is known.

Archimedes produced his extraordinary number of the total grains of sand in the universe based on these assumptions. Scholarship in the philosophy of science has emphasized that experiments are reproducible and transparent to readers; so much is true for Archimedes' experiments as well. He selected a small seed, accessible in Greek markets throughout the Mediterranean. Poppy, its latex and its seeds, had been known in the Greek Mediterranean since Homer and were widely used in Hellenistic medicine.[79] No special capability other than basic able-bodiedness is expected of the experimenter. Anyone can line poppy seeds against a ruler and count them. The mundane, everyday aspect of counting poppy seeds conveys

---

76 Archimedes *Arenarius* 222.11–14 Heiberg: τὸ μὲν οὖν ἀκριβὲς λαβεῖν οὐκ εὐχερές ἐστι διὰ τὸ μήτε τὰν ὄψιν μήτε τὰς χεῖρας μήτε τὰ ὄργανα, δι' ὧν δεῖ λαβεῖν, ἀξιόπιστα εἶμεν τὸ ἀκριβὲς ἀποφαίνεθσαι.
77 Archimedes *Arenarius* 222.16–17 Heiberg: ἀποχρῆ δέ μοι ἐς τὰν ἀπόδειξιν τοῦ προκειμένου γωνίαν λαβεῖν.
78 Archimedes *Arenarius* 236.3–16 Heiberg.
79 Scarborough 1995.

believability to Archimedes' larger claim about the extraordinary sets of numbers he discusses. With his experiments measuring the angle of the sun's rays and the diameter of poppy seeds, Archimedes makes a seamless transition from the ordinary to the extraordinary, from the smallest to the largest. If subject matter is entertaining, the experimenter's self provides both a model for education in the form of imitation and the presence of authority. Archimedes' verbal description of his experiments subordinates image to text and does so in a didactic, entertaining fashion.

Authorial *Personae*: Expertise

Archimedes' authority in the manipulation of his experiments reminds us that a scientist may possess social credibility in a number of ways: from experiential knowledge, such as comes from personal experience or from demonstration of competence and skill; or from appeals to authority such as comes from manifest textual authenticity; or from power and authority vested in accepted social graces. The social dynamic of a court science treatise is a social inferior addressing a social superior to be a guide to a scientific subject. The social inferior must prove his credentials to speak on the subject at hand in order to be a trustworthy guide to this area of knowledge. He must validate his presented knowledge. Yet, although the scientist will be trustworthy if he is knowledgeable, the social superior who must trust him lacks the technical expertise to evaluate his competence of this knowledge. The potential gap between a scientist's claimed competence and performed competence perhaps seems larger and more dangerous to us than it did to ancient society. We moderns have well-developed social procedures where knowledgeable peers review, authenticate, and certify professional knowledge. But ancient society perhaps took the asymmetry in stride; after all, employers of scientists faced this problem throughout in antiquity.

Apollonius emphasizes the problem of professional competence in his introduction. Other physicians, he claims, have not undertaken the task to elucidate Hippocrates because of their inexperience or hesitation to prove themselves, in spite of the fact that the joint injuries discussed occur in wrestling-yard. For his own part Apollonius claims to have competence in the subject.

> Some of these I refitted myself, some I saw in Alexandria while apprentice to Zopyrus. Posidonius, who lived with the same doctor, can witness for us that the named man treated [patients] mostly by following Hippocrates in the case of surgery for bone breaks and joint dislocations.[80]

---

[80] Apollonius 12.1–5 K-K: ὧν τινὰς μὲν καὶ αὐτὸς κατήρτικα, τινὰς δὲ καὶ Ζωπύρῳ παρηδρευκὼς ἐν Ἀλεξανδρείᾳ τεθεώρηκα. ὅτι δὲ ὁ ῥηθεὶς ἀνὴρ ἐπί τε τῶν καταγμάτων καὶ ἐπὶ τῆς τῶν ἐξαρθρήσεων χειρουργίας κατὰ τὸ πλεῖστον Ἱπποκράτει κατακολουθῶν ἐθεράπευεν, μαρτύρήσειεν ἂν ἡμῖν Ποσειδώνιος τῷ αὐτῷ συνδιατετριφὼς ἰατρῷ.

Apollonius claims that he has personal experience with joint dislocation, names his teacher Zopyrus, and offers a further witness, Posidonius, of his teacher's competence and theoretical orientation. Some commentators have drawn attention to the name Posidonius and suggested that this man was the famous Stoic philosopher, Posidonius of Apamea. But this seems unlikely, for two factors suggest that Posidonius was a doctor. First, Apollonius refers to a person competent enough to understand multiple therapeutic orientations to joint dislocations so that the Hippocratic approach is chosen instead of another. Second, Apollonius says that Posidonius "has lived with" Zopyrus: Posidonius was Zopyrus' student. In the apprenticeship model of medical education in antiquity, many Hellenistic physicians are said to have lived with or been part of the household of their teacher.[81] Apollonius' mention of Posidonius is meant to bolster the image of Zopyrus, a theoretical physician with multiple students. Apollonius' medical competence, as he presents it, depends on his association with Zopyrus. Zopyrus was probably a physician of some prominence, working in Alexandria, offering a famous antidote to an unknown King Ptolemy, and serving as a medical advisor-by-distance to the Hellenistic king Mithridates VI.[82] Apollonius' invocation of his teacher and his own experience as witnesses to his medical competence parallel the two witnesses which applicants for the position of public physician would adduce, their teacher and their patients. Therefore Apollonius' introduction situates himself within the specific social setting of a physician applying for the job of public physician. Apollonius writes to King Ptolemy as if he is previously unknown to Ptolemy.

Biton's treatise, like Apollonius', seeks to emphasize his own expertise. Biton always records the inventor of the six mechanical inventions for war-making that comprise his treatise. The reader's initial impression is that Biton is either paraphrasing and compressing the original inventor's account or is drawing his knowledge from a handbook of war-machines. Although *prima facie* Biton seems to have contributed nothing of his own, Biton announces his own competence in an important passage.

> Following those [previous machines] we describe for you the construction of the city-sacker (ἑλεπόλεως) which Posidonius the Macedonian designed for Alexander son of Philip. The employment of the woods is varied. That for long parts and for planking is either fir or silver-fir or pine. That for the axles and wheels is oak or ash, and the same for the beams and supports. You must know in advance that the magnitudes of the city-sackers ought to be constructed in relation to the assaults on the walls and that the city-sackers ought to be superior in magnitude. There is a theory for this approach, which I have discussed in my *Optics*. For I am competent in the category of surveying.[83]

---

81 Berrey 2014b: 433–436.
82 *EANS* 851.
83 Biton 52–53.2 W: ἑχομένως δὲ τούτων ἑλεπόλεώς σοι κατασκευὴν ὑποτάσσομεν, ἣν ἠρχιτεκτόνευσε Ποσειδώνιος ὁ Μακεδὼν Ἀλεξάνδρῳ τῷ Φιλίππου. ἔστι δὲ ἡ τῶν ξύλων κατεργασία παντοδαπή. ὅσα γὰρ εἰς τοὺς ἄξονας καὶ τροχούς, δρύϊνα ἢ μελέϊνα, τὰ δὲ αὐτὰ καὶ εἰς τοὺς κανόνας

Biton's reference to his own work and competence matters within its context. Mobile siege engines, such as the city-sacker (ἑλεπόλεως), were supposed to be rolled up to city walls so that soldiers could pass from the siege engine to the wall at height. It is necessary for the attacker to calculate the height of walls accurately, a calculation done by (more or less) triangulating surveying equipment at a distance. Biton's competence in surveying proves essential to the building and employment of certain war-machines.

Biton gives further reason for confidence in his experience when discussing how to build these machines. Biton's discussion of the types of wood to be employed for the city-sacker parallels his initial advice to Attalus about adaptability in measurement and appropriate construction materials.

> Try to use general science. For it is necessary to use both the measurements and even proportions of the materials here set down. When it comes to woods attempt to construct them for use from ash woods, for these would most suitable.[84]

The end of the treatise passses from discussion of construction materials to the flexibility of the designs themselves.

> I have copied especially those machines I thought fitted you. For I am persuaded that you will invent similar ones through these. Do not be disturbed that, because I have used established measurements, it will be necessary for you to use the same measurements. For if you wish to construct bigger or even smaller engines, do so; only try to preserve the proportion.[85]

Biton's authority comes directly from the practical implementation of these devices. Unlike his fellow mechanician **no. 139** Philon Biton never mentions his practical experience nor his association with fellow engineers and builders. Although he never uses the word "experience" Biton's expertise appears to lie in the adaption and building of the war machines his text describes. Roby reads the text as demonstrating Biton's craftsmanship rather than his control of theoretical detail.[86] His

---

καὶ ὑποστυλώματα. δεῖ δέ σε προειδέναι ὅτι πρὸς τὰς προσβολὰς τῶν τειχῶν καὶ τὰ μεγέθη τῶν ἑλεπόλεων δεῖ κατασκευάζειν, καὶ ὑπεραίρειν τοῖς μεγέθεσι τὰς ἑλεπόλεις. ἔστι δὲ καὶ τοῦτο μεθοδικὴ θεωρία, ἣν διείλεγμαι ἐν τοῖς Ὀπτικοῖς· ἔγκειται γάρ μοι τὸ γένος τοῦ διοπτρικοῦ. I follow Marsden's 1971 identifications of the woods.

84 Biton 44.3–6 W: πειρῶ δὲ ταῖς ἐπιστήμαις χρῆσθαι. χρὴ γὰρ χρῆσθαι καὶ τοῖς μέτροις καὶ ἔτι τοῖς ῥυθμοῖς τῶν προβεβλημένων. πειρῶ δέ, ὅσα μὲν ἂν ᾖ ξύλινα, κατασκευάζειν εἰς τὴν χρείαν διά τε τῶν μελεΐνων ξύλων· ταῦτα γὰρ ἁρμόσειεν ἂν μάλιστα.

85 Biton 67.5–9 W: ὅσα μὲν οὖν μάλιστα ἐνομίζομέν σοι ἁρμόζειν ἀνεγράψαμεν. πεπείσμεθα γὰρ ὅτι διὰ τούτων τὰ ὁμοιοειδῆ ἐξευρήσεις. μὴ παραταραχθῇς δὲ ὅτι ἱσταμένοις μέτροις κεχρήμεθα, μήποτε καὶ σὲ δεήσῃ τοῖς αὐτοῖς μέτροις κεχρῆσθαι. ἐάν τε γὰρ βούλῃ μείζονα κατασκευάζειν, ἐπιτέλει, ἐάν τε ἐλάσσονα· μόνον πειρῶ τὴν ἀναλογίαν φυλάττειν.

86 Roby 2016: 45, 216–221 (45): "Biton describes how the wood functions in a catapult to establish the artifact as part of a system of craft-knowledge, and Biton himself as the arbiter of that knowledge."

selection of designs to copy is motivated by a criterion of adaptability. In a passage describing a second stone-throwing catapult, Biton advises: "For often the positions of locales do not admit the same kind of engines."[87] So even Biton's seeming repetition of subject matter in fact is motivated by the diverse contextual possibilities of their use. A reader intending to learn how to use war-machines comes to appreciate Biton's understated expertise.

Apollonius and Biton are at pains to define the nature of their expertise and defend it. The anxiety suggests that they write to their addressees without previous interaction. By contrast, Eratosthenes and Archimedes offer no account of their experience, as if their expertise was already known to their addressees.

Eratosthenes' letter to King Ptolemy opens blandly: "Eratosthenes to King Ptolemy, greetings".[88] The letter transitions immediately to a history of the mathematical problem at hand, with no further qualifications of Eratosthenes' listed. Finally, after a moderate list of the problem and various solutions proposed with light criticism, Eratosthenes announces his own solution: "An easy instrumental solution has been devised by us, by which we will find not only two means of two given lines, but as many as one commands."[89] Eratosthenes' mathematical competence, then, emerges from his own solution to the problem. He has no need to cite others' appraisals of his work, such as that of Archimedes.[90] Rather, as we might expect from his social situation, Eratosthenes' qualifications were already known to Euergetes; he is not writing him for the first time on the subject of mathematics. The proof of his expertise lies in the evidence of his solution.

Like Eratosthenes' address to King Ptolemy, Archimedes' letter to King Gelon opens without comment on Archimedes' abilities as a mathematician. Netz has shown that Archimedes' literary technique favors narrative shifts that surprise the reader and exhibits a playful, authoritative presence.[91] Netz has suggested that Archimedes' authorial presence comes through most strongly in the *Sand-Reckoner*.[92] Certainly Archimedes' ludic, authoritative *persona* rapidly develops in the treatise. If the text itself were not enough evidence of Archimedes' mathematical capabilities, Archimedes mentions that the *Sand-Reckoner* has a relationship to another book he wrote: "I suppose it useful that the naming of the numbers be

---

[87] Biton 48–49.1 W: πολλάκι γὰρ αἱ τῶν τόπων θέσεις οὐκ ἐπιδέχονται τὰ αὐτὰ τῶν ὀργάνων.
[88] Eutocius *In Archimedis de Sphaera et Cylindro Libros II* 88.4 Heiberg: Βασιλεῖ Πτολεμαίῳ Ἐρατοσθένης χαίρειν.
[89] Eutocius *In Archimedis de Sphaera et Cylindro Libros II* 90.11–13 Heiberg: ἐπινενόηται δέ τις ὑφ' ἡμῶν ὀργανικὴ λῆψις ῥᾳδία, δι' ἧς εὑρήσομεν δύο τῶν δοθεισῶν οὐ μόνον δύο μέσας, ἀλλ' ὅσας ἄν τις ἐπιτάξῃ.
[90] Archimedes *Methodus* 71.col1.33–71.col2.8 Netz et al. = 428.18–21 Heiberg: "Seeing that you, just as I said, are learned, remarkably preeminent in philosophy, and honoring any science that passes under your eyes, I thought it good to write to you."
[91] Netz 2009a: 66–107.
[92] Netz 2009a: 105–106.

said, so that even those readers who do not come upon my writing *To Zeuxippus* do not imagine that nothing has been said earlier about the naming in this book."[93] Scholarship has argued that *To Zeuxippus* was the formal, technical counterpart to the *Sand-Reckoner*'s generalist account. Even granting that distinction, Archimedes does not list his experiential qualifications for Gelon. Like Eratosthenes, Archimedes too is not demonstrating his mathematical expertise to his addressed audience for the first time.

Authorial *Personae*: Courtiership

Yet scientific authority does not only derive from experience. As we have seen, empirical knowledge-making is a form of truth-telling; scientific authority does not necessarily follow from epistemic originality. Since the *persona* of the ancient courtier aspired to virtues of truth-telling, *parrhēsia*, *paideia*, and elite friendship within the culture of the symposium, the scientific practitioner at court adopted the sympotic *persona* with epistemic virtues good for empirical knowledge-making. In chapter 2 I grounded the discursive self of Hellenistic courtiers in an account of court friendship: the *personae* of court friendship are the truth-teller or the flatterer. Authorial self-presentation in extant court science treatise corresponds to this description. As suggested, each author presents his material as an account of *paideia*. Apollonius and Archimedes invoke the language of truth-telling directly, figuring themselves as *parrhēsiasts*. Both Apollonius and Archimedes use the language of friendship to address their social superiors: through the written text, the kings become "friends-of-science". Eratosthenes deftly presents himself as an equal and, with his skillful handling of criticism and praise of kingship, comes near to being a flatterer. But Biton's muted authorial voice does not fit the *persona* of flatterer or *parrhēsiast*, making his treatise a weak attempt at securing royal patronage.

Apollonius' text is addressed to King Ptolemy who has directed Apollonius to write the commentary. Apollonius opens his text with praise for the king: "I observe that you are disposed in a way friendly toward doctors (φιλιάτρως), King Ptolemy."[94] Apollonius praises Ptolemy with the unique adverb φιλιάτρως, a *hapax legomenon* occurring only here in all of Greek literature.[95] Apollonius uses the cognate verb φιλιατρέω twice in his text, once clearly referring to Ptolemy, another time referring to τοὺς φιλιατροῦντας. These terms denote a social relationship be-

---

93 Archimedes *Arenarius* 236.17–22 Heiberg.
94 Apollonius 10.3–4 K-K.
95 A search in *Thesaurus Linguae Graecae* <http://stephanus.tlg.uci.edu> for "φιλιάτρως" returns only this passage.

tween Apollonius and his patrons. Apollonius twice attacks adherents of an opposing medical sect, the Herophileans.[96] Apollonius accuses the Herophileans of confusing and leading astray the "friends-of-learning" (τῶν φιλομαθούντων) and "friends-of-doctors" (τοὺς φιλιατροῦντας) from true understanding of the subject matter. While Apollonius and his Herophilean opponents are medical specialists, Apollonius' audience are social elites with an interest in science but no practice. Apollonius' attack on the Herophileans is the rhetorical move of the protagonist within the group presenting "insider information" to a confidante outside the group. By communicating this insider information publicly, the scientific author both claims true understanding of the information of the scientific field and offers himself as a guide to the field for the lay public: he is poised to mediate appropriately between inside and outside.

In attacking the Herophileans Apollonius is making an epistemological claim and a social claim too. Apollonius turns the scientific dispute between Herophileans and Empiricists into a social issue: the Herophileans will confuse and lead astray patrons who sponsor medicine (τοὺς φιλιατροῦντας), like Ptolemy. In the dichotomy of social relationships between the scientific inside and royal outside, Apollonius is internally trying to gain traction in his Empiricist dispute with the Herophileans by bringing Ptolemy's superior social standing to bear against Hegetor and the Herophileans. For the royal patrons on the outside, Apollonius' aim was to place the Herophileans outside the knowledgeable scientific community and thereby to denigrate the social standing of the Herophileans as possible clients for royal patrons.

Apollonius writes as a *parrhēsiast*. Since Apollonius positions his work as the application for position of court physician, it is possible to see Apollonius' quest for authority in light of court friendship. Apollonius' book represents his own usefulness as a skilled surgeon. Apollonius offers his book to Ptolemy as an *eranos*, the sympotic gift brought by the guest to the host.[97] Furthermore, friendship and truth-telling are important themes to Apollonius' authorial *persona* from his frequent conflation of social and epistemological claims against the Herophileans when invoking Ptolemy as a friend-of-medicine. Finally, Apollonius displays his *paideia* or intellectual culture by his elucidation of Hippocrates' text. Grammatical and linguistic terms developed by the Hellenistic grammarians are widespread in Apollonius' text.[98] Apollonius elucidates a confusing term in the Hippocratic text

---

96 Apollonius 16.10–11 K-K: "How could one not say that he, constrained by his inexperience, has twisted (διεστροφέναι) the understanding of doctors and friends-of-learning (τῶν φιλομαθούντων)?" Apollonius 80.15–16 K-K "Hegetor not only raves but has confused (διέστρεφεν) friends-of-doctors (τοὺς φιλιατροῦντας) as much as he can." Berrey 2014a: 162–167, Berrey 2015: 479–484 discuss Apollonius' attacks on Hegetor.
97 Apollonius 10.17 K-K.
98 Cf. Kollesch and Kudlien's 1965 index *s. v.* ἑρμηνεία, λέξις, λόγος, ὁ. Roselli 1998 has analyzed Apollonius' grammatical knowledge in detail.

by determining its meaning in Hippocrates' original Coan dialect.[99] Apollonius' display of *paideia* takes in contemporary book culture. He critiques earlier lexicographers' efforts to understand or (in his view) disparage Hippocrates, reifies the Hellenistic focus on geographical differences in language and register, and produces an Empiricist reading of Hippocrates that unites the ethical and biographical. Apollonius' authorial stances plays on the court theme of the absent advisor, the written work as a substitute for the live person. He inverts Demetrius of Phalerum's dictum: "What friends do not dare to tell kings is written in books."

Archimedes too writes as a *parrhēsiast*. A more powerful writer than Apollonius, he cloaks himself with the virtues of courtiership: *paideia*, truth-telling, friendship. Archimedes recognizes the difficulty of his subject matter but also concedes to Gelon that his previous knowledge will allow him to understand it.

> I will try to show you by geometric proofs which you will follow closely that, of those numbers named by me and published in *To Zeuxippus*, some exceed not only the number of sand equal in magnitude to the filled earth, just as I said, but also the number equal in magnitude to the universe. You understand that the universe is called by most astronomers the sphere whose center is the center of the earth and whose radius is equal to the straight line between the center of the sun and the center of the earth. For you have read these things in the proofs written by astronomers.[100]

Gelon apparently has had a good education about the properties of the physical universe. Yet Mendell points out the strangeness of Archimedes' order of objects in the universe: Archimedes uses the antiquated cosmology of the philosopher Anaximander over the more contemporary cosmology of Aristotle.[101] Is Archimedes conceding an antiquated arrangement of the universe for Gelon's sake? It is difficult to believe that Gelon's education has encompassed Anaximander and the early Presocratic philosophers but not yet the more recent developments of fourth-century BCE astronomy. A better explanation lies in a literary reading of the passage: Archimedes sets up an (older) theory of the universe in which the sun lies

---

99 Apollonius 28.1–16 K-K.
100 Archimedes *Arenarius* 216.15–218.7 Heiberg: ἐγὼ δὲ πειρασοῦμαί τοι δεικνύειν δι' ἀποδειξίων γεμετρικᾶν αἷς παρακολουθήσεις, ὅτι τῶν ὑφ' ἁμῶν κατωνομασμένων ἀριθμῶν καὶ ἐκδεδομένων ἐν τοῖς ποτὶ Ζεύξιππον γεγραμμένοις ὑπερβάλλοντί τινες οὐ μόνον τὸν ἀριθμὸν τοῦ ψάμμου τοῦ μέγεθος ἔχοντος ἴσον τᾷ γᾷ πεπληρωμένα, καθάπερ εἴπαμες, ἀλλὰ καὶ τὸν τοῦ μέγεθος ἴσον ἔχοντος τῷ κόσμῳ. κατέχεις δέ, διότι καλεῖται κόσμος ὑπὸ μὲν τῶν πλείστων ἀστρολόγων ἁ σφαῖρα, ἇς ἐστι κέντρον μὲν τὸ τᾶς γᾶς κέντρον, ἁ δὲ ἐκ τοῦ κέντρου ἴσα τᾷ εὐθείᾳ τᾷ μεταξὺ τοῦ κέντρου τοῦ ἁλίου καὶ τοῦ κέντρου τᾶς γᾶς· ταῦτα γὰρ ἐν ταῖς γραφομέναις παρὰ τῶν ἀστρολόγων δείξεσι διάκουσας.
101 Mendell 2016: "One would think that the whole world is the sphere of the fixed stars and everything within and that the sun is lower than the fixed stars, as Aristotle argues, and not the cosmology of Anaximander, who does place the sun as the outermost object. Instead, Archimedes seems to place the sun as the outermost, since 'world' (kosmos) should encompass everything, and he is aiming to give as large a universe as possible on each of the two rival theories."

outermost; he now moves to a recent hypothesis by the astronomer Aristarchus of Samos in which the sun lies at the center of the universe. Archimedes takes Aristarchus' heliocentric universe as the basis of his subsequent calculations.[102] Archimedes establishes an antithesis between a universe with a sun at the edge and a universe with the sun at its center. A literary reading of Archimedes' *Sand-Reckoner* gains more evidence from Archimedes' list of astronomers who gave astronomical distances, from Eudoxus, Archimedes' father Pheidias, and Aristarchus.[103] Apart from Aristarchus, whom Archimedes introduced, both Eudoxus and Pheidias are not introduced, as if they were already known to Gelon. Given Archimedes' close association with the royal house of Syracuse, it would not be surprising that Gelon knew Pheidias.[104] Archimedes obliquely refers to his family's court friendship with Gelon and Gelon's father, Hiero.

Gelon had as much familiarity with astronomers as other king of his age. Some, perhaps such as Pheidias and Archimedes, he knew personally or by repute; others, such as Anaximander or Eudoxus, perhaps he knew by reading. After all, Archimedes reports that Gelon had read astronomy books: "For you have read these things in the proofs written by astronomers."[105] Archimedes includes Gelon explicitly among the friends of science at the end of the *Sand-Reckoner*: "to those have understood and given thought about the distances and the sizes of the earth, the sun, the moon, and the entire universe, these matters [sc. Archimedes' calculations] will be believable on account of the proof".[106] Above I read this statement as an example of Gelon's explicit inclusion among those knowledgeable about the natural world thanks to Archimedes' teaching. Doubtless that was Archimedes' intended praise of Gelon and self-promotion. But perhaps Gelon's previous knowledge of astronomy was not so minimal as a cursory reading suggests. Gelon was invested in scientific knowledge through *paideia*.

---

102 Archimedes *Arenarius* 218.31–220.7 Heiberg: "I say, if the sphere of sand were such in magnitude as Aristarchus supposes the sphere of the fixed stars is, that thus certain of the beginning numbers that have a name [τῶν ἐν ἀρχᾷ ἀριθμῶν τῶν κατονομαξίαν ἐχόντων Heiberg: τῶν ἐν Ἀρχαῖς ἀριθμῶν τῶν κατονομαξιῶν ἐχόντων mss.] exceed in multitude the number of the sand that has a magnitude equal to the aforementioned sphere, when the following are supposed." Since Netz 2003 has shown the fundamental thematics of naming to the treatise, I believe Heiberg's printed text of this disputed passage is sound. Archimedes concludes the *Arenarius* 258.1–5 Heiberg by naming the number of sand in the sphere of the universe based on Aristarchus' model: "It is clear that the multitude of sand that has a magnitude equal to the sphere of the fixed stars, which Aristarchus supposes, is less than one thousand myriads of the eighth numbers."
103 Archimedes *Arenarius* 220.20–25 Heiberg.
104 Plutarch *Marcellus* 14.12: Ἀρχιμήδης Ἱέρωνι τῷ βασιλεῖ συγγενὴς ὢν καὶ φίλος "Archimedes was a kinsman and friend to Hiero the king".
105 Archimedes *Arenarius* 218.5–7 Heiberg: ταῦτα γὰρ ἐν ταῖς γραφομέναις παρὰ τῶν ἀστρολόγων δείξεσι διάκουσας.
106 Archimedes *Arenarius* 258.7–10 Heiberg.

Archimedes presents himself in the categories of a *parrhēsiast*. He offers Gelon the discourse of court friendship. Archimedes' experiments, described earlier, establish his authority as a paradigm for his handling and manipulation of instruments. Since the experiments are transparent and reproducible to readers, Archimedes claims the category of truth-telling. At the same time, he shares his knowledge about astronomy and reminds Gelon of his reading in *paideia*. Netz's persuasive reading of the *Sand-Reckoner*, with its characteristically Hellenistic literary themes – the mix of the large and small, the poetic and prosaic, the wonder of display from the natural world – reminds us that Archimedes presents his *paideia* with cultural sophistication. And what sort of question *is* the motivating issue of the *Sand-Reckoner*? The task of naming a number of the grains of sand in the universe, while pursued seriously, is a type of entertainment. To name a number greater than the grains of sand filling the universe seems a sort of trivializing game. Clearchus described the natural world in sympotic "inquires": "For [the ancients] proposed inquiries to their fellow drinkers not just as people now ask questions, what sex position or which and what kind of fish is sweetest and most seasonable, or what is particularly good to eat after Arcturus or the Pleiades or after the Dog-star."[107] Sympotic inquiry about the natural world begins with its potential for human benefit. Archimedes poses the inquiry about the multitude of sand as a progression beginning from Gelon's immediate context: "I mean not only that around Syracuse and the rest of Sicily, but even all the land, both inhabited and uninhabited."[108] Archimedes' *parrhēsia* in court friendship serves as a ready advisor to King Gelon on the ludic questions arising in sympotic discussion.

If Apollonius and Archimedes openly represent themselves as *parrhēsiasts* on a spectrum of court friendship, Eratosthenes writes so skillfully that, even after extensive reflection, I am unable to decide whether his self-figuration as a *parrhēsiast* is in fact closer to a flatterer, the antonymic figure of court friendship. Eratosthenes' self-presentation in the letter comes closer to flattery than we might expect, especially when we reflect on Isocrates' nod towards "skillful flattery".[109] The theme of Eratosthenes' letter is Eratosthenes' own contribution toward the goal of good kingship. Eratosthenes' cautious and deft criticism of tragedy's representation of Minos avoids direct criticism of the king. The majority of the letter is professional and technical. Eratosthenes uses the terminology of "friends-of-science" only once, when urging an impersonal professional attention to mechanical detail to assure a good mathematical result.[110] But more direct praise of the king occurs in the final epigram explicitly addressed to Ptolemy: "O Ptolemy, happy that as

---

107 Clearchus fr. 63 Wehrli = Athenaeus 10.457d.
108 Archimedes *Arenarius* 216.3–6.
109 Isocrates *Ad Nicoclem* 28.
110 Eutocius *In Archimedis de Sphaera et Cylindro Libros II* 94.4–7 Heiberg: "For more accuracy one must be a friend-of-*technē* (φιλοτεχνητέον) by seizing the lines in order that, in drawing the plates together, all remain parallel, without division, and equally in contact with each other."

father coming into your youth with your son you yourself gave all that is friendly to the Muses and Kings!"[111] I continue the analysis from above that these lines begin a praise of the Ptolemaic vision of dynastic kingship in a particular mixture of Greek and Egyptian traditions. On one hand, Eratosthenes replicates the royal ideology of gift-giving that portrays the king as the ideal friend of the kingdom.[112] Yet the king has already performed his act of giving. The king's acts have preceded Eratosthenes' own contribution. It is not that the king does not belong among the friends-of-science but rather that Eratosthenes' text does not help move the king toward this status. Eratosthenes seems to recognize Ptolemy's prior scientific competence. Eratosthenes plays the role not as advisor of scientific knowledge but as a laudator of the king's prior knowledge. Perhaps Ptolemy, unlike Minos in Eratosthenes' quoted tragedy, knows how to select a builder who can properly triple a volumetric measurement.[113]

While Eratosthenes consistently avoids bringing moral reproach against the ideology of kingship and always praises the current king Ptolemy, he criticizes only his scientific rivals by name. Four lines of the epigram replicate the earlier criticism of the solutions of the earlier mathematicians Archytas, Menaechmus, and Eudoxus.[114] Plutarch admonished that the flatterer thinks invective and *isogēria* are a substitute for *parrhēsia*.[115] Eratosthenes does not engage in personal invective against previous mathematicians but only their solutions; his criticism is professional, not personal. Eratosthenes' flattery is limited to praise of the current monarch. Eratosthenes' authorial voice employs some of the techniques of the flatterer while avoiding others; he is far from representing himself straightforwardly as a *parrhēsiast*. In addition to his role as a mathematical innovator Eratosthenes in his capacity as a courtier plays a flatterer, praising existing royal knowledge and power.

Biton's authorial voice is far more muted than the other court scientists. Biton writes with very little artifice. His text emphasizes the usefulness of the inventions as well as their adaptability. Astonishingly, Biton polemicizes against no one in his brief treatise.[116] In the agonistic tradition of Greek science authors polemicize against their predecessors as a means of expressing their own authority and asserting truth about the natural world. Biton's treatise contains egotism primarily in his

---

111 Eutocius *In Archimedis de Sphaera et Cylindro Libros II* 96.22–24 Heiberg.
112 Diodorus Siculus 1.70.6; Ps.-Aristeas 286.
113 In chapter 4 I argue that the epigram (and the *stele* as a whole) is addressed to the Ptolemaic bureaucracy of the *dioiketes* and subordinates.
114 Eutocius *In Archimedis de Sphaera et Cylindro Libros II* Heiberg 90.4–11, 96.16–19.
115 Plutarch *Quomodo Adulator ab Amico Internoscatur* 66D.
116 Biton's refusal to advance criticism is not due to the material under discussion. Even Philon in *Belopoeica* 72.26–27 Th advances criticism against Ctesibius at one point whom he has otherwise praised: "Ctesibius seems to me to have erred in this".

act of writing, not his invention or conception of the material.¹¹⁷ Biton presents himself as an organizer and dispensator of textual material; Roby writes that Biton "casts himself in the role of technologically savvy anthologist".¹¹⁸ His authority develops from his deployment of *historia* from the writers of war-machines that fit different situations. As an organizer of *historia* Biton seems to share the literary strategies of the **no. 110, 113,** and **119** paradoxographers. Biton's discussion of past technologies offered the author an opportunity to display book learning and the fruits of wide reading; it invested readers in the curiosities derived from study of the natural world. As discussed above, Biton's own competence and expertise comes from his reading selection, his knowledge of construction, and means of developing these war machines. His work can be read as the combination of pratical experience and learned knowledge, although Biton does not foreground his practical experience.

Does Biton write for Attalus' patronage? Again, the absences in the treatise are notable. Biton never uses the language of court friendship nor describes his own background or social standing. Since he does not polemicize against other authors, he does not offer readers a canon of truth-telling. All the same Biton's treatise does invoke a criterion of utility. At the opening of his treatise he explains the purpose of these machines to Attalus: "I am convinced by them that you will easily overturn those engines of your enemies in their attacks, when you take the field against them with the underwritten methods."¹¹⁹ These devices are useful for Attalus to defeat his enemies. Meißner has pointed out that Biton, unlike **no. 139** Philon, does not aspire to a level of generality in his mechanical treatment of war-machines; Philon treats war-machines within the context of general mechanical principles.¹²⁰ Biton instead offers specific mechanical examples directed toward the king's military interests: four kinds of offensive weapons for overcoming defensive walls and two kinds of defensive weapons for repulsing besiegers. Attalus' war-

---

117 Biton 43 W: ἐπιβέβλημαι γράψαι; 44.6 W: ἀρξόμεθα καταβαλέσθαι; 48 W: ὑπογράψομεν; 52.1 W: ὑποτάσσομεν; 53.4 W: ἔστω πρῶτον ἡμῖν ὁ λόγος; 55.1 W: μὴ λανθανέτω δέ σε ὅτι ὑπὲρ μιᾶς ἐπιφανείας διελέχθην; 57 W: ἐχομένως δὲ τῶν προγεγραμμένων ὑπογράψομεν; 57 W: ὑπογράψω δέ σοι; 61 W: προκεχείρισμαι ἀναγράψαι; 65 W: ὑπογράψομεν; 67.5 W: ὅσα μὲν οὖν μάλιστα ἐνομίζομέν σοι ἁρμόζειν ἀνεγράψαμεν.
118 Roby 2016: 221.
119 Biton 44.1–3 W: δι' ὧν [sc. ὀργάνων] πέπεισμαι ὅτι ταῦτα τὰ κατὰ τὰς προσβολὰς τῶν πολεμίων ὄργανα ῥᾳδίως ἀναστρέψεις, ἀντιστρατευόμενος ταῖς ὑπογεγραμμέναις μεθόδοις. I find this sentence extremely puzzling. With four of his six machines Biton describes offensive weapons for protracted sieges, not defensive weapons intended to destroy besiegers' equipment. Rehm and Schramm 1929: 6–7 had already seen the difficulty and argued that the surviving text is an excerpt of a longer work. Marsden 1971: 62 rejected the view that the surviving text is lacunose but does not explain the passage at hand. Meißner 1999: 97 takes the sentence at face-value, placing it within a tradition of military machines for defense. Biton's criterion of utility is at least clear.
120 Meißner 1999: 164.

making, one of the chief practices of Hellenistic kingship, will be bettered through Biton's expert knowledge. Yet Biton does not seem to adapt a stance of equality with the king nor include a profession of his loyalty. The categories of *parrhēsiast* or flatterer seem inappropriate descriptions of Biton's ethical self-presentation. If Biton did seek Attalus' patronage, Biton's subservience, pedesterian style, hidden references to his own experience, and muted authorial *persona* made this treatise an especially weak and likely insufficient attempt to attract a royal reader.

Conclusions

A close reading of Hellenistic treatises written from Archimedes, Eratosthenes, Biton, and Apollonius to kings has shown generic similarities in belatedness and praise, text and image, and expertise and courtiership in authorial *personae*. The authors use this variety of techniques to master the asymmetry between expert and lay person. All court science authors situate their works as a form of *paideia*. Their technical expertise is either introduced by a quotation from high literature or structured as an exposition on previous technical texts. All authors use the text-image complex to instruct their lay readers, while privileging verbal expression as the primary vehicle for carrying meaning. Archimedes, Eratosthenes, and Apollonius invoke the language of court friendship, praising the king as a friend-of-science and figuring the king as the ideal friend of the kingdom. The *persona* of the king as the named addressee governs the authors' intended literary features examined here. For it is the king as an addressee that gives rise to the author's need to offer an ethical *persona*, articulate his expertise, and frame his work in a didactic and entertaining way that overcomes the asymmetry of expertise to a lay reader.

When we view these works as the literary remains of the gift economy, we can see how the treatises function as substitutes for absent advisors. In accord with the thick description of court science articulated in this and previous chapters, the literary remains of the gift-exchange in antiquity show that scientific gift-exchange was initiated by the scientist, a courtier socially inferior to the king. Scientists in their *personae* as courtiers sought to persuade the *persona* of the king through the authorial strategies outlined here that they could serve the king and kingdom as advisors about their area of specialized expertise. While Apollonius and Biton strive to prove their technical expertise, Archimedes and Eratosthenes write to kings to whom their expertise is already known. The authors adopt ethical *personae* of courtiers designed to secure patronage for themselves. Since Archimedes and Apollonius write as *parrhēsiasts*, their scientific selves combine the epistemic virtues of truth-telling, *paideia*, and loyal friendship with specialized expertise useful for the kingdom. As an ambiguous ethical self nearing the skillful flatterer, Eratosthenes combined truth-telling and flattering royal praise as a complement to royal entertainment. Biton's dim authorial presence offers no polemical opportunity to claim the mantle of truth-teller but claims only the mantle of *paideia*.

4 Technology and Performance in Eratosthenes and Andreas

This chapter culminates the specific version of the argument for science at the court of Ptolemy III Euergetes and Ptolemy IV Philopator that began in the Introduction. Having given a descriptive prosopography of the court, the social dynamics of court patronage recounting the virtues of king and courtiers, and a close reading of the genre of extant court science treatises, I now move to individual case studies. Here I examine two specific works of court science from the court of Ptolemy 246–205 BCE: **no. 136** Eratosthenes' instrument and **no. 141** Andreas' machine. I aim to contextualize these scientific products within the thick description of court science. My argument is that these technological products of cross-disciplinary scientific investigation gained social currency first through their entertainment value as court science.

Eratosthenes

On the 7th of March in the year 238 BCE the Greek-Egyptian pharaoh Ptolemy III Euergetes promulgated a decree in three languages, one of the great multilingual decrees of the Ptolemaic house whose most famous representative is the Rosetta Stone. The Canopus Decree recognized various honors for Euergetes and his sister-wife Berenice and set up a calendar reform that added a leap-year day to the 365-day Egyptian calendar every 4 years. Although the leap-day did not last beyond Euergetes' reign, the decree instituted the extra day as a festival day in celebration of the regnal titulature of Ptolemy and Berenice, as chapter 2 showed. It was a gift to the god-kings of Ptolemaic Egypt. And what a royal gift, the creation of entire new day! Even longer life to the royal couple! Did **no. 136** Eratosthenes stand beside Euergetes on that March day as the intellectual responsible for the calendar reform?

A creative novelist would give a more satisfying answer to that question than a historian, since we have only the Canopus Decree itself and fragments of Eratosthenes' life as a basis for our answers. Yet examining our extant evidence in a different light leads to new, less biographical questions appropriate for historical investigation. For the Canopus Decree suggests a type of court science interested in materiality, hybrid Greco-Egyptian culture, and the ideology of Ptolemaic kingship. The *stele*'s top frieze shows Thoth, Egyptian god of writing and knowledge, whom the Greek calls Hermes, standing behind Euergetes and Berenice. Stephens and others have shown the ability of Ptolemaic court poets to "see double" by manipulating the symbols and narratives from classical Egyptian ideology of kingship to portray the bicephalous Ptolemaic monarchy.[1] Scientific historians have

---

[1] Koenen 1993, Selden 1998, Stephens 2003, Berti 2009b.

contributed to this scholarly conversation, although infrequently.[2] I will argue that the intercultural model of Stephens applies also to science produced in Alexandria for the Ptolemaic court. As chapter 3 discussed, natural scientists also dedicated treatises to Ptolemaic rulers and participated in the discourses of the court, including Eratosthenes. Eratosthenes' poem *Hermes*, interwoven with science, might have told of Thoth's invention of letters, since *interpretatio Graeca* equates Thoth with Hermes.[3] Might his court science have seen double too?

Instead of the Canopus Decree I focus on Eratosthenes' *Letter to Ptolemy*, included in chapter 3 as one of court science treatises.[4] This text on the doubling of the cube is ascribed to Eratosthenes and preserved in Eutocius' commentary on Archimedes. At five complete pages of Greek in Heiberg's critical edition, it is Eratosthenes' most extensive piece of extant science.[5] I accept the *opinio communis* of scholarship that the text is authentic.[6] Netz has shown that the text evidences a characteristically Alexandrian concern for belatedness.[7] The text's belatedness pro-

---

[2] On medicine see von Staden 1989: 1–31, Jouanna 2012, Lang 2012. Lang's 2012: 101–239 sophisticated overview of Egyptian and Greek medicine under the Ptolemies concludes that there was little exchange of theoretical views between Greek and Egyptian medical traditions but a greater cultural exchange in physicians' self-presentation and in the therapeutic options available to all patients, both Egyptian and Greek. Lang 2012: 243–266 sees Alexandria as a world apart from the *chora*, with the elite practitioners of the capital expressing themselves within a tradition exclusively of Greek thought and competing with each other in Greek agonistic fashion.

[3] *SH* 397.

[4] Throughout I will refer to Eratosthenes' text as preserved in Eutocius by this title. Eutocius himself gives no other title than "As Eratosthenes". Pappus 3.21 records the title *Mesolabos* "Meanstaker" when describing Eratosthenes' instrumental solution. Pappus' description of Eratosthenes' solution differs in several crucial respects from that recorded by Eutocius. See footnote 46.

[5] Eutocius *In Archimedis de Sphaera et Cylindro Libros II* 88.3–96 Heiberg. The extant *Catasterismoi*, although based on Eratosthenes' work in some way, is of unclear origin.

[6] Knorr 1989: 63–153 must be read first, since all subsequent scholarship depends on his careful arguments. Eratosthenes' *historia* of the development of the problem was Knorr's motivation to declare the entire text authentic: Knorr 1986 notably relied on this passage centrally as evidence for the pre-Euclidean development of certain curves. Geus 2002 and Taub 2008 accept the text as authentic; Vitrac 2008 believes it is authentic but somewhat revised by a later editor, perhaps Eutocius. Implicitly accepting the text's authenticity, Leventhal 2017 argues that the entire text presents a Ptolemaic vision of literary empire. Eratosthenes' text is embedded in Eutocius' record of solutions to the problem of the two mean proportionals: some of the other texts, such as Plato's solution, are clearly inauthentic while still others of doubtful form, such as Menaechmus'. The air of suspicion that pervades Eutocius' catalogue has been partially responsible for the slow acceptance of the authenticity of Eratosthenes' text. I thank Max Leventhal for sharing his preliminary and completed work with me and hope that our remaining disagreements stimulate further dialogue about a fascinating text.

[7] Netz 2009a: 132, 160–164. Netz reads the text in a dialectic between poetry and mathematics; he is therefore concerned with predecessor models for Eratosthenes' literariness. In his view, the mythicizing of the text undercuts its mathematical and practical content (162): "One notes a certain tension in Eratosthenes' rhetoric. In the following technical account, Eratosthenes would refer to the (genuine) practical applications of the problems of in the construction of war-engines, as well

vides insight into the Ptolemaic court scientist's appropriation of Egyptian symbols. I argue that the text shows Egyptian symbols only of contemporary Ptolemaic Egypt because of the agonistic context of Hellenistic Greek science.

The text details Eratosthenes' solution to a classic problem of proportionality that first arose as an equivalent problem to the doubling of the cube. "Doubling the cube" was one of the classical problems of ancient Greek geometry: the mathematical task is to produce a cube with double the volume of any given cube. From a modern algebraic viewpoint, a cube whose volume is a product of its side produces a double volume when that initial side is multiplied by the cube root of 2. Ancient mathematicians were able to reduce the problem of volume in doubling the cube to the relationship between the sides of the two cubes. They expressed this relationship as a continuous proportion. Known as the problem of two mean proportions, it required (in modern algebraic terms) to find x and y such that a : x :: x : y :: y : b. Eratosthenes' contribution was to build a novel machine to produce mathematically the mean proportionals x and y.

Eratosthenes' *Letter to Ptolemy* presents itself as double, comprised separately of a letter and an inscription from a dedicatory stele: the letter begins with an address to Ptolemy, followed by a history of the problem, the usefulness of Eratosthenes' instrumental solution,[8] and a geometric proof; the dedication (including

---

as in solid measurement in general. However the tendency of the mythical introduction is to undercut radically any practical motivation of the problem: it is first suggested by a tragic speech and then interest flourishes as a consequence of an oracle, the duplication in both cases motivated by purely *ad hoc* architectural requests." My interpretative strategy is the opposite of Netz's: I am concerned with the practicality of the device and the denigration of past mathematicians. But I agree with Netz 2009, Taub 2008, and Leventhal 2017 that the text at least partially functions as entertainment.

8 Throughout I will refer to Eratosthenes' instrument without name as "device" or "instrument", since Eratosthenes himself gives no other name to this machine than ὄργανον. Less careful scholarship has conflated the name of the instrument described in Eutocius' text with Pappus' 3.21 = 54.22–56.2 Hultsch description of Eratosthenes' accomplishment: "Attacking the problem by means of instruments in an amazing way they brought it to implementation and construction, as is possible to see from the treatises attributed to them, I mean the *Mesolabos* of Eratosthenes and the *Mechanics* or *Catapults* of Philon and Heron (ὡς ἔστιν ἰδεῖν ἀπὸ τῶν φερομένων αὐτοῖς συνταγμάτων, λέγω δ' ἐν τῷ Ἐρατοσθένους μεσολάβῳ καὶ τοῖς Φίλωνος καὶ Ἥρωνος μηχανικοῖς ἢ καταπαλτικοῖς)". Vitruvius 9.*pr*.14 understands the name of the device as *mesolabos*. But Knorr 1989: 75n.40 was correct to point out that the term *mesolabos* used in Pappus refers to a book-title, not a machine name. Nor is the epigram's term μεσόγραφα in Eutocius *In Archimedis de Sphaera et Cylindro Libros II* 96.20 Heiberg the name for the machine. I cannot share the view of Knorr 1989: 151n.9, who seems to understand the text this way. Netz 2004: 288 translates μεσόγραφα μυρία in line 20 of the epigram as "thousands of means." He has followed the lead of Heiberg, who translates in his Latin "sexcentas medias proportionales". Apart from the mistaken identification of the machine as *mesolabos*, LSJ's entry *s. v.* μεσόγραφος is correct; the LSJ's citation of the use of the word in Plutarch *Marcellus* 14 establishes that the μεσόγραφα μυρία are the products of the machine, not the device itself.

an 18-line epigram) reverses the order. Read as halves, a difference in the utility of Eratosthenes' innovation stands out: the letter highlights preserving the proportions of altars and temples and enlarging war machines.

> Once this (instrument) is discovered, we will be able to set in general a given solid bound by parallelograms into a cube, or change the shape from one (solid) into another (solid), and do likewise to even enlarging [a given solid] by preserving its likeness, such as altars and temples. We will also be able to set both wet and dry measures into a cube, I mean a *metretes* or a *medimnos*, and by the side of this (instrument) to measure the volumes of these receiving vessels. The invention will also be useful for those wishing to enlarge catapults and stone-throwing engines. For it is necessary to enlarge proportionally all their thicknesses, sizes, slits, washers, and tensile strings, if the throw is to be enlarged proportionally, since these are not possible without the discovery of the means.[9]

The activities of building altars, temples, and catapults are particularly associated with the Ptolemies in their roles as pharaohs. Building altars and temples are adumbrated as kingly duties in Callimachus' *Hymn to Apollo* and were mainstays of pharaonic self-presentation, as in the enormous religious complex at Philae initiated by Philadelphus and continued by Euergetes and Philopator.[10] As previous chapters have described, **no. 139** Philon's *Artillery Constructions* shows that the Ptolemaic monarchs gave financial support for technological developments in Hellenistic catapult technology.

On the other hand, Eratosthenes' epigram highlights preserving the measurements of animal-pens, granaries, and wells in its first sentence:

> If, friend, you plan to fashion a doubled cube from a little (cube) or to change any solid's shape into another, this is possible for you, even if you measure a cattle-byre or a grain-pit or the wide hollow of a water-cistern, when by the tips of the frames you seize the means run-together inside the double rulers.[11]

---

9 Eutocius *In Archimedis de Sphaera et Cylindro Libros II* 90.13–27 Heiberg: τούτου δὲ εὑρισκομένου δυνησόμεθα καθόλου τὸ δοθὲν στερεὸν παραλληλογράμμοις περιεχόμενον εἰς κύβον καθιστάναι ἢ ἐξ ἑτέρου εἰς ἕτερον μετασχηματίζειν καὶ ὅμοιον ποιεῖν καὶ ἐπαύξειν διατηροῦντας τὴν ὁμοιότητα, ὥστε καὶ βωμοὺς καὶ ναούς· δυνησόμεθα δὲ καὶ τὰ τῶν ὑγρῶν μέτρα καὶ ξηρῶν, λέγω δὲ οἷον μετρητὴν ἢ μέδιμνον, εἰς κύβον καθίστασθαι καὶ διὰ τῆς τούτου πλευρᾶς ἀναμετρεῖν τὰ τούτων δεκτικὰ ἀγγεῖα, πόσον χωρεῖ. χρήσιμον δὲ ἔσται τὸ ἐπινόημα καὶ τοῖς βουλομένοις ἐπαύξειν καταπαλτικὰ καὶ λιθόβολα ὄργανα· δεῖ γὰρ ἀνάλογον ἅπαντα αὐξηθῆναι καὶ τὰ πάχη καὶ τὰ μεγέθη καὶ τὰς κατατρήσεις καὶ τὰς χοινικίδας καὶ τὰ ἐμβαλλόμενα νεῦρα, εἰ μέλλει καὶ βολὴ ἀνάλογον ἐπαυξηθῆναι, ταῦτα δὲ οὐ δυνατὰ γενέσθαι ἄνευ τῆς τῶν μέσων εὑρέσεως. For the translation of χοινικίδας as "washers" in contemporary catapults, see Marsden 1971: 53 and Philon *Belopoeica* 53.10 Th, 57.25 Th.
10 Callimachus *In Apollinem* 61–69; on Euergetes at Philae see Bingen 2007: 31–43.
11 Eutocius *In Archimedis de Sphaera et Cylindro Libros II* 96.10–15 Heiberg: εἰ κύβον ἐξ ὀλίγου διπλήσιον, ὦθαγέ, τεύχειν / φράζεαι ἢ στερεὴν πᾶσαν ἐς ἄλλο φύσιν / εὖ μεταμορφῶσαι, τόδε τοι πάρα, κἂν σύ γε μάνδρην / ἢ σιρὸν ἢ κοίλου φρείατος εὐρὺ κύτος / τῇδ' ἀναμετρήσαιο, μέσας ὅτε τέρμασιν ἄκροις / συνδρομάδας δισσῶν ἐντὸς ἕλῃς κανόνων.

Every Egyptian village has its cattle-pen, its granary, and its water-channels to irrigate fields; these epigram lines describe the local village economy. Over the course of Eratosthenes' text then we have moved from Alexandria, in the letter to the king, to the local village, in the dedicatory stele.

We should take seriously the explicit generic self-categorization of the text as separately a letter to Ptolemy and a transcription of the previously placed *stele*. The hybrid nature of the text comes from its composition in two broad chunks, for two separate audiences. The letter's audience is the king; the dedication *stele*'s audience are the Greco-Egyptian scribes of the native village economy subservient to the *dioiketes*. The text's recognition of separate audiences should perhaps not be surprising, for the introductory letter is meant to describe to the king what exists on the *stele*. Presumably then the *stele* has been placed in a context available for Egyptians to read and examine somewhere outside of Alexandria. There is reason for confidence that Eratosthenes' *stele* was a historical object, not merely a literary invention. We have at least two ancient parallels for mathematical *stelai*. First is the *stele* of Archimedes' tomb as discovered by the Roman politician Cicero: the *stele* had a verse inscription and an inscribed diagram.[12] The second *stele* is Ptolemy's *Canopus Inscription*, an extant text that records star sightings. While preserved in manuscript the text is prefaced by the phrase "as on the *stele* in Canopus".[13] The parallels suggest that Eratosthenes' *stele* was a real, material object. Eratosthenes' *stele* existed somewhere outside of Alexandria whose audience was the Greco-Egyptian scribes who worked in the traditional village economy.

There are further clues beyond the social environment that the intended audience of the stele is a mixed Greco-Egyptian audience. These come, perhaps surprisingly, from the mathematics itself. I discuss first the practical application of the proportional solution and second the proof style. The Rhind mathematical papyrus from dynastic Egypt in the Middle Kingdom contains problems on measuring produce for granaries of cylindrical and orthogonal shapes and problems on distributing them.[14] Consider the example of the Egyptian mathematician's calculation of volume from a cylindrical granary.

> Example of working out a circular container of diameter 9 and height 10. You are to subtract a ninth of 9, namely 1; remainder 8. Multiply 8 eight times, result 64. You are to multiply 64 ten times; it becomes 640. Its half is now added to it; it becomes 960. This is its content in *khar*. You are to take a twentieth of 960, namely 48. This is the amount which will go into it in quadruple-*heket*, namely 48 hundreds of quadruple-*heket* corn.[15]

---

12 Cicero *Tusculanae Disputationes* 5.64–65.
13 Ptolemy *Inscriptio Canobi* capitulum, 149.1 Heiberg ὡς ἐν τῇ ἐν Κανώβῳ στήλῃ.
14 These are problem numbers 41–46.
15 *P.Rhind* no. 41; I cite the translation from Peet 1923.

Past mathematical historians have focused on the scribe's approximation of the formula for the volume of a cylinder, namely $\pi r^2 h$: here the scribe seems to use an approximation of π as $(16/9)^2$, a relatively accurate approximation.[16] More recently, however, Imhausen has stressed the functional purpose of Egyptian calculations and their embeddedness in daily-life practices.[17] As an example of the functional embeddedness of Egyptian mathematics she counts six surviving problems about grain containers, twenty-one surviving problems about the production of beer and bread, twelves problems about the divisions of grain rations, and four problems about the work production of grain.[18] Following Imhausen, we may therefore read the Rhind papyrus' calculation of the cylinder's volume as a problem about the contextual production of grain in the Egyptian economy, in which *khar* and *heket*, the quantifying units of measurement, are cubic measurements of larger and smaller sizes. The *khar* is useful to quantify the large quantities of the grain stores from a given area; the *heket* on the other hand is a unit appropriate for distribution of grain to workers or for production to beer or bread.[19] Agricultural volumetric calculations were a mainstay of Egyptian mathematics, the very application highlighted in Eratosthenes' epigram.

The second clue that the audience intended for Eratosthenes' *stele* is a mixed Greco-Egyptian audience comes from the proof style of the mathematics. Both the letter and the dedicatory stele each contain two different styles of mathematical proof, a classical Euclidean proof and a mechanically-derived proof from Eratosthenes' invention. An important, recent historiography of the development of mathematical proof has focused on the comparative role of the rhetoric of mathematical persuasion within a given social context.[20] Imhausen too has suggested that the procedural style is important for contextualizing Egyptian mathematics. All Egyptian mathematical problems contain a stylized working out of the problem by manipulating certain numbers in columns in a particular way. Just such a manipulation of numbers follows the above problem from the Rhind papyrus. Eratosthenes' mechanical instrument depends for its working on a two-step movement of the rulers of the machine as they slide along a track. There is a procedural element in the positioning of the ruler. Thus the proceduralism of Eratosthenes' mechanical proof is comparable to the algorithmic style of certain Egyptian proofs, so perhaps Eratosthenes' proof styles may vary to persuade different social audiences.

As we have seen, the use of a proportional tool to measure cattle-pens, granaries, and water-channels is the mathematical challenge of the Egyptian economy.

---

16 Peet 1923 *ad loc.*
17 Imhausen 2003 is most important but she has continued to argue her thesis in Imhausen 2010a, 2010b.
18 Imhausen 2010b: 3. Imhausen 2010b closely reads two tombs scenes to show that scribes are an integral part of the artistic presentation of tomb scenes involving accumulation, cleaning, measurement, transportation, and summation of grain stores.
19 On these terms see Robins and Schute 1987: 14.
20 Chemla 2012.

Scholarship on the Ptolemaic economy has varied, from the previous view that it was a centrally-controlled and planned economy to the more recent views of Manning, who sees a mixture of *ad hoc* centralization and local techniques.[21] I caution against seeing in the lines of Eratosthenes' epigram about cattle, grain, and water notions of a planned centralized economy with a corresponding planned increase of output. Rather, the reader's attention is drawn to the notions of measurement and calculation.

The centrality of cattle, granaries, and water-channels to the Egyptian economy was surprisingly long-standing. The Theban tomb of Rekhmire, the vizier of the pharaohs Thutmosis III and Ahmenhotep II of the 18th dynasty (fifteenth century BCE), contains lengthy hieroglyphic inscriptions about the duties of the vizier, the dynastic Egyptian official in charge of the judiciary, economy, and some aspects of the military. The inscriptions concern Rekhmire's responsibilities for ensuring the irrigation channels are clear and recording and measuring the tax tributes of grain and cattle. Rostovtzeff, the social and economic historian of Ptolemaic Egypt, recognized in the instructions to the vizier Rekhmire an ancestor of Ptolemaic bureaucratic instructions on the management of the economy found in papyrus Tebtunis 703. This late third century BCE papyrus contains instructions probably from the Alexandrian *dioiketes* to his subordinate. The papyrus enumerates among the other duties of the subordinate his responsibility to check the irrigation channels, measure grain for transport to Alexandria, and record and manage the cattle herds. I present the relevant portions of the two texts side-by-side.

Tomb of Rekhmire[22]	P.Tebt. 703[23]
Water: §698 "It is he who dispatches the official staff, to attend to the water-supply in the whole land." §707 "It is he who inspects the water-supply on the first of every ten-day period [*lacuna*] concerning every matter of the judgment hall."	**Water:** recto col. ii., lines 29–40: "… [check] the channels (ὑδρ]αγωγούς) through the plains, whether the inlets of water into them have the set depth and the receiver in them is effective: from these the farmers are accustomed to introduce the water onto the land each sows. Likewise [check] the mentioned canals (διώρυγας) from which the inlets come into the aforementioned channels, whether they are dug, whether the inflows from the river are best clean, and whether they function properly."[24]

---

21 Manning 2003, Manning and Morris 2005, Kehoe 2010.
22 I cite the translation of Breasted 1906.
23 See the original publication in Hunt and Smyly 1933: 66–102 (commentary written by Rostovtzeff).
24 *P.Tebt.* 703 recto col. ii., lines 29–40: τούς τε διὰ τ[ῶν πε]δίων | ἠγμέν[ους ὑδρ]αγωγούς, εἰ τὰ συν[τ]αχ[θ]έντα | βάθη ἔχου[σιν] αἱ εἰς αὐτοὺς ἐπιρύσεις τοῦ ὕδα[τος καὶ ἐκ[ποιοῦ]σα ὑποδοχὴ ἐν αὐτοῖς ὑπάρ|χει ἀφ' ὧν ε[ἰσάγει]ν εἰώθασιν οἱ γεωργοὶ τὸ ὕδωρ | εἰς ἣν γῆν ἕ[κα]στος κατασπείρει·

Tomb of Rekhmire	P.Tebt. 703
Grain: §§748–749 are images of Rekhmire receiving and inspecting the offerings of grain to the temple of Amun, and its subsequent threshing, milling, and baking.	**Grain:** recto col.iii, lines 70–87: "Let it be a care for you that the present grain in the nomes except that spent in the locales for seed and unsold [*letters missing*] is carried down. In this way it is easy to load it into the first ships that arrive; give yourself wholly to this task. For if the ship captains [*letters missing*] their own [*letters missing*] in each locale [*letters missing*] spending time. Let it be a concern for you that the contracted provisions of grain sail down to Alexandria, whose list we are sending you: send them at the right times, not only in the right order but also proofed and fit for use."[25]
Cattle: §§718–728, §§730–745 are images of Rekhmire inspecting and counting the tax duties of various enumerated cities, districts, and their officials. The primary products are gold, cloth, and cattle.	**Cattle:** verso col. i, lines 183–191: "Take care to see also to the cattle-pens (τὰ μοσχοτροφ[ῖ]α) and make every effort that feed in them is present from when it is green, and that the written *per diem* on the cows be spent and [*letters missing*] be sold regularly, the part from the locales themselves and, if there is need of further income, those from the villages."[26]

The stability of the Egyptian economy's agricultural products and its administrative bureaucracy over the *longue durée* from the fifteenth to the third centuries BCE is striking. It is obvious that there was need in Ptolemaic Egypt to measure and calculate the proportional dimensions of, in Eratosthenes' words, "a cattle-byre or a

---

ὁμοίως δὲ καὶ τὰς [δηλ]ουμένας διώρυγας ἀφ' ὧν | εἰς τοὺς προγεγραμμένους ὑδραγωγοὺς αἱ ἐπιρρύσεις γίνονται, εἰ αὐταί τε ὠχύρων|ται καὶ εἰ ἀπὸ τοῦ βελτίστου αἱ ἐμβολαὶ | ἀπὸ τοῦ ποταμ[οῦ καθ]αριῶνται ⟨καὶ⟩ εἰ ἄλλως | πως ἐν ἀσφαλεία[ι εἰσί]ν.

25 *P.Tebt.* 703 recto col.iii, lines 70–87: ἐπι[μελ]ὲς δέ σοι γινέσθω καὶ ὅπως [[καὶ]] | ὁ ὑπάρχων σῖτος ἐν τοῖς νομοῖς πλὴν | τοῦ ἐν αὐτοῖς τοῖς τ[ό]ποις δαπανω|μέν[ου] [εἰς τὰ σ]πέρ[μ]ατα [κ]αὶ τοῦ ἀπλώ|του ..[.......]ς κατάγηται· οὕτω δὲ | ἐμβα[λεῖν εἰ]ς τὰ πρῶτα παριστά|μενα [πλοῖα ῥαίδιο]ν, καὶ πρὸς τὸ τοι|οῦτον [μήποτε] παρ[έ]ργως σαυτὸν δί|δου. εἰ[... γ]ὰρ οἱ ναύκληροι τὰς ἰδί|ας ολ.[......]οις ἐφ' ἑκάστων τῶν | τόπ[ων.. δ]ιατριβόντων. ἐπιμελὲς | δέ σοι ἔστω καὶ ἵνα αἱ διαγεγραμμέ|ναι ἀγοραὶ κατάγ[ωντ]αι εἰς Ἀλεξάν|δρειαν ὧν σοι καὶ [τ]ὴν γραφὴν ἐπιστέλ|λομε[ν ἀπ]οστέλλων κα[ὶ] κατὰ τοὺς | καιρο[ύς, μὴ μ]όνον ἀριθμὸν ἔχουσαι | ἀλλὰ κα[ὶ δ]ε[δο]κιμασμέναι καὶ ἐπι|τήδεοι πρὸς τὰς χρείας.

26 *P.Tebt.* 703 verso col. i, lines 183–191: ἐπιμέλ[ου δ]ὲ ἐπισκοπ[εῖ]ν καὶ τὰ μοσχοτρο|φ[ῖ]α, καὶ τὴ[ν] πλείστην σπουδὴν ποιοῦ ὅ|πως ὅ τε σῖτος ἐν αὐτοῖς παρε[σχη]μένος | ᾖ[ι] μέχρι τῶν χλωρῶ[ν], καὶ εἰς [τ]οὺς μόσχους | ἀναλίσκηται ὁ διαγ[ε]γραμμένος καθ' ἡ|[μ]έραν, καὶ τ[ὸ .....]ρο[.] εὐτάκτως ἀποδί|[δ]ωται, τό τ' ἐξ αὐτῶν τῶν τόπων καί, ἐὰν | [π]ροσδέωνται τοῦ προσανακομιζομέ[ν]ου, | [κ]αὶ ἐξ ἄλλων κωμῶν.

grain-pit or the wide hollow of a water-cistern." Eratosthenes' *sphragis* epigram is clearly referencing the native Egyptian economy and products of native village life.

Yet is Eratosthenes really referencing dynastic Egyptian mathematics? Stephens and others have shown the ability of Ptolemaic poets to manipulate dynastic Egyptian images to portray royal Ptolemaic power. It is difficult to construe Eratosthenes' mathematics as a clear reference to the past. Rather, the contextually-Egyptian references of his letter and dedication concern the present and future. On the other hand, Eratosthenes denigrates the past and its knowledge in both the letter and dedication. His opening, for instance, spurns past lay knowledge:

> They say that one of the old tragedians brought on Minos, who was preparing a tomb for Glaucus, and upon learning that it would altogether one hundred feet said: "You have named a small enclosure for a royal tomb: let it be double, and, that it not fail in its beauty, speedily double each side of the tomb." He seems to have erred. For, if the sides are doubled, a plane-figure becomes four-times greater and a solid eight-times greater.[27]

As I argued in chapter 3, Eratosthenes' criticism ambiguously falls either on the king or the tragedian, where the remainder of the text avoid criticism of the king. In either case, the criticism is directed toward past lay knowledge about mathematics. Later Eratosthenes denigrates the technical knowledge of past mathematicians.

> The result was that they all [sc. geometers in Plato's Academy] wrote a demonstrative solution but to employ it and bring it to use was difficult except in short that of Menaechmus and his only with trouble.[28]

Why denigrate the technical knowledge of the past? As Lloyd has recognized, Greek scientists often include first-person egotistic claims touting the superiority of their own discoveries over previous authorities in the famous agonistic context of Greek science.[29] Eratosthenes' text is filled with such egotistic claims. His criticism of rivals helps define a canon of truth-telling in his quest to speak as a courtier with *parrhēsia*. Furthermore, as Lang has shown in a different context, Greeks

---

27 Eutocius *In Archimedis de Sphaera et Cylindro Libros II* 88.5–13 Heiberg: τῶν ἀρχαίων τινὰ τραγῳδοποιῶν φασιν εἰσαγαγεῖν τὸν Μίνω τῷ Γλαύκῳ κατασκευάζοντα τάφον, πυθόμενον δέ, ὅτι πανταχοῦ ἑκατόμπεδος εἴη, εἰπεῖν· Μίκρον γ' ἔλεξας βασιλικοῦ σηκὸν τάφου / διπλάσιος ἔστω, τοῦ καλοῦ δὲ μὴ σφαλεὶς / δίπλαζ' ἕκαστον κῶλον ἐν τάχει τάφου. ἐδόκει δὲ διημαρτηκέναι· τῶν γὰρ πλευρῶν διπλασιασθεισῶν τὸ μὲν ἐπίπεδον γίνεται τετραπλάσιον, τὸ δὲ στερεὸν ὀκταπλάσιον.
28 Eutocius *In Archimedis de Sphaera et Cylindro Libros II* 90.8–11 Heiberg: συμβέβηκε δὲ πᾶσιν αὐτοῖς ἀποδεικτικῶς γεγραφέναι, χειρουργῆσαι δὲ καὶ εἰς χρείαν πεσεῖν μὴ δύνασθαι πλὴν ἐπὶ βραχὺ ταῦτα Μεναίχμου [ἐπὶ βραχὺ ταῦτα Μεναίχμου *scripsi*: ἐπὶ βραχυτητι του Μενεχμου *mss.*: ἐπὶ βραχύ τι τὸν Μέναιχμον Heiberg] καὶ ταῦτα δυσχερῶς. Netz 2004: 295 follows the manuscripts. The fact that none of the other mathematicians named in the text has an article in front of the name tells against the manuscript text and Heiberg's emendation.
29 Lloyd 1987: 50–108.

thought Egyptian science fixed and nearly unchangeable.[30] Eratosthenes' egotistic and agonistic denigration of past technical knowledge is directed indiscriminatingly against the past, both Greek and dynastic Egyptian. The superiority of Eratosthenes' mathematical invention is a triumph of the present world over the Greek and Egyptian past.

The temporal progression of Eratosthenes' *sphragis* epigram confirms a presentist reading of Eratosthenes' invention.

> If, friend, you plan to fashion a doubled cube from a little ⟨cube⟩ or to change any solid's shape into another, this is possible for you, even if you measure a cattle-byre or a grain-pit or the wide hollow of a water-cistern, when by the tips of the frames you grasp the means run-together inside the double rulers. Do not seek the mechanically-difficult works of Archytas' cylinders nor cut the cones of the Menaechmean triads, not even if some curved shape of divine Eudoxus is recorded in the lines. For you may easily fashion myriad mean-drawings in these plates by beginning from a small base. O Ptolemy, happy that as a father coming into your youth with your son you yourself gave all that is friendly to the Muses and Kings! Later, o Heavenly Zeus, may he grasp the scepters from your hand. May these come to pass and may someone looking at this dedication say that it was the work of Eratosthenes of Cyrene.[31]

The poem begins in the present of the native Egyptian economy, as we have seen. Next comes the present superiority of Eratosthenes' innovation: "this is possible for you." In the next lines past mathematicians' solutions to the problem are rejected for their difficulty: Eratosthenes tells readers to avoid the work of the three Greek mathematicians Archytas, Menaechmus, and Eudoxus. In the following lines we return to the present superiority of Eratosthenes' invention, which is said to be "easy" and capable of "myriad" results. At last we transition from the present to the future in a rhetorical triad: Eratosthenes praises the current king Ptolemy, wishes for a successful crowning of the next monarch, Ptolemy's son, and looks forward to a future day in which a future reader recognizes the entire dedicatory *stele* as the work of Eratosthenes. The *sphragis* epigram is imbued with the egotistic and agonistic context of Greek science. The present is seen as the apex of knowledge and the future is imagined as a continuation of the present. There is no attempt here to praise or manipulate the mathematics of the dynastic Egyptian past.

---

30 Lang 2012: 128–134.
31 Eutocius *In Archimedis de Sphaera et Cylindro Libros II* 96.10–27 Heiberg: εἰ κύβον ἐξ ὀλίγου διπλήσιον, ὤθαγέ, τεύχειν / φράζεαι ἢ στερεὴν πᾶσαν ἐς ἄλλο φύσιν / εὖ μεταμορφῶσαι, τόδε τοι πάρα, κἂν σύ γε μάνδρην / ἢ σιρὸν ἢ κοίλου φρείατος εὐρὺ κύτος / τῇδ' ἀναμετρήσαιο, μέσας ὅτε τέρμασιν ἄκροις / συνδρομάδας δισσῶν ἐντὸς ἕλῃς κανόνων. / μηδὲ σύ γ' Ἀρχύτεω δυσμήχανα ἔργα κυλίνδρων / μηδὲ Μεναιχμείους κωνοτομεῖν τριάδας / διζήσῃ, μηδ' εἴ τι θεουδέος Εὐδόξοιο / καμπύλον ἐν γραμμαῖς εἶδος ἀναγράφεται. / τοῖσδε γὰρ ἐν πινάκεσσι μεσόγραφα μυρία τεύχοις / ῥεῖά κεν ἐκ παύρου πυθμένος ἀρχόμενος. / εὐαίων, Πτολεμαῖε, πατὴρ ὅτι παιδὶ συνηβῶν / πάνθ' ὅσα καὶ Μούσαις καὶ βασιλεῦσι φίλα, / αὐτὸς ἐδωρήσω· τὸ δ' ὕστερον, οὐράνιε Ζεῦ, / καὶ σκήπτρων ἐκ σῆς ἀντιάσειε χερός. / καὶ τὰ μὲν ὣς τελέοιτο, λέγοι δέ τις ἄνθεμα λεύσσων / τοῦ Κυρηναίου τοῦτ' Ἐρατοσθένεος.

Eratosthenes' science references contemporary Ptolemaic Egypt, not its classical Egyptian past. Despite the presentism of Eratosthenes' text, there is a clear attempt to author a text appropriate for both a Greek and Egyptian audience in both the mathematics and social context. On the one hand, Eratosthenes references the classical problems and proof style of Classical Greek mathematics; on the other hand, Eratosthenes seems to reference a variety of proof styles, including the proceduralism of native Egyptian mathematics. We know from demotic papyri that these native Egyptian mathematics continued into the Ptolemaic period. The social context of Eratosthenes' text is also deeply embedded in both the economy of the native Egyptian village and the dynastic functions of pharaonic kingship. The Eratosthenic mathematician sees double.

The Greco-Egyptian context of Eratosthenes' *Letter* suggests that the primary audience of the *stele* is the royal bureaucracy. For the *dioiketes* and his subordinate Greco-Egyptian scribes, with professional interest in both the agricultural products of native village life and the dynastic functions of pharaonic kingship, best represent Eratosthenes' mixed audience of the present of Ptolemaic Egypt. While it is speculative to associate the *Letter to Ptolemy* with any of the named *dioiketeis* **nos. 19–22**, the conclusion that the *Letter*'s primary audience is the royal bureaucracy gains strength from the comparandum of Eratosthenes' work *On the Measurement of the Earth*. That treatise's use of royal bematists and title associates it as a product of the Ptolemaic bureaucracy concerned with taxes and land.[32] Such a bureaucracy is effectively the office of the *dioiketes* and his subordinates. Therefore Eratosthenes' letter to Ptolemy can be best contextualized within a political ideology of the monarch's care for his current subjects, a presentist scientific contribution of the Ptolemaic court scientist. Although based in Alexandria, the Ptolemaic

---

[32] Eratosthenes' *On the Measurement of the Earth* (περὶ ἀναμετρήσεως τῆς γῆς) was a remarkable treatise; cf. Geus 2002: 223–259, Roller 2010: 263–267. Its main scientific contribution was to yield an informed estimation of the circumference of the earth, 252,000 stades. Scholars have universally hailed the ingenuity of Eratosthenes' solution to the problem and, with varying guesses about an equivalent modern value for his stade, credited his estimation within .8%–17% accuracy of the modern value. Yet Eratosthenes' mathematical ingenuity would not have been possible without basic geographical information that came from his access to Ptolemaic royal power. Greeks identified the origins of geometry in the Egyptian practice of land measurement. The legendary Egyptian king Sesotritis "sent people to investigate and measure (ἀναμετρήσοντας)" how much of a peasant's allotted land had been lost due to the action of the flooding of the river Nile; Herodotus 2.109. The term ἀναμέτρησις "measurement" appears in documentary papyri, usually in reference to landholdings (e.g. *P.Teub.* 3.793). Thus Eratosthenes' Greek title περὶ ἀναμετρήσεως τῆς γῆς resembles the paper trail of the Ptolemaic bureaucracy for tax purposes. More explicitly, Martianus Capella 6.598 witnesses that Eratosthenes was dependent on the Ptolemaic step-measurers: "Indeed, Eratosthenes, made knowledgeable about the number of stades from Syene to Meroe by the royal measurers of Ptolemy, learning how much that extent of the earth was and multiplying by the fraction of the whole, determined without delay how many thousands of stades the great circle embraces the measure of the earth."

court scientist's presentism leads him to be interested in the intercultural world of Ptolemaic society outside the capitol.

The *Letter to Ptolemy* thus echoes the Canopus Decree in its materiality, Greco-Egyptian hybridity, and the ideology of Ptolemaic kingship. But the treatise also speaks the language of Ptolemaic intellectuals and court science in other ways. The text is shot through with notions of utility. Eratosthenes criticizes Minos for failing to build a doubled tomb and the geometers of the Academy for the difficulty of putting their solutions into practice.[33] His own solution, by contrast, is capable, useful, and easy to do.[34] In its claims for the utility Eratosthenes echoes the royal expectation of Hellenistic sympotic contributions. As analyzed in chapter 2, the king ought to invite to symposia to serve as his advisors "friends-of-learning and men capable of suggesting what is useful to the kingdom and to the lives of its subjects".[35] Eratosthenes' invention benefits royal activities (building altars, temples, measuring wet or dry measures, or enlarging catapults) as well as the more economical and plebian activities of Egyptian village life (measuring cattle-byres, a grain-pits, or water-cisterns).[36] Eratosthenes' invention is potentially useful to the kingdom and its inhabitants' lives, to the greater glory of the king.

Furthermore, Eratosthenes' *Letter to Ptolemy* espouses the aesthetic features of Ptolemaic court literature: Greco-Egyptian context, hybridity, performance. I have already explored its potential to "see double" within its Egyptian social and literary context, a theme found extensively in other Ptolemaic court authors discussed as **nos. 109–111, 117–119.** Hybridity is quite literally inscribed in the text as a mixture of a formal letter and an inscription; its formal opening and conclusion are themselves genres of tragedy and epigram.[37] Beyond the formal property of the text the mathematical machine and its product is a hybrid of different sciences.

Eratosthenes mixes mechanics and mathematics in his approach to determine two mean proportionals.[38] Cross-disciplinary science mixes different domains of

---

[33] Eutocius *In Archimedis de Sphaera et Cylindro Libros II* 88.11–13; 90.9–11 Heiberg.
[34] Eutocius *In Archimedis de Sphaera et Cylindro Libros II* δυνησόμεθα 90.14, 90.18, δυνατά 90.27, πάρα 96.12; χρήσιμον 90.21; ῥᾳδία 90.12, ῥεῖα 96.21 Heiberg. Leventhal 2017 emphasizes the thematic of "handiness" in the text.
[35] Ps.-Aristeas 286.
[36] Eutocius *In Archimedis de Sphaera et Cylindro Libros II* 90.17–26; 96.12–13 Heiberg.
[37] Should we read the generic features with which Eratosthenes opens and closes the text as hybridity or belatedness? That is, does the epigram replace tragedy as the genre that presents a contemporary view of intellectual Greek life? Consider Dioscorides of Nicopolis' **no. 111** epigrams on previous tragedians or Philopator's **no. 8** imitation of the master-stylist of the tragic genre, Euripides. Leventhal 2017 opts to read the similar themes between the tragic fragment and the epigram as hybridity.
[38] Berryman 2009 has underscored how multifaceted ancient mechanics could be. Its most restricted sense concerned only the principles of leverage. But the discipline comes to embraces principles of mutual contact, force upon bodies, and automated systems of continuous feedback. I follow Berryman's 2009: 54–130, 146–167 argument that a well-defined sense of mechanics as a

scientific knowledge. Eratosthenes' solution is unique in contrast to previous attempts. After Hippocrates of Chios first recognized the problem, three mathematicians solved the problem through the intersection of different mathematical curves. Archytas manipulated the surfaces of half-cylinders and Eudoxus particular curved lines. Menaechmus' contribution has been more difficult to elucidate, although the most obvious interpretations remains the use of the three conic sections.[39] At any rate these are all strictly mathematical solutions that, while feats of ingenuity, are subject to Eratosthenes' just criticism that they are difficult to implement. Geometrical solutions are given that do not readily yield expressible magnitudes.

An evidently experimental strategy for solving the problem of the two mean proportionals comes from Eratosthenes' contemporary **no. 139** Philon. Philon's *Artillery Constructions* correctly states that solving the mathematical problem of the two mean proportionals allows for corresponding scaling of artillery machines. The hole holding the tension-spring is the magnitude determinative of the machine's capabilities. Philon first describes that earlier engineers experimented to find correct proportions for this hole; later Alexandrian engineers, financially supported by the Ptolemaic pharaohs, made more systemic experiments.[40] Philon himself reports that he will transmit the method "just as I learned after long association with those craftsmen in Alexandria who concern themselves with such matters and after being known to not a few engineers in Rhodes."[41] He proceeds to give a rough table of cubic roots with instructions on how to derive them.

> Reduce to units the weight of the stone in reference to which the machine must be constructed. From the reduced amount, however many units its cube root possesses, make the diameter of the hole so many dactyls while adding a tenth of the obtained cube. If the weight does not have a cube expressible in fractions, take the nearest and if it exceeds try to lessen the tenth proportionally. If it falls short, fill it out by adding a tenth.[42]

---

discipline dates to the third century BCE. Eratosthenes' work would have been considered cross-disciplinary by his elite audience.

39 Eratosthenes' epigram describes "the Menaechmean triads" (Eutocius *In Archimedis de Sphaera et Cylindro Libros II* 90.16 Heiberg) and Eutocius himself preserves a text entitled "As Menaechmus" (Eutocius *In Archimedis de Sphaera et Cylindro Libros II* 78.13–80.24) in solution to the problem of the two mean proportionals. Yet Eutocius' text "As Menaechmus" has been heavily rewritten in the post-Apollonian language of conic sections, a chronological impossibility for the pre-Apollonian Menaechmus. "The Menaechmean triads" might most obviously be the three conics sections, viz. hyperbola, parabola, and ellipse, but Knorr 1986: 61–67 has rejected this interpretation.
40 Philon *Belopoeica* 50.14–29 Th.
41 Philon *Belopoeica* 51.10–12 Th: καθότι καὶ αὐτοὶ παρειλήφαμεν ἔν τε Ἀλεξανδρείᾳ συσταθέντες ἐπὶ πλεῖον τοῖς περὶ τὰ τοιαῦτα καταγινομένοις τεχνίταις, καὶ ἐν Ῥόδῳ γνωσθέντες οὐκ ὀλίγοις ἀρχιτέκτοσι.
42 Philon *Belopoeica* 51.15–20 Th: τὸ τοῦ λίθου βάρος, πρὸς ὃ ἂν δέῃ τὸ ὄργανον συστήσασθαι, εἰς μονάδας ἀγαγεῖν καὶ τοῦ συναχθέντος πλήθους ὅσων μονάδων ἡ ⟨κυβικὴ⟩ πλευρά, τοσούτων δακτύλων τὴν τοῦ τρήματος διάμετρον ποιεῖν προσθέντα ἔτι τὸ δέκατον μέρος τῆς εὑρεθείσης πλευρᾶς. ἐὰν δὲ μὴ ἔχῃ ῥητὴν πλευρὰν τὸ βάρος, ὡς ἔγγιστα λαμβάνειν, καὶ ἐὰν μὲν ὑπεράγῃ,

Philon follows this guide with a table of weights for the stone shot and corresponding diameter for the hole of the tension spring. Advancing from weights of 10 minae to 3 talents by 5 minae, his table is probably the product of physical experimentation. The length of the diameter of the hole for the tension-spring are relatively large, from 11 up to 27 dactyls. Perhaps that is why, as Philon himself explains, it is not precise but rather adjusted by rounding. Philon's experimental method yields rough solutions to the problem of two mean proportionals, but only in the context of its application to artillery machines.[43]

By contrast Eratosthenes' machine produces geometrically accurate results that are precise and expressible magnitudes. He adapts the certainty of geometric methods to yield expressible magnitudes in the fashion of experimental mechanics.[44] We can see Eratosthenes' mixture of the mechanical and mathematical in the machine itself, as well as its products. Like the *stele* it too is a material object. Eratosthenes describes his as bronze and soldered to the top of the *stele*. He describes its construction cursorily in the letter:

> A frame in wood, ivory, or bronze has three equal plates fashioned as thin as possible. One is fixed in the middle, the other two are pushed in the grooves, each persuading themselves by size and measure.[45]

On the text inscribed on the *stele*, where this three-plate exemplar is soldered, Eratosthenes adds that more plates can be installed if more than two mean proportionals are sought. Eratosthenes saw the possibility of an iterated mechanical solution, not a general mathematical claim.[46] He repeats this idea in the epi-

---

τὸ δέκατον μέρος ἐλασσοῦν πειρᾶσθαι τῷ κατὰ λόγον, ἐὰν δὲ προσλείπῃ, προστιθέντα τὸ δέκατον προσαναπληροῦν.

43 Philon goes on to give a geometrical proof, stressing that he is providing a shortened version of a geometric proof in an earlier and now lost book of the *Mechanical Collection*. Eutocius *In Archimedis de Sphaera et Cylindro Libros II* preserves an account of a geometric proof from Philon that ought to be compared to the extant text.

44 Several results in Eutocius' catalogue of solutions are mechanical solutions: Eratosthenes', the solution ascribed to Plato, and Nicomedes'. The others should all be dated subsequent to Eratosthenes'.

45 Eutocius *In Archimedis de Sphaera et Cylindro Libros II* 92.27–94.3 Heiberg: διαπήγνυται πλινθίον ξύλινον ἢ ἐλεφάντινον ἢ χαλκοῦν ἔχον τρεῖς πινακίσκους ἴσους ὡς λεπτοτάτους, ὧν ὁ μὲν μέσος ἐνήρμοσται, οἱ δὲ δύο ἐπωστοί εἰσιν ἐν χολέδραις, τοῖς δὲ μεγέθεσιν καὶ ταῖς συμμετρίαις ὡς ἕκαστοι ἑαυτοὺς πείθουσιν.

46 Eutocius *In Archimedis de Sphaera et Cylindro Libros II* 96.6–8 Heiberg. In this brief comment may lie the genesis of the astounding insight of René Descartes' *La Géométrie* 2. Pappus 3.23 records Eratosthenes' instrumental solution in a geometric fashion. However Pappus does not include Eratosthenes' statement that it possible to include more plates to obtain further means. Descartes *La Géométrie* 2 (pp. 315–319 in the 1637 edition of *Discours de la Methode*; I use the facsimile of that edition in Descartes 1954) imitates Pappus' 3.20–22 classification of plane, solid, and linear problems and uses Pappus' 3.23 presentation of Eratosthenes' instrument to produce higher order solutions. Yet Descartes goes beyond Pappus' presentation of cubic equations to equations of degrees 4, 5, and 6 by introducing, essentially in imitation of Eratosthenes' comment, more parallel-producing

gram.⁴⁷ Eratosthenes' machine is a set of physical sliding rectangles that move parallel to each other along grooves. When they move past each other the imaginary mathematical diagonals that cross between the sides of the plates remain parallel to each other. The sought mean proportionals are the heights to the diagonals.

And here something extraordinary happens. The mean proportionals sought exist both as a physical object and as a mathematically conceptual equivalent. The reader must use his fingers to grasp the measurement offered by the moving plates. The epigram describes the moment of discovery: "when by the tips of the frames you grasp the means run-together inside the double rulers".⁴⁸ Netz has suggested that this was a hunting metaphor.⁴⁹ But I believe that Eratosthenes' description is the image of picking up numbers off the counting board.⁵⁰ The flat counting board was widespread in the ancient world and would have been a part of the armamentarium of the Ptolemaic *dioiketes* and his subordinates, charged with taxes, land measurement, and keeping track of the products of agriculture. The tokens on the counting board, manipulated from point to point as in a game of backgammon, are simply grasped as a physical object by the scribe's opposed digits. Netz summarized the implication: "the Greeks imagined [numbers] as an entity grasped between the thumb and the finger."⁵¹ The counting token's numerical significance is a function of its initial placement on the board. Eratosthenes imagines the user of his machine acting with an extra set of fingers and scooping up the two mean lines, grasped between opposed digits, just as pebble counters off a counting

---

rectangular plates. Three mathematical projects intertwine in a triple parentage: Eratosthenes', Pappus', and Descartes'. It seems to have been Pappus' contribution to recognize the general problem of classification and Descartes' to present linear curves of higher degree polynomials as a yet more general and abstract contribution to the problem of classification. From Eratosthenes' off-hand comment on the *stele* about the possibility of further means, a comment unelaborated in the letter, it is hard to believe that he saw the problem in the general terms that Pappus or Descartes framed it. That fact alone tells us that Eratosthenes saw the problem of mean proportionals in radically different terms from the frame of analytical geometry. Hence Pappus's 7.3 statement that Eratosthenes' *On Means* was part of the so-called *analymenos* or *loci*-related analyses is the anachronistic reading of Pappus' own concerns for mathematical generality. So was Descartes was aware of the crucial differences between Pappus and Eutocius' presentation of Eratosthenes' solution? The answer is not yet clear. Modern scholarship to Descartes' *La Géométrie* has concentrated on Pappus' presentation over Eutocius': for example, Bos's 2001 treatment, otherwise essential, does not recognize the difference.

47 Eutocius *In Archimedis de Sphaera et Cylindro Libros II* 96.20–21 Heiberg.
48 Eutocius *In Archimedis de Sphaera et Cylindro Libros II* 96.14–15 Heiberg.
49 Netz 2004: 298.n183.
50 Netz 2002, a paper of unparalleled brilliance, is the best discussion how Greeks imagined numbers with counting boards; Cuomo 2001: 12–13 has archaeological evidence.
51 Netz 2002: 9.

board. The mean proportionals are the physical objects of the space between the user's fingers whose numerical function is a product of the geometrical parallel lines preserved by the rectangular plates. They are drawn out by the machine and called μεσογράφα "mean-drawings".[52] Eratosthenes' machine produces mean proportionals that are both materialist and have mathematical magnitude.

The user's manipulation of the machine, or performance, is therefore an essential aesthetic feature of Eratosthenes' text. Literary texts highlighting motion and sound were thick at the contemporary Ptolemaic court, as in **nos. 110** Callimachus and **111** Dioscorides. The machine itself is a source of spectacle: the rectangular plates move along a track, clanking in their bronze rails. The machine, the user's interaction with the machine, and the text with its complex mix of genres – embedded tragedy, monumental inscription, and *sphragis* epigram – act as entertainment.

In conclusion Eratosthenes sees no contradiction between utility and entertainment. Court science set expert advice in the context of the king's governing the kingdom. Court science served an aesthetic utility for the pleasure of elite court society. Such science shared the aesthetic features of contemporary court poetry, which drew on earlier written literature, mixed genres and cultures, and highlighted motion and sound. Eratosthenes' machine is both useful for the lives of the kingdom and shares the aesthetic features of contemporary court literature. Eratosthenes uses the discourse of friendship to refer to the user of his machine. "For more accuracy one must be a friend-of-*technē* by seizing the lines in order that, in drawing the plates together, all remain parallel, without division, and equally in contact with each other."[53] Proper use of the machine, technical accuracy, and the social dynamics of court patronage run together here. Eratosthenes' *Letter to Ptolemy* is court science in the intellectual Greek tradition, capped with a poem on a material *stele*, with a physical machine dedicated to improving the village economy of the Ptolemaic kingdom. I suggest that Eratosthenes' *stele* might have been placed at a village not far from Alexandria, perhaps at Canopus. The temple at the Canopic branch of the Nile, where the *theoi Euergetai* shared the shrine with Osiris, is mentioned in Euergetes' Canopus Decree as a spot at which the Egyptian priests met; this temple is further said to be point from which the bark of Osiris sails to Alexandria, a mythological account that might be referenced in Eratosthenes' *Catasterisms*.[54] Many Egyptian temples contained personal dedications inside the outer courtyard forming the temple complex. Building on the previous argument about scientific egotism, we should see Eratosthenes' *stele* possibly placed within the outer courtyard of the Canopic temple complex as a personal

---

52 Eutocius *In Archimedis de Sphaera et Cylindro Libros II* 96.20 Heiberg.
53 Eutocius *In Archimedis de Sphaera et Cylindro Libros II* 94.4–7 Heiberg: πρὸς δὲ τὸ ἀκριβέστερον λαμβάνεσθαι τὰς γραμμὰς φιλοτεχνητέον, ἵνα ἐν τῷ συνάγεσθαι τοὺς πινακίσκους παράλληλα διαμένῃ πάντα καὶ ἄσχαστα καὶ ὁμαλῶς συναπτόμενα ἀλλήλοις.
54 Canopus Decree *OGIS* 56; bark of Osiris Geus 2002: 219.n50, 221.n62.

monument celebrating himself, his deeds, and his relationship to the reigning monarch, in a Ptolemaic Greek equivalent of an Egyptian model such as the famous naophorous statue of the Egyptian courtier Udjahorresne.[55] Without a doubt Eratosthenes set himself and his science to the service of his pharaoh; court science spoke the language of Ptolemaic poetry and entertainment just as much as it spoke the language of Ptolemaic power. Novel cross-disciplinary scientific knowledge emerged from the court matrix of utility and entertainment.

Andreas[56]

Eratosthenes' technology raises questions about how we recognize the discourses in the effects produced by material artifacts. For how should we reconstruct lost scientific machines? The answer seems deceptively simple – we should reconstruct them as scientific devices. We simply need to piece together the disparate parts to function as designed. Imagine reassembling Eratosthenes' instrument: we need to polish and shape some bronze into rectangles that slide within fixed rails. Here then the commonsense answer assumes that "the scientific machine" is a stable transhistorical category whose defining characteristic consists in accurate mechanical functioning with due allowance for historically appropriate mechanical parts and error tolerances. Reconstructing a historical machine according to the commonsense understanding of scientific devices assumes a relatively uncritical and unproblematic transfer between contemporary mechanical knowledge and historically-limited mechanical knowledge: the commonsense understanding implies that we moderns can always reconstruct the mechanically-limited machines of the past. Rather than rejecting this understanding entirely, I wish to problematize it. The commonsense position corresponds only to our contemporary understanding of mechanical functioning. The commonsense position ignores historical actors' contextual understanding of the machine and reduces their knowledge to the machine's rote mechanical function. Reconstructing a historical machine means more than reconstructing its mechanical functioning by presenting the working machine to a modern audience; it also means recapturing what the lost machine meant to historical actors in its discursive context. What might sliding rectangles have meant to Eratosthenes and his contemporaries? My reading of his machine sug-

---

55 Now preserved in the Vatican museum, a statue of Udjahorresne stands holding a small temple as representative of his rebuilding Egyptian temples for the new Persian pharaoh Cambyses. The statue recounts Udjahorresne's life and praises his relation to political power. Translation and commentary in Lichtheim 2006: 36–41.
56 An earlier version of the following analysis was published as Berrey 2017.

gests that they did not understand the machine's motion and sound in the same way we do.⁵⁷

Enlarging the meaning of a lost machine's reconstruction to include what the machine meant to historical actors has historiographical benefits. For not all lost scientific machines exist as physically disparate pieces, ready to be reassembled according to a textual description. Most lost ancient machines are known now only thanks to texts describing them. Sometimes these texts themselves are fragmentary. Such is the case with a surgical machine for the reduction of joints devised by a now-fragmentary author, **no. 141** Andreas of Carystus, the court physician to the pharaoh Ptolemy IV Philopator.⁵⁸ (A reminder of terminology: I will be discussing surgical devices that acted as levers supplying force to return joints to their sockets; the technical medical procedure for the resetting of a dislocated joint is a reduction.) The extant description of Andreas' machine prompts the historiographical question: what is recoverable about a now-lost machine described in fragmentary testimonia? My thesis is that reconstructing lost objects sometimes means reconstructing context as well as content. In the particular case of Andreas' machine I suggest that we can recover more information by understanding what its performance meant to historical actors. Andreas' machine reflects similar aesthetic qualities in contemporary court poetry because they shared a similar audience.

What was technology in antiquity? The modern definition of the word "technology" is a nineteenth century coinage, emphasizing in particular the devices of the Industrial Revolution and their products – in a word, artifacts.⁵⁹ The ancient equivalent of the modern term is best expressed in the transliterated ancient Greek word *technē*, which denotes not only products of knowledge but the discipline that produces those products as well. The semantic field of the ancient Greek word *technē* was vast.⁶⁰ Medicine was a *technē*, as was mathematics, but so were disciplines like rhetoric, sculpting, and cooking. *Technē* was identified variously with other words, such as *sophia* (knowledge/wisdom), *epistēmē* (rigorous, stable knowledge), and *dynamis* (power). Technology in antiquity was thus more widespread than our conception of machines and devices, and in this widespread conception was more integrated into intellectual and practical life.

---

57 Archaeologists have interrogated the categories involved in contextualizing ancient scientific objects; Baker 2013: 85–109.
58 For previous scholarship on Andreas see von Staden 1989; von Staden 1998; von Staden 1999a; Massar 2005.
59 See *OED s.v.* 'technology', especially 4a and 4b. The transliterated term *technologia* exists in ancient Greek but means the comprehensive nature of a *technē*; cf. Mansfeld 1998: 82.n29. Schatzberg 2012 attempts to disentangle the discourses of fine arts, industrialism, and science in the eighteenth and nineteenth centuries as the modern sense of technology emerged.
60 I echo the excellent account of Cuomo 2007: 7–12.

Historians of science and historians of technology have not always lived happily together.[61] The reasons are complex but are partially due to an artificial antithesis between working against nature and working with nature, derived from Galileo. Studies in the history of technology in Greco-Roman antiquity have shown this to be a false and anachronistic dichotomy.[62] Technology was not conceived in opposition to science but in combination with it. The integration of material artifacts into scientific investigation was strongest during the third century in the Hellenistic period, when Andreas lived. This was a period that saw the invention of screws, pistons, steam-driven toys, and numerous developments in the size and scope of weaponry.

The strong integration of material artifacts into ancient scientific study and its conceptual similarity to the use of modern technology in science invites us to apply the conclusions of modern studies of technology to ancient material. One main argument of contemporary sociology of science is that machines are inscription devices.[63] They write and produce paper; it is the writing on this paper (technically, the inscription) that is important for the advancement of scientific concepts and theories. Sometimes therapy is derived from this inscription, as in an X-ray machine; sometimes the charts and data produced by the machine are incorporated into a scientific paper. Ancient machines, by contrast, never produced paper trails. They do not "inscribe" their effects.

Instead, I suggest that we think about ancient machines as a type of performance. The term "performance" indicates a repeatable visual, aural, or tactile activity that provokes a reaction for the watching audience.[64] Speaking of technology as a performance ought not to invoke the metaphor of actors upon a stage, playing roles – as if the people, props, and spoken lines are temporary identities of their real selves, which hide somewhere behind the curtain. No, the technology and its medical use were real; the doctor's successful reduction of dislocated joints by machine was real. Yet as in all ancient medicine, the medical care created a spectacle for a crowd at which a doctor might undertake an epideictic display. The performance context for ancient medicine was important. The picture of everyday healing in Greco-Roman antiquity, drawn primarily from texts not written by physicians, shows a public event without privacy, the doctor performing his medical knowledge and practice for curious and gaping on-lookers.[65] The doctor's reputation was at stake in the public performance of his science. Further, the performance of tech-

---

61 Alexander 2012 is a helpful overview of recent controversies.
62 See Cuomo 2007; Schiefsky 2007; Berryman 2009: 43–49, 236–249.
63 Classic statement of this thesis in Latour and Woolgar 1986.
64 See Mol 2003 for a performative analysis of contemporary medical knowledge.
65 Nutton 2012: 270–271. For the epideictic display of medicine in Hippocrates see Jouanna 1999: 75–111 and in Galen see von Staden 1995, Gleason 2009.

nology established an emotive connection between the doctor and his audience and embellished the doctor's reputation in front of an audience.

The emotion aroused in the audience by watching machines was *thauma*, "wonder". The treatises of the mechanician Hero provide explicit evidence for this phenomenon.[66] Hippocratic medical writings juxtapose, rather than oppose, nature and *technē*. The Hippocratic surgical text *On Fractures* expects performance from the physician, especially regarding machines: "it is shameful and untechnical not to devise how to work devices."[67] The Hippocratic treatise *On Joints* describes the construction of the Hippocratic bench on which the patient's limbs were stretched to reduce dislocated joints.[68] Since there was no institutionalized system of healthcare in antiquity, the medical marketplace provided patients with access to a range of healers, including magicians and village wise-people, as well as physicians trained in the teacher-pupil lineage typical of Hippocratic and subsequent medical philosophies. The ancient physician's ability to attract patients to himself and away from his competitors depended on his reputation as a healer. In a large city the individual physician faced greater competition from other healing traditions and other physicians as well. A large machine and the medicine effected on it may have filled patients with wonder and allowed the physician to offer a unique service, distinguishing himself from his competition. Technology could thus have served the physician as an aid to persuasion.

The interchange of "performance" among artifacts, discourses, and effects in ancient *technē* and modern technology provides a useful term for analysis. Yet the unity of the term performance and its ease of application between ancient *technē* and modern technology obscure the loose footings of the discursive category. Performance is itself not a stable transhistorical notion productive of a continuous genealogy from the ancient dynasties to the multipolar nation-states of the twenty-first century. To think so would be to commit the same error that plagues the commonsense understanding of machine reconstruction: it ignores historical actors' contextual understanding of the machine. Rather, technological performance must be a historically-situated concept capturing an audience-specific response to a machine's motion, sound, and appearance. An explanatory appeal to "technological performance" is an appeal to the discursive contextual understanding of historical

---

[66] Berryman 2009: 50–55, Tybjerg 2003.
[67] Hippocrates *De Fracturis* 30 = 3.524L with translation by Cuomo 2007: 16: αἰσχρὸν γὰρ καὶ ἄτεχνον μηχανοποιέοντα ἀμηχανοποιέεσθαι. Hippocratic texts explicitly recognized the epideictic potential of machines: Hippocrates *De Fracturis* 13 = 3.466L, repeated with an important alteration in Hippocrates *De Articulis* 72 = 4.296L: ἄριστον δέ, ὅστις ἐν πόλει μεγάλῃ ἰητρεύει, κεκτῆσθαι ἐσκευασμένον ξύλον "Best is for whoever practices medicine in a city with many people to acquire a fashioned machine." Cf. the discussion in chapter 3 footnote 56.
[68] Hippocrates *De Articulis* 72 = 4.296–300L.

actors. Technological performance can offer a historically sensitive account of Andreas' machine.

The textual fragments of Andreas' machine (its name is not preserved) are actually testimonia.[69] The most detailed information about the machine is preserved in the *Medical Collection* of Oribasius, a doctor of Late Antiquity, who is himself quoting or summarizing Heliodorus, a doctor dating from the second century CE.[70] Testimonia about Andreas' machine are embedded in Oribasius' complex reorganization of mechanical knowledge. Oribasius reviews surgical machines from Hippocrates' onward; all these machines were intended to exert strong applied force to the muscles and joints of the body in order to reset broken limbs or restore dislocated joints. Oribasius describes most machines not individually but by describing a part, then attributing such a part to this or that device. From Oribasius' summary we learn fifteen named parts of Andreas' machine, although not all these parts are distinct.[71] Despite these names and knowledge of their functions, we do not have enough information to attempt a physical reconstruction of Andreas' machine. We lack information about the frame, as well as the size and position of several parts. No scholar has attempted a physical reconstruction of Andreas' device.

What is the relation between Oribasius' textual description and Andreas' physical machine? Although a basic historiographical issue, the question has yet not been previously posed. No easy and wholly satisfactory answer can be given. It is not even clear from other testimonia that Andreas ever wrote a description of his machine. Since Celsus and Galen testify that Andreas' machine was known to them through books, it appears that Andreas' machine entered textual circulation in some kind of doxographical listing of surgical machines.[72] Later physicians' knowledge implies that at some point during the Hellenistic period a contemporary recorded his understanding of the machine's parts, its functioning, and its movement. Importantly the evidence surviving in Oribasius represents a kind of reader-response from a historical contemporary of Andreas. I have argued above that a

---

69 Nine testimonia (*An.* 11–17 vS) concern Andreas' machine. I have located one additional testimonium about Andreas' machine beyond these: Oribasius *Collectiones Medicae* 49.4.53–55 = 4.9–10 Raeder.
70 Oribasius was the doctor to the Roman emperor Julian (reigned 361–363 CE), who is the dedicatee of Oribasius' collection. On Heliodorus see Marganne 2006.
71 ἀγκῶνες "elbows"; βάσις "base"; διάπηγμα "cross-bar"; ἐλασμάτιον χαλκοῦν "hammered bronze plate"; κοχλίας "screw"; ὀδούς/ὀδοντάριον "tooth/toothlet"; παρασκέλια "side-legs"; περόνη "pin"; πλινθίον "frame"; πτέρναι "heels"; πτερώσεις "feathers"; σκέλη "legs"; σπάθη "spade"; στρομώματα "hinges"; χελώνη "tortoise". The names come from Oribasius *Collectiones Medicae* 49.4 = 6–10 Raeder; cf. von Staden 1998: 158.
72 *An.* 11 vS = Celsus *De Medicina* 8.20.4; *An.* 13 vS = Galen *In Hippocratis De Articulis* 18A.338–339K, especially "For the machine is described by countless of the recent machinists in the books they have written about instruments".

reconstruction of the machine ought to embrace historical actors' contextual understanding of the machine's performance; this is the very historiographical category of the surviving evidence about Andreas' machine.

So while our lack of knowledge prevents a physical reconstruction of Andreas' machine, our evidence captures the machine's functioning by situating it within the tradition of ancient mechanics. By taking Andreas' contemporary **no. 139** Philon to represent the discourse of Hellenistic mechanics, von Staden has shown that all named parts of Andreas' machine are drawn from mechanics.[73] The named parts of Andreas' machine parts come from contemporary war machines. Given the complex transmission of testimonia about Andreas' machine, it is unclear whether the precise terminology given by Oribasius goes back to Andreas or depends on some intermediary source. Nevertheless, since Oribasius' evidence represents a historical actor's contextual understanding of Andreas' machine, the mechanical vocabulary shows that Andreas' contemporaries understood that the machine depended on the pieces of Hellenistic war machines. Andreas' machine depended on screw and gearing mechanisms adapted from contemporary war machines developed in Alexandria.

Rather than list the individual parts of the machine and describe how they relate to contemporary mechanics, I instead focus on how the machine moved. According to one testimonium from Oribasius, all parts of Andreas' machine were joined together.

> Let there be a construction made just as in the instrument of Andreas: the tortoises. Each tortoise is bored through and through the hole the screw is placed, and inside the hole an iron or hammered bronze plate is nailed to the tortoise. The hammered plate is called a tooth. This tooth of the tortoise is keyed into the screw-thread of the screw. The final result is that, by twisting depending on the direction of the screw, the hammered plate (the so-called little tooth), held continuously in the screw-thread furrow, moves the tortoise. Some parts of these very square screw-threads are simple, some are double. The simple screw cut with one spiral moves one tortoise. The double kind of screw is cut with two spirals and moves two tortoises. Such is the screw in the large frame of Andreas' instrument. For the wood piece between the cross-bars is cut from its midpoint by opposite screw-threads, so that, by a different directional turn of the screw, the tortoises either move from the midpoint to the cross bars or meet up from the cross-bars to the middle.[74]

---

73 von Staden 1998.
74 *An.* 17 vS = Oribasius *Collectiones Medicae* 49.5.1–5 = 10.12–27 Raeder: ἔστω οὕτως ἔχουσα ἡ κατασκευὴ ὥσπερ ἐν τῷ τοῦ Ἀνδρέου ὀργάνῳ χελῶναι· ἑκάστη δ' αὐτῶν τέτρηται, καὶ διὰ τοῦ τρήματος διενήνεκται ὁ κοχλίας, ἔνδοθεν δὲ κατὰ τὸ τρῆμα σιδηροῦν ἢ χαλκοῦν ἐλασμάτιον προσήλωται τῇ χελώνῃ. τὸ δ' ἐλασμάτιον τοῦτο κέκληται ὀδούς. οὗτος ἤδη ὁ τῆς χελώνης ὀδοὺς εἰς τὴν τοῦ κοχλίου ἕλικα κατακέκλεισται· λοιπὸν δὴ συμβαίνει τῇ ποιᾷ τοῦ κοχλίου συστροφῇ εἰλούμενον τὸ ἐλασμάτιον τὸ λεγόμενον ὀδοντάριον περὶ τὸν κοχλίαν ἐν αὐτῇ τῇ κοίλῃ ἕλικι συνεχόμενον κινεῖν τὴν χελώνην. αὐτῶν δὲ τούτων τῶν τετραγώνων οἱ μέν εἰσιν ἁπλοῖ, οἱ δὲ διπλοῖ. ἁπλοῦς ἐστι κοχλίας ὁ μιᾷ ἕλικι συντετμημένος καὶ μίαν κινῶν χελώνην· διπλοῦς δὲ τοιοῦτός ἐστι κοχλίας ὁ δυσὶν ἕλιξι διατετμημένος καὶ δύο κινῶν χελώνας. τοιοῦτος δ' ἐστι κοχλίας ἐν τῷ μεγάλῳ πλινθίῳ τῷ Ἀνδρέου ὀργάνῳ· ἀπὸ γὰρ τῆς μεσότητος τὸ ξύλον ἐκ τῶν διαπηγμάτων ἀντιθέτοις ἕλιξι

As the screw turned, the teeth of the plate caught. Since the plate was embedded in the tortoise (so-called because it was a slow-moving part of the machine), the tortoise followed the movement of the teeth around the furrows of the screw and moved. The rotational motion of the screw was translated into the rectilinear motion of the tortoise. One tortoise climbed and caused the so-called elbows to rise while the other tortoise descended. The patient's limbs were probably positioned over the tortoises, in either a vertical or horizontal position (depending on whether the machine's side-legs were engaged). The movement of the tortoises, either apart or toward each other, accomplished different types of reductions. Andreas' machine was capable of accomplishing the classic types of reductions in ancient surgical treatments.[75] The gearing and screw mechanisms of Andreas' machine allowed it to control the force and its location that the machine could apply to the patient's body. Since older surgical machines such as the Hippocratic bench did not move nor used screws, Andreas' machine was able to exert more force with greater mobility and greater control than older devices. The possibility of more force with greater mobility and control translated to greater therapeutic effectiveness. For the patient Andreas' machine offered therapeutic precision, refinement, relative gentleness, and efficacy.

Yet to understand the machine only as a mechanically-advanced therapeutic device ignores the motivation of its author, embedded in court society. As we saw in chapter 1, Andreas feuded with contemporary court poets and intellectuals, named plants after Serapis (god of the Ptolemies), and had a patronage relationship with Sosibius **no. 15**, the king's most powerful *philos*.[76] Andreas clearly worked within a sophisticated social milieu; the interpretation of his machine might be understood within that social context. For the operation of the machine is not only a matter of physical functioning but the contextual understanding for that functioning, or how the authors' contemporaries would have understood his meaning. As argued above, this is the discursive category of performance. What did the movement of Andreas' machine mean to his contemporaries?

τέμηται, ὥστε κατὰ ποιὰν τοῦ κοχλίου συστροφὴν ἤτοι ἀπὸ τῆς μεσότητος ἐπὶ τὰ διαπήγματα ὁρμᾶν τὰς χελώνας ἢ ἀπὸ τῶν διαπηγμάτων εἰς τὸν μέσον τόπον συντρέχειν.

75 See τὸν καθ' ὑπεραιώρησιν καταρτισμόν "reduction by lifting-over" Oribasius *Collectiones* 49.4.5 = 6.19 Raeder; πρὸς τοὺς κατὰ ἀνάτασιν καταριστμοί "resettings by extension" *Collectiones* 49.4.8 = 6.27–28 Raeder; τὴν κατ' ἐξελκυσμὸν μοχλείαν "the lever-bar in an extension by pulling-out" *Collectiones* 49.4.12 = 6.32 Raeder; τὴν κατὰ περίσφαλσιν μοχλείαν "the lever-bar in a reduction by slippage" *Collectiones* 49.4.19 = 7.17 Raeder. The surgical writings of the Hippocratic Corpus had previously discussed these technical movements in reduction.

76 Polemics *An.* 2a vS = *Etymologicum Magnum* 198.12–21 *s. v.* βιβλιαίγισθος, *An.* 44 vS = Σ.Aristophanem *Aves* 266, *An.* 45 vS = Athenaeus 7.312e, *An.* 46 vS = Σ.Nicandrum *Theriaca* 823; Serapis *An.* 37 vS = Dioscorides *De Materia Medica* 3.126–127; treatise dedicated to Sosibius, *philos* of Philopator *An.* 9 vS = Soranus *Gynaecia* 4.1 = 131.4–7 Ilberg. On doctors and patronage relations at court see Massar 2005: 51–63, 184–193.

Andreas' machine is a performative use of technology. The machine moves itself in complex ways through its so-called tortoises. Its therapeutics aim to elaborate on the anatomical knowledge of the human body developed from the investigations of Andreas' medical teacher, Herophilus. Andreas' machine integrates contemporary learned medicine and mechanical knowledge; this is cutting-edge cross-disciplinary science. The integration of these different sciences in a deliberate mixture I call hybridization in a conscious attempt to apply Kroll's famous term for the generic strategy of Hellenistic poetry to Hellenistic science as well.[77] The machine's self-performance and motion, its agonistic position in the tradition of ancient surgical machines, its hybridity of mechanical and anatomical knowledge are aesthetic objects of wonder. It is possible to understand the machine's movement and hybridization within the performance context of court aesthetics. The proper contextualization for the performance of Andreas' machine is court society, where Andreas practiced.

The generic strategy of hybridization applies equally to Hellenistic poetry and science. The wonders of Andreas' machine are similar to the interests in performance, belatedness, hybridity, and paradoxography of Andreas' court contemporaries **no. 111** Dioscorides, **no. 113** Archelaus, and **no. 117** Istrus. Archelaus and Istrus are known for their compilations of natural wonders, in continuation of **no. 110** Callimachus' collection of marvels.[78] Istrus wrote in prose; Archelaus at least occasionally in poetry. Archelaus' paradoxography read before a royal Ptolemy focused on the hybridity of the natural world: scorpions emerge from dead crocodiles, wasps from horse corpses.[79] Istrus' wonders include curious reports about the spring on the Athenian acropolis called the Klepsydra. Istrus and Dioscorides had an interest in belated antiquarianism: Istrus wrote at least one multi-volume work specifically on Athenian history; Dioscorides wrote a series of epigrams on Archaic and Classical poets.[80] Some of Dioscorides' other epigrams show a fascination with performance and movement in dance and song: singing the Trojan war, dancing and drums for the goddess Cybele.[81] The aesthetic discourses of these three authors and Andreas himself are part of the continued influence of Callimachean aesthetics in the later third century BCE.

Why does it matter that Andreas' machine exhibits novelty, surprise, and hybridity, the same qualities as the literary work of Archelaus, Istrus, and Dioscorides?

---

77 Kroll 1924. Fantuzzi and Hunter 2004: vii–viii caution against an uncritical adoption of Kroll's phrase.
78 Callimachus frr. 407–410 Pfeiffer. On the influence of Callimachean prose see Benedetto 2011 and the still useful survey of Fraser 1971: 1.770–784.
79 Antigonus of Carystus *De Mirabilibus* 19; Archelaus 125–129 SH.
80 *BNJ* 334 (Istros) F1–16; Dioscorides of Nicopolis 17–24 *HE*. Both the scientific disparagement of the past and Callimachean aetiology fall under "antiquarianism" in the sense of a conscious literary belatedness on the part of the author.
81 Dioscorides of Nicopolis 2, 16, 35, 36 *HE*.

Since Andreas' machine evoked wonder in its audience, the audience was receptive to the machine's aesthetic qualities as an epideictic display. The aesthetic qualities shared by Andreas' machine and the texts of his contemporary Archelaus, Istrus, and Dioscorides imply that these authors and Andreas shared the same audience: the society of Philopator's court.

Identifying court society as the relevant causal explanation for these aesthetic qualities is a prelude to unpacking the complex phenomenon of court society. As argued in chapter 1, I see a moderately closed court society in the third century BCE with multiple families of hereditary aristocrats drawn from the Macedonian aristocracy and from the families of Greek settlers and colonists who emigrated in the early part of the third century; the court society of these families loyal to the dynasts was augmented by exiles and individuals (including natives of Egypt or Persia) who successfully pursued promotion among the priesthood or military. I see the assemblage of these court actors as a cultural space for *epideixis* or display, as chapter 2 argued. Performance in court society is the opportunity for entertaining display in front of these historical actors.[82] In symposia fit for his dignity the king feasted with his *philoi* and generals, while page boys, cup bearers, jesters, and lovers circulated. The king hosted distinguished visitors: there were Jewish priests, as in Ps.-Aristeas; there were philosophers, as in anecdotes about the Stoic philosopher **no. 124** Sphaerus.[83] Scientists too adopted the sympotic *personae* of courtiers. While we know no other physicians other than Andreas under Philopator, we know three physicians **nos. 132, 133, 135** at the court of Euergetes, his predecessor. The political and diplomatic intrigues of Apollophanes of Seleucia, described in chapter 2, are perhaps better known than Euergetes' physicians. Philopator's court astronomer and correspondent of Archimedes, Dositheus **no. 138**, wrote a book partially about resident poets and scientists at Hellenistic courts. Andreas himself met his death sleeping within the king's tent at the battle of Raphia.[84] While there were the usual hedonistic excesses of the Hellenistic monarchies at these symposia, the examples from Ps.-Aristeas and Sphaerus show monarchs carefully attuned to sophisticated intellectual questions of knowledge and philosophy. As chapter 2 argued, tradition endowed the king with idealized prudence and social tact, to be sure, but we have specific evidence that Hellenistic court society was curious about contemporary scientific knowledge. The *Lives of Aratus* records that the Hellenistic king Antigonus Gonatas urged the poet Aratus to translate Eudoxus' astronomical *Mirror* into poetry; the physician Apollonius of Citium was urged to write and illustrate a commentary on Hippocrates by a "King

---

[82] Strootman 2014: 46–49 makes the more general argument that Hellenistic court society knew itself to be a display.
[83] Ps-Aristeas 187–300, Diogenes Laertius 7.177.
[84] *An.*1 vS = Polybius 5.81.

Ptolemy".⁸⁵ Rather than denigrating Hellenistic monarchs for reducing scientific knowledge to mere entertainment in these examples, we ought instead to recognize the particular historical form that scientific knowledge took within ancient court culture. The evidence suggests that Hellenistic court society appreciated technical scientific achievements and arguments presented within the generic forms and aesthetic codes of contemporary sympotic entertainment, namely philosophical dialogue, poetry, and performance.

In summary, Andreas' machine participated in contemporary courtly discourses. The machine moved and was unexpectedly capable of transforming itself into different shapes. It was built using the novel contemporary technology of siege engines, applied to the body, for the doctor to wage war against the diseases of bones and joint dislocations. It combined traditional medical techniques of reduction with the artifact pieces of sophisticated weaponry. The aesthetic qualities Andreas' machine exhibited – performative movement, an aesthetic belatedness toward antiquarian surgical machines, and a hybridization of anatomical and mechanical knowledge – were aesthetic qualities shared by contemporary court writers Archelaus, Dioscorides, and Istrus. These similarities are meaningful; they indicate a similar causal explanation. Andreas' machine was designed to exhibit these qualities because it was intended for an audience that respected these qualities in their sympotic entertainment. Andreas' machine was a piece of technology designed to appeal to patients who wanted novelty and refinement. It was designed to appeal socially to elite patients. The machine of Andreas belonged within the context of Philopator's court.

Andreas' machine fits into the larger pattern of Andreas' claims to authority at court in his *technē*. He authored several books and was the author of the surgical machine.⁸⁶ But his authority as a court physician did not come from his originality in authoring texts and physical materials. The court polymath Eratosthenes called Andreas a book-Aegisthus or book-seducer and accused him of rewriting Eratosthenes' own works.⁸⁷ Perhaps this polemic conceals Andreas' authorial technique of hybridizing prior discourses and knowledge. Even while integrating prior knowledge into his own productions, Andreas rejected the authority of previous experts: he accused Hippocrates of burning down the archives in Cnidus, he scorned Archelaus' claims that sea-eels mate with desert vipers.⁸⁸ Agonism strengthened Andreas' own claim to competence in his *technē*. By establishing a canon of truth-

---

85 *Vita Arati 1*; Apollonius 10 K-K.
86 Surviving titles are *Casket, To Sosibius, On Medical Genealogy, On Poisonous Animals, On False Beliefs*; von Staden 1989: 473.
87 *An.* 2a vS = *Etymologicum Magnum* 198.12–21 s. v. βιβλιαίγισθος.
88 *An.* 47 vS = Soranus *Vita Hippocratis* 4; *An.* 46 vS = Σ.Nicandrum *Theriaca* 823. The relation between Andreas' genealogy of Hippocrates and contemporary biographies has been discussed in chapter 1.

telling against his contemporaries and previous authorities Andreas positioned himself within the discourse of court friendship as a *parrhēsiast*.[89] Andreas' cultural production and authority within the court life of his age endured to culminate in Ps.-Aristeas' story of the Septuagint translation from Hebrew into Greek. There the fictitious bodyguard of Ptolemy named 'Andreas' forms an intimate member of court as courtier and diplomat; Ps.-Aristeas probably intended his audience to recall the historical Andreas, courtier and physician.[90] The parallels between Andreas' textual and material production offer insight into his authorial technique. They illustrate once again the resonant interchange of performance as a discursive category among artifacts, discourses, and effects in ancient *technē* at an elite court.

So how should we reconstruct lost scientific objects? The simple answer "as rebuilt scientific objects" is insufficient. If analysis of Andreas' machine has left "the scientific" undertheorized with respect to the content of science, it has at least shown that the scientific cannot be separated from its discursive contexts of production. If we cannot recover a physical reconstruction of the machine, we can sometimes recover the contextual meaning of its operation. Differently said: if we cannot recover what the scientific means to us – the physical working of the machine – we can perhaps recover what the scientific meant to historical actors. Reconstructing lost objects sometimes means reconstructing context in addition to content. In the case of Andreas' machine, we must recognize how the social context of court science determined the emergence of an entertaining, cross-disciplinary technological artifact.

---

89 It is beyond our evidence to make biographical claims about Philopator's reaction toward his feuding clients. Biagioli's 1993: 73–90 remarks on the attitude of Baroque Italian patrons toward their scientific champions suggest that patrons aimed less for a finality to debate than for the notion of aristocratic 'good sport', with appropriate give and take in the manner of chivalric duels. Galileo's patrons, the Medici, and other sixteenth and seventeenth century CE Italian aristocrats were caught up in Baroque codes of comportment, an attitude not easily applicable to Hellenistic monarchs. As for the ancient context, Strootman 2014: 175–184 notes how the asymmetry of power between the Hellenistic king and the well-established loyal families at court who expected to be rewarded increased over time. Family loyalty was not at issue here, since Eratosthenes, Archelaus, and Andreas were immigrants to Alexandria. I for one am doubtful that the Hellenistic monarch demanded 'fair play' from his feuding clients; rather, I would point to the equality and respect demanded of the king's *persona*: the monarch was supposed to be capable of being a fair referee and judge, as Ps.-Aristeas 227 claims. After all, what grounds would the monarch have for judging a dispute between two rival spokesmen for the natural world? Certainly Philopator was not in a position to judge the dispute on its merits. The asymmetry of knowledge prevents the king from judging which is the truer fit to *physis*. It would be a historiographical mistake to expect the king to close debate on a scientific subject.
90 Ps.-Aristeas 12, 40; von Staden 1989: 475.n11.

Technology and Performance

Why did technology and performance go hand in hand? Medicine for one developed certain domains as sites for public consumption of medical knowledge during the Hellenistic era: dietetics, pharmacology, and exegesis of Hippocratic texts.[91] Dietetics thematized symposia and royal banquets, as we have seen; kings received pharmaceutical recipes from physicians, planted herb gardens, and experimented with poisons; Apollonius of Citium dedicated his exegesis of Hippocrates to King Ptolemy. The example of Andreas' machine shows that technology too became a site for the consumption of medical knowledge.

Yet to pose the investigation within the terms of one field of science neglects the broader picture. From Eratosthenes' machine to Andreas', technology became a site for cross-disciplinary science and an object of court spectacle. The production of artifacts and their performative effects appealed to Ptolemaic court society for multiple reasons. As in the case of Eratosthenes' machine, artifacts provided a useful contribution to the lives of the kingdom's inhabitants within the ideal parameters of court friendship. As in the case of Andreas' machine, sophisticated artifact pieces echoed the aesthetics of contemporary court society: they emphasized motion and sound, mixed different scientific fields, and suggested the intellectual triumph of the present moment while drawing on generic expectations of previous accounts. Technology enabled court scientists to present novel scientific ideas in the discourse of court society, the socially-legitimizing social codes of Ptolemaic intellectual life. In accord with the thesis presented in the Introduction, the social codes of Ptolemaic court society shaped the emergence of new cross-disciplinary scientific knowledge. New cross-disciplinary science in the Hellenistic period gained social currency and subsequent scientific success first through its entertainment value as court science.

---

91 Massar 2005: 203–273.

5 Herophilus' Pulse and Archimedes' Mechanized Mathematics

The previous chapters have considered the full context of court science produced at the court of Ptolemy 246–205 BCE. We have listed the historical actors at the court, explained the discourse of friendship and the places of exchange, examined the authorial strategies in extant works to appeal for patronage, and considered in case studies the court science of Eratosthenes and Andreas. While this chapter continues the investigation through two case studies of court science, I move away from the court society under Euergetes and Philopator 246–205 BCE. I show how court science can be located elsewhere in the Hellenistic world, as well as the limits to the reading of court science.

Herophilus' Pulse

In this section I examine the fragmentary evidence about Herophilus' discovery of the pulse. I show that Herophilus' pulse carries the aesthetic qualities found in other Hellenistic court science. I argue that we can work backwards from the evidence of the science itself to deduce the social context of its origin. As a case study in the history of court science, this chapter studies the historical genesis of how an ancient scientific idea came to be. The Introduction described this genesis as scientific emergence. Emergence concerns the historical development of actors' beliefs, that is, how historical actors bring epistemological concepts into a holistic account. Despite its possible appeal for the Aristotelian world of coming-to-be, with features about emergence, salience, robustness, and the intellectual productivity of scientific objects, classicists have rarely thematized the coming-to-be of scientific ideas. Rather than speculate on the reasons for silence I move to tell the biography of a scientific idea and analyze the story of its birth.

Born in Bithynia and trained on Cos, Herophilus moved to Alexandria as a colonist and was a practicing physician in Alexandria during the first quarter of the third century BCE under Ptolemy I Soter and Ptolemy II Philadelphus.[1] He is most famous for conducting the first anatomical studies of the human body in

---

[1] Any study of Herophilus is deeply indebted to von Staden 1989, whose analysis remains compelling even when I understand the evidence differently. See von Staden 1989: 26–29, 36–50 for a biography of Herophilus. I cite the fragments of Herophilus as vS after von Staden 1989. I use the term 'fragment' in a loose sense: von Staden's 1989 collection assembles testimonia concerning Herophilus' writings and doctrines and only a few testimonia quote verbatim Herophilus' own writings.

global history. Through his anatomical studies, which probably involved vivisecting criminals condemned by the Ptolemies, he discovered the ventricles of the brain, distinguished between the sensory and motor nerves, discovered and named the four membranes of the eye, gave the first classic description of the liver, investigated the pancreas and named the duodenum, discovered and named the epididymis and the two vasa deferentia in men and the Fallopian tubes and ovaries in women, and he investigated the auricles and valves of the heart while giving an anatomical basis to the difference between arteries and veins.[2] In addition to this outstanding record of anatomical discovery, Herophilus also 'discovered' the pulse, a word I will defend momentarily.[3]

The most important difference between ancient and modern ideas about pulsation concerns the role of the heart. Modern medical dramas on television have given the false impression that pulsation can be graphically shown on an EKG monitor. Rather the peaks and troughs represented on the monitor are the electric firings of the heart. Ancient physicians believed that the pulse was connected to the function of the heart in some contested fashion but they had no idea of the circulation of blood. Instead, at the center of Herophilus' pulse theory was a vision of the pulsating artery. Herophilus distinguished between two arterial motions. The artery moved outward toward the physician's touch in a motion called διαστολή "dilation". It moves inward again away from the touch in a motion called συστολή "contraction". Herophilus thought only of the artery, not the heart. Contemporary Western biomedicine uses Herophilus' transliterated terminology of *diastole* and *systole* in relation to the movement of the artery but uses the translated meanings of *dilation* and *contraction* in relation to the heart's pumping motion. In accord with Herophilus' theory, I will use the English translations of the Greek terms "dilation" and "contraction" in reference to the artery's motion.

A brief contrast of ancient and modern ideas of pulsation hits at the central philosophical problem. Herophilus' pulse was certainly a salient and intellectually productive premodern scientific object. As the originator of Greek theories about the pulse, Herophilus was the intellectual parent of Greek medicine's pulse lore. Several thousand pages of ancient pulse lore in both Greek and Latin texts survive from the physician Galen and other ancient authors.[4] Galen's pulse treatises in turn were taught in the medical schools of Late Antiquity, and thus passed to the medieval and early modern periods as sophisticated elaborations of Herophilus' basic idea. Galen's pulse charts were hung and memorized in some European medi-

---

[2] Discussion of all these in von Staden 1989: 155–181, 250–259.
[3] Herophilus frr. 144–188b vS contain doctrines on the physiology of the vascular system and pulse theory.
[4] Scholarship on ancient pulse theory is limited. See Lewis 2015, Lewis 2016 for a listing of current bibliography. Lewis 2017 arrived too late for my use.

cal schools even up to the early nineteenth century.⁵ The longevity and elaboration of Herophilus' pulse indicates that the work it did in premodern medicine was valuable to practitioners. But salience, intellectual productivity, and cultural success may seem to miss the point. I have so far avoided the terms perhaps more familiar to debates about premodern scientific objects, realism and relativism. Modern scholarly debates about ancient pulse lore in terms of realism and relativism have followed the shifting waves of broader intellectual currents. Deichgräber, a leading twentieth century scholar of ancient medicine and empiricism, tried to correlate Galen's complex lore with modern understandings of blood pressure in the arterial vessel.⁶ In the forward march of scholarship realism gave to relativism. More recent scholarship has made the strong claim that premodern pulse theory is a cultural construction with no empirical basis, and hence no comparison between the mutually unintelligible scientific paradigms of ancient and modern is possible.⁷

But the debate between realism and relativism is better left to the Science Wars of the 1990s, smacks of a certain teleological history dismissed in the Introduction, and most importantly distracts us from a better understanding of Herophilus' pulse. I hold that Herophilus in an important sense 'discovered' the pulse and gave it empirical content, but I also hold that any comparativist project between ancient and modern conceptions is unsustainable. A more productive analysis of Herophilus' pulse offers the opportunity to see how a premodern scientific object comes to be, and separates both the analytic categories that compose it into transhistorical empirical phenomena and historically specific cultural categories. Such an analysis also offers the opportunity to understand emergent holism with a case 'on the ground'. The holistic idea contains properties that are not reducible to the properties of the original components. Premodern scientific objects can be materially complex, therapeutically useful, novel in their intellectual genealogy, and the basis of further elaboration in an intellectual tradition. I suggest that Herophilus' pulse offers similar strengths. To begin we should analyze the categories that compose Herophilus' pulse into transhistorical empirical phenomena and historically specific cultural categories.

Herophilus' pulse did not arise in a conceptual vacuum. To speak of historical actors' recognition of the concept of the pulse requires identifying the categories that are combined into an emergent holistic account. We need to recognize the pulse as a natural, non-pathological motion within an individual body. The pulse is a regular, involuntary biological faculty that is born with us and dies with us. We further need to discriminate that faculty to one body part, namely the arteries. While for us the distinction between arteries and veins is routine, Herophilus' teacher Praxagoras had been the first to distinguish between them.

---

5 Deichgräber 1984.
6 Deichgräber 1984.
7 Kuriyama 1999, Barton 1994.

The Greek verb Herophilus used, σφύζω "to pulse", originally indicated violent motion and pathological symptoms associated with fear and fever, not a normal vital function.[8] Throughout classical antiquity the verb retained a pathological sense in addition to its newfound medical meaning. Herophilus' teacher Praxagoras was the first Greek to state that only arteries pulse. Praxagoras conflated several types of involuntary movement, claiming that "tremor", "spasm", and "palpitation" as well as "pulse" were likewise involuntary movements of the arteries and differed only by size, not kind.[9] Herophilus refined Praxagoras' discovery and attributed Praxagoras' other involuntary motions to the muscular and nervous systems, the other material sites of Herophilus' anatomical investigations. For Herophilus "tremor", "spasm", and "palpitation" became involuntary pathological motions of the muscular and nervous systems. Consequently Herophilus was the first to identify pulse as an involuntary and natural movement unique to the arteries. Herophilus was therefore the first Greek physician to determine that arteries alone pulsate naturally.

Herophilus' pulse presents a natural motion of the body that is the faculty of a unique somatic material. Empirical and conceptual categories are required for a scientific idea to "emerge" as a distinct concept, isolated from other natural phenomena. I now turn to the analytical categories of description for this concept.

The physician Galen preserves an important quotation from Herophilus' treatise *On Pulses*, in which Herophilus describes the analytical categories he applied to the pulse.

> First introducing the subject [Herophilus] says as follows – for I will write down the entire passage ... "In general pulse seems to differ from pulse in amount, size, speed, vehemence, and rhythm. From their differences in these respects pulse at times appears proper and [at times] not proper. One pulse seems to differ and be recognized generally as different from another, as was said, in rhythm, size, speed, vehemence. If in the same rhythm one pulse seems to differ from another in speed, size, and vehemence."[10]

While later doctors added more *differentiae* of the pulse to Herophilus' initial assessment, it appears that Herophilus analyzed the movement of the artery in terms of these categories. But did Herophilus mean four or five *differentiae* of the arteries? We come at once to a conundrum in Herophilus' pulse theory. Herophilus' first

---

8 LSJ *s.v.* σφύζω 3.
9 Von Staden 1989: 271.
10 Herophilus fr. 162.77–86 vS = Galen *De Dignoscendis Pulsibus* 8.959–960K: πρῶτον μὲν ἐπιφέρων εὐθύς φησιν ὡδί – γράψω γὰρ τὴν ῥῆσιν ὅλην ... καθ' ὅλου μὲν οὖν δοκεῖ διαφέρειν σφυγμὸς σφυγμοῦ πλήθει, μεγέθει, τάχει, σφοδρότητι, ῥυθμῷ. ἐκ τοῦ κατὰ ταῦτα διαφέρειν φανερὸς γίνεται ἐνίοτε ὅ τε οἰκεῖος καὶ οὐκ οἰκεῖος. φαίνεται δὲ διαφέρειν καὶ ἐπιγινώσκεσθαι καθόλου μὲν ἕτερος ἑτέρου σφυγμός, ὡς εἴρηται, ῥυθμῷ, μεγέθει, τάχει, σφοδρότητι. εἰ δὲ ἐν τῷ αὐτῷ ῥυθμῷ φαίνεται διαφέρειν ἕτερος ἑτέρου σφυγμὸς σφυγμοῦ τάχει, μεγέθει, σφοδρότητι.

sentence lists five *differentiae* but two sentences later he lists only four *differentiae*: size, speed, vehemence and rhythm. Galen believed that Herophilus used only these four.[11] However, I will argue below that Herophilus intended the fifth *differentia*, "amount", to refer to frequency, which he measured with a water-clock.

Beyond these explicit *differentiae* Heophilus also described pulses by age-groups. Herophilus developed normative pulses for four different age-groups: newborns, children, adults, and the elderly.[12] The age-groups determine the normative *differentiae* of the pulse. Herophilus' *differentiae* of pulses are limited to his classification of age-groups. A newborn has a normative pulse described in terms of size, speed, vehemence, and rhythm; a child has a normative pulse described in separate terms of size, speed, vehemence, and rhythm; and so on for each group. The Herophilean physician first placed the patient into an age-group before determining the generic characteristics of the patient's pulse.

Given that Herophilus believed pulse to be composed of two arterial motions, dilation and contraction, which did he mean to be measured by his *differentiae*? His procedure for constructing a normative pulse rhythm begins from the dilating artery.[13] So I proceed by assuming that he thought the dilation was the most important movement. We imagine Herophilus feeling the forearm of a patient, already placed in one of the four age-groups, in order to classify the four *differentiae* of the patient's pulse. He felt how much the artery moves outward to him; this is the *size* of the pulse. He classified pulse sizes as "sufficient", "good-sized", and "remarkable".[14] Herophilus felt how quickly the artery dilates toward his touch; this is the *speed* of the pulse. He classified pulse speeds as "regular", "fast", and "slow".[15]

---

11 In Herophilus frr. 163a, 163b Galen transmits the view of Archigenes, a leading physician *fl.* 100 CE, that Herophilus uses other specific *differentiae* between pulses – regularity, irregularity, evenness, and unevenness – although without treating them as generic differences between pulses as size, speed, vehemence, and rhythm.
12 Herophilus fr. 177 vS.
13 Herophilus fr. 183.8–9 vS: "Supposing that the perceptible time-unit in which he found the artery dilating was primary ..." Yet Galen was not entirely sure whether Herophilus thought dilation or contraction was more important: in Herophilus frr. 157, 158 vS Galen at last decides that Herophilus meant contraction to be the active motion. Herophilus fr. 183 vS shows that Galen is likely mistaken.
14 See von Staden 1989: 285–286. Three different testimonia show these descriptions. (1) Herophilus fr. 184 vS = Galen *De Dignoscendis Pulsibus* 8.869K: "Well they at any rate call the pulse of a child small, but Herophilus never called it small, but sometimes named it sufficient in size, sometimes remarkable, or something similar." (2) Herophilus fr. 180 vS = Galen *Synopsis de Pulsibus* 9.453K: "The pulse of the child at any rate Herophilus says is sufficient in size, but Archigenes [says it is] small." (3) Herophilus fr. 181 vS = Galen *De Dignoscendis Pulsibus* 8.853K "Herophilus at any rate calls this pulse [sc. a child's pulse exceeding Galen's moderate pulse] good-sized."
15 Von Staden 1989: 284–285. Herophilus fr. 186 vS = Pliny *Historia Naturalis* 11.89.219: "The pulse of the arteries is clear chiefly at the end of the limbs; it is nearly a marker of diseases and was divided into fixed meters and metrical laws by age, ⟨as⟩ regular or fast or slow, by Herophilus a prophet of medicine with wondrous skill."

Herophilus felt the strength of the artery's motion in dilation; this is the *vehemence* of the pulse. There is no evidence how he classified pulse vehemences, although later physicians classified them as "strong" or "weak".[16] Finally, Herophilus felt the length of time of the dilation and compared it to the length of time of the contraction; this is the *rhythm*.

Clearly the most important factor in Herophilus' analysis of the pulse was his ability to perceive by touch. The Herophilean physician must use his fingers alone to observe and to analyze four specific characteristics in the motion of the pulsating artery, as well as to distinguish between the artery's dilation and contraction. The tactile ability requires time and training. The skill necessary in the doctor's touch was almost certainly part of the repertoire of clinical procedures and knowledge taught *in situ* by ancient physicians to their students: there is no record in the surviving fragments of Herophilus' *On Pulses* of written instructions for feeling the pulsating artery.[17] It remains unclear how Herophilus himself acquired his sophisticated ability to perceive arterial characteristics by touch.

Aristoxenus and Rhythm

Since the *differentiae* of size, vehemence, and speed are given subjective measurements and descriptions, the more objective criterion in Herophilus' treatment of the pulse is the timing of arterial pulsation. Herophilus appears to have approached the measurement of pulse timing in two different ways: through appropriating Aristoxenus' theory of musical rhythms and by employing a water-clock to measure pulse frequency. I will show that Herophilus' timing of the pulse – whether analyzed by appropriated Aristoxenian rhythms or by water-clock – employs a concept of time that measures an event by discrete units of elapsed time. First I discuss Herophilus' appropriation of Aristoxenus' musical theory for his description of the pulse' rhythm.

Galen remarks that Herophilus' statements on pulse rhythms are difficult to understand because they presuppose the readers' acquaintance with rhythm in musical contexts. To understand Herophilus' medical rhythms one needs to understand the technical language of both medicine and music. Herophilus compared dilation and contraction to the up-beat (ἄρσις) and down-beat (θέσις) of musical

---

[16] Von Staden 1989: 274–275. See also Galen's description of vehemence at *De Differentiis Pulsibus* 8.501K: "It is further necessary that it have tension, either weak and faint, or readily and strongly active."

[17] See Lewis 2015 for the ancient clinical procedure of feeling a patient's pulse as recorded by Marcellinus. Galen *De Pulsibus ad Tirones* 8.454K states that the doctor best determines a patient's pulse by feeling the arteries in the temples, wrist, or the instep of the foot. Skills of touching were almost certainly passed from teacher to pupil in bedside training.

rhythm.[18] The artery rises toward the physician's touch in dilation like the up-beat, the artery falls away from the physician's touch in contraction like the down-beat. The new conceptual *differentia* that Herophilus was able to specify by using the terminology of musicians is the ratio of dilation to contraction, or the rhythm of the pulse. The category of pulse rhythm was founded on an analogy with the terminology of musical rhythm so as to describe as exactly as possible the new concept.

Underlying Herophilus' project of categorization is the standard of measurement. While Herophilus now had physical objects, arterial dilation and contraction, to analogize to the elements of musical rhythm, *arsis* and *thesis*, he lacked a conceptual tool by which to measure the time of dilation and contraction. The theory of rhythm by the Peripatetic Aristoxenus of Tarentum offered the conceptual apparatus to solve this problem.[19] Central to Herophilus' appropriation was Aristoxenus' concept of the "primary time-unit."

Aristoxenus in his *Elements of Rhythm* aimed to provide an analysis of rhythm on the basis of empirical phenomena.[20] In his terminology, rhythm is composed of time-lengths (χρόνοι) and elements that are capable of assuming rhythm are called ῥυθμιζόμενα, for whose translation Pearson has coined the English neologism "rhythmizables". In rhythm and rhythmizables Aristoxenus contrasts the notional and material properties of rhythm. The smallest unit of time into which rhythmizables can be broken down is the πρῶτος χρόνος, "primary time-unit". While the first book of Aristoxenus' *Elements of Rhythm* is lost, the second book preserves an account of the primary time-unit.

> Time is divided by rhythmizables in each of its parts. There are three rhythmizables: speech, song, and bodily motion. ... Let be called a primary of time-units that capable of being divided by none of the rhymthizables ... The meaning of the primary time-unit must be understood in the following way. One of the appearances that presents itself readily to perception is that the speeds of motions do not increase to an unlimited degree, but are somehow fixed in arranged time-units, in which the parts of things that are set in motion are made. By things-set-in-motion I mean how the voice moves in speaking and singing and [how] the body moves in signifying, dancing, and making the rest of such sort of movements.[21]

---

18 Herophilus fr. 183.3–7 vS.
19 It is helpful to restate the case for Herophilus' appropriation, since von Staden 1989: 278–279 is ambivalent about the possibility.
20 Most scholarship on Aristoxenus concerns his *Elementa Harmonica*, the major surviving work. Pearson 1990 collects evidence relating to Aristoxenus' rhythmical theory. See further accounts of Aristoxenus' rhythmical theory in West 1992: 224–225 and Gibson 2005: 77–98.
21 Aristoxenus *Elementa Rhythmica* 2.9–11, 6–8 Pearson: διαιρεῖται δὲ ὁ χρόνος ὑπὸ τῶν ῥυθμιζομένων τοῖς ἑκάστου αὐτῶν μέρεσιν. ἔστι δὲ τὰ ῥυθμιζόμενα τρία· λέξις, μέλος, κίνησις σωματική ... καλείσθω δὲ πρῶτος μὲν τῶν χρόνων ὁ ὑπὸ μηδενὸς τῶν ῥυθμιζομένων δυνατὸς ὢν διαιρεθῆναι ... τὴν δὲ τοῦ πρώτου δύναμιν πειρᾶσθαι δεῖ καταμανθάνειν τόνδε τὸν τρόπον. τῶν σφόδρα φαινομένων ἐστὶ τῇ αἰσθήσει τὸ μὴ λαμβάνειν εἰς ἄπειρον ἐπίτασιν τὰς τῶν κινήσεων ταχύτητας, ἀλλ' ἵστασθαί που συναγομένους τοὺς χρόνους, ἐν οἷς τίθεται τὰ μέρη τῶν κινουμένων· λέγω δὲ τῶν οὕτω κινουμένων, ὡς ἥ τε φωνὴ κινεῖται λέγουσά τε καὶ μελῳδοῦσα καὶ τὸ σῶμα σημαῖνόν τε καὶ ὀρχούμενον καὶ τὰς λοιπὰς τῶν τοιούτων κινήσεων κινούμενον.

A brief explanation of Aristoxenus' account of the primary time-unit (πρῶτος χρόνος) will be helpful. The primary time-unit cannot further divided by any element assuming rhythm. One of those elements capable of assuming rhythm is bodily movement. Aristoxenus gives the example of dance; we might equally think of the twisting artery. The primary time-unit is the amount of time coequal with the shortest elements within any rhythmizable. As such, it remains whole and unable to be divided into further rational units. In Aristoxenus' system it is not an absolute measurement: it will vary with the tempo applied to the medium of rhythm. The primary time-unit is like the beat of the tempo. It can be shorter if the dancer dances quickly or longer if the dancer dances more slowly.[22] So the primary time-unit is not an absolute length of time, only the time taken for the smallest divisible element of the rhythmizable.

We ought to distinguish between two different types of temporal measurements, since the distinction will facilitate our understanding of Herophilus' multiple attempts to time the pulse. I call normative time when we measure an event by the duration of the primary time-unit. In Aristoxenus' example of the bodily movement, the dancer's individual foot-motions function as the normative measurement by which other temporal phenomena are counted. The duration of the primary time-unit of the foot-motion establishes the norm against which the dance is measured. By contrast, the length of passing time the dancer takes to dance the movement I call standardized time. Suppose then that two dancers dance the same movement. Each dancer will use a foot-motion as the same primary time-unit but one dancer will express the dance faster than the other: there is one normative time, the primary time-unit, and thus one rhythm, but there are two different standardized times, the passing duration of their performances, and thus two tempos.

Herophilus used Aristoxenus' primary time-unit in his pulse theory. Several Herophilean fragments employ the term of primary time-unit.[23] The case for Herophilus' appropriation of Aristoxenus rests on more than similar terminology. Herophilus appears to have actively employed Aristoxenus' theory of the primary time-unit in his writings on pulse rhythm. Herophilus used a unit of measurement to measure all parts of arterial pulsation.[24] Herophilus constructed some time-unit,

---

22 Pearson 1990: 76: "A performer establishes the *tempo* of what he is going to play by deciding the length of the primary *chronos*, just as a musician today may fix the metronome setting at [quarter note] = 120, deciding that each [quarter note] will take half a second, a hundred and twenty to the minute." A modern musician can equally establish a tempo at half the speed: he sets the metronome so that each quarter note takes a second. The primary time-unit remains the same but the tempo has changed.
23 Herophilus frr. 178.2, 183.25 vS.
24 Herophilus fr. 174.1–5 vS = Galen *De Dignoscendis Pulsibus* 8.913K "How therefore was Herophilus first to establish some time-unit in relation to sense-perception, by which he, in measuring the other ⟨time-periods⟩, claimed that they consist either of two or three or more [of these units], or [that these units] are both perfect and not-subject-to-increase, as they themselves call them, or decreased a little or a great degree or the greatest degree?"

identified as a "primary time-unit" later in the passage,[25] and used it to measure the time of dilation and contraction. A Herophilean pulse rhythm therefore might consist of two or three primary time-units. There is a common temporal measurement to all parts of the cycle of arterial dilation and contraction. The time length of further dilations and contractions measured in relation to the initial time-unit produce proportions of varying complications; they yield the rhythm of arterial dilation to contraction.

Galen records an important testimonium about Herophilus' implementation of the primary time-unit to explain his empirical identification of pulse rhythms, "a time-unit in relation to sense-perception".[26]

> So the time-units with dilation and contraction have been written also by Herophilus, since he drew ratio into rhythms for the sake of his age-groups. For just as musicians establish those ⟨rhythms⟩ in certain defined arrangements of time-units by comparing the up-beat and down-beat with each other, so Herophilus, supposing that the dilation of the artery was analogous to the up-beat and that the contraction of the artery was analogous to the down-beat, made his observation beginning from a newly born child. Supposing that a perceptible time-unit in which he found the artery dilating was primary, he claims that the ⟨time-unit⟩ of the contraction was at least equal to it, not quite distinguishing about either of the rests.[27]

Herophilus took a newborn to measure its pulse rhythm in dilation and contraction. He felt the dilation of an artery and called its temporal duration the primary time-unit. The time of dilation of the artery of the newborn became the common temporal measurement by which other sphygmological phenomena are measured. Galen remarks that Herophilus claimed that the length of time of arterial contraction is the same as that of the dilation in a newborn.[28] The ratio of dilation to

---

25 Galen *De Dignoscendis Pulsibus* 8.915K.
26 Herophilus fr. 174.1–2 vS.
27 Herophilus fr. 183.1–11 vS = Galen *Synopsis de Pulsibus* 9.463–465K: γέγραπται μὲν οὖν καὶ Ἡροφίλῳ τὰ κατὰ τοὺς χρόνους μετὰ τῆς διαστολῆς τε καὶ συστολῆς, ἕνεκα τῶν ἡλικιῶν εἰς ῥυθμοὺς ἀνάγοντι τὸν λόγον. ὥσπερ γὰρ ἐκείνους οἱ μουσικοὶ κατά τινας ὡρισμένας χρόνων τάξεις συνιστῶσι παραβάλλοντες ἀλλήλαις ἄρσιν καὶ θέσιν, οὕτως καὶ Ἡρόφιλος ἀνάλογον μὲν ἄρσει τὴν διαστολὴν ὑποθέμενος, ἀνάλογον δὲ θέσει τὴν συστολὴν τῆς ἀρτηρίας, ἀρξάμενος ἀπὸ τοῦ νεογενοῦς παιδίου τὴν τήρησιν ἐποιήσατο, πρῶτον χρόνον αἰσθητὸν ὑποθέμενος ἐν ᾧ διαστελλομένην εὕρισκε τὴν ἀρτηρίαν, ἴσον δ' αὐτῇ καὶ τὸν τῆς συστολῆς εἶναι φησιν, οὐ πάνυ τι διοριζόμενος ὑπὲρ ἑκατέρας τῶν ἡσυχιῶν. While Herophilus evidently defined only two parts of the arterial cycle, namely dilation and contraction, Galen recognized a cycle of four parts: dilation, rest, contraction, rest.
28 The text is loose with language here. In Herophilus fr. 183.9–10 vS ἴσον δ' αὐτῇ καὶ τὸν τῆς συστολῆς εἶναι φησιν equivalence is between the time-unit of contraction (τὸν τῆς συστολῆς) and dilation itself (αὐτῇ), when it needs to be between the time-unit of contraction and the *time-unit* of dilation, which would be τῷ αὐτῆς in Greek. Yet it is clear that the equivalence between time-units is what is meant, since Herophilus fr. 183.22 vS says that the rhythm of the pulse in newborns is "proportional" (δι' ἴσου).

contraction, or the rhythm of the infant's pulse, is a primary-time unit to a primary time-unit. Later in Galen's testimonium Herophilus calls this pulse rhythm "proportional".

Herophilus established a normative pulse rhythm not only for infants: each age-group has its own pulse rhythm. A significant passage for our understanding of Herophilus' pulse rhythms reveals how Herophilus qualified pulse rhythm for all other age-groups; it is easy summarized.[29] Children, as we have just seen, have a pulse rhythm two primary time-units long, one for dilation and one for contraction. Youths have a pulse three time-units in length, two for dilation and one for contraction. Adults have a pulse four time-units long, two for dilation and two for contraction. The elderly have a pulse three time-units in length, one for dilation and two for contraction.

The text shows Herophilus establishing normative rhythms for each age-group. Although the term "primary time-unit" does not appear here, Herophilus employed some similar notion, for the passage talks in terms of time-units. What is the basis of the primary time-unit that forms these time-units? We must infer that Herophilus employed the time of dilation of the artery of the newborn as the primary time-unit for all age-groups, for there is some comparison across age-groups. If there is no descriptive *mensurandum* common to all age-groups, it is difficult to see how Herophilus intended to compare pulse rhythms.[30]

In turn, Herophilus' solution that the infant's pulse is the primary time-unit of all age-groups leads to two possibilities, both problematic. The first possibility is that Herophilus described the rhythm of the infant's dilation to contraction generically so that any member of the infantile age-group has a one-to-one pulse rhythm. Such a solution would imply that the primary time-unit Herophilus identifies is generic and typologized to some degree. The identified time of the primary time-unit, identified by Herophilus' perception, is taken to stand for the primary time-unit of all other infants. Further, this primary time-unit is the basis of the temporal measurements of the rhythms of all other age-groups. The procedure would have allowed Herophilus to compare pulse rhythms across age-groups. But in so doing Herophilus would have moved the primary time-unit closer to a set time and thus negated the flexibility of a measurement that can vary with each individual. The second possibility is that each individual patient has his or her own pulse rhythm that can be measured according to a unique primary time-unit derived from the

---

29 Herophilus fr. 177.12–30 vS = Rufus of Ephesus *Synopsis de pulsibus* 4 = 223–5 Daremberg/Ruelle.
30 Herophilus fr. 177.12–30 vS = Rufus of Ephesus *Synopsis de pulsibus* 4 = 223–5 Daremberg/Ruelle seems to indicate that there is a comparison of time-units across age-groups: the short syllables of the infant's pyrrhic pulse rhythm are comparable to the short syllables of the teenager's trochaic pulse rhythm. Herophilus fr. 174.1–10 vS = Galen *De Dignoscendis Pulsibus* 8.913K claims that Herophilus establishes "some time-unit in relation to sense-perception, by which he, in measuring the other ⟨time-periods⟩, claimed that they consist either of two or three or more [of these units]".

observation of the individual. Herophilus identified a unique primary time-unit with each individual patient; as the individual grows into new age-groups, the patient's two-to-two rhythm of adulthood is measured on the basis of his or her childhood primary time-unit. The second possibility, while conforming better to Aristoxenus' sense of the flexibility of the temporal measurement of the primary time-unit and allowing a comparison of pulse rhythms across age-groups per individual, seems to destroy the clinical efficacy of the measurement of pulse rhythm. A doctor without knowledge of an individual patient's childhood primary time-unit is unable to use the pulse rhythm for diagnosis and prognosis of the same individual as an adult or as a member of any other age-group.

And yet Herophilus clearly appropriated Aristoxenus' theory of the primary time-unit. If we interpret Herophilus' identification of the infant's primary time-unit in the sense that each infant is unique, the more our interpretation will assert Herophilus' appropriation of Aristoxenus in all respects of Aristoxenus' theory and yet Herophilus' pulse theory will be less medically coherent as well as less useful for diagnosis and prognosis. If we interpret Herophilus' identification of the infant's primary time-unit in the sense that the infant's pulse is typologized, the more our interpretation will lessen the effectiveness of Herophilus' appropriation of Aristoxenus' primary time-unit and yet will give medical purpose to Herophilus' comparison of pulse rhythms. I prefer to give an interpretation that assumes the medical purposiveness of Herophilus' theory at the expense of his appropriation of Aristoxenus in all respects.

Therefore I argue that we should understand the infant's one-to-one pulse according to the initial interpretation that the infant's pulse is a typologized and generic measurement. The infant's pulse serves as the basis for the temporal phenomena of the pulse rhythms of other age-groups. When comparing the infantile age-group's one-to-one pulse rhythm to the adult age-group's two-to-two rhythm, infants and adults have the same ratio of dilation to contraction but different rhythms. Since the infantile primary time-unit is the temporal measurement common to age-groups, the infant's cycle of dilation to contraction moves twice as quickly as an adult's pulse rhythm. The primary time-unit of the infant is a normative time for the infantile age groups, but a standardized time for the adult age group. Within this schema the patient's classification into an age-group, not the individual patient's uniquely determined primary time-unit, has become the determinative factor.

In summary, Herophilus employs a primary time-unit to serve as the standard *mensurandum* of pulse phenomena. He identifies this *mensurandum* with his own ability to determine the time of the dilation of the infant's artery. Herophilus' adaption of Aristoxenus' primary time-unit to measure pulse phenomena utilizes the descriptive attributes of nature in one field, music, and applies that description to another aspect of nature, the expansion of the infant's dilating artery. The naturalizing link is made at one point: Herophilus' identification of the time of the infant's dilating artery with Aristoxenus' primary time-unit.

Herophilus' Water-Clock

The other objective measurement Herophilus attempted for the pulse was frequency, as in the sole testimonium about this event.

> Herophilus declared that a patient had fever whenever the pulse became more frequent, larger, and more vehement with much internal heat. If therefore [the pulse] should lessen its vehemence and size, the fever is in remission. He says that the frequency of the pulses becomes primary when fevers begin and remains so until their final resolution. The story goes that Herophilus was so encouraged in using the frequency of pulse as a secure sign that he constructed a water-clock holding an expressed measurement (χωρητικὴν ἀριθμοῦ ῥητοῦ) for the natural pulses of each age-group and that on entering in to the patient and setting down the water-clock felt the patient [for his pulse]. By as much as the greater movements of pulses overshot (παρέλθοιεν) the ⟨magnitude⟩ natural for the filling-out (ἐκπλήρωσιν) of the water-clock, by so much did he reveal the pulse to be more frequent, that is, either more or less feverish.[31]

Marcellinus recounts that Herophilus constructed a water-clock to quantify the extent of a patient's fever. The clock contained measurements, likely lines drawn inside the bowl, set for each age-group. Frequency, the *differentia* measured by Herophilus' water-clock, is different from the *differentia* of rhythm measured by Aristoxenus' primary time-unit.[32] I suggest that the *differentia* of frequency mentioned by Marcellinus corresponds to the amount, the fifth *differentia* listed at the

---

31 Herophilus fr. 182 vS = Marcellinus *De Pulsibus* 255–267 Schöne: ὁ δὲ Ἡρόφιλος πυρέσσειν ἀπεφήνατο τὸν ἄνθρωπον, ὁπόταν πυκνότερος καὶ μείζων καὶ σφοδρότερος ὁ σφυγμὸς γένηται μετὰ πολλῆς θερμασίας ἔνδον. εἰ μὲν οὖν προαπαλλάξειε τὴν σφοδρότητα καὶ τὸ μέγεθος, ἔνδοσιν τοῦ πυρετοῦ λαμβάνοντος· τὴν δὲ πυκνότητα τῶν σφυγμῶν ἀρχομένων τε τῶν πυρετῶν πρώτην συνίστασθαι καὶ συμπαραμένειν μέχρι τῆς τελείας αὐτῶν λύσεως λέγει. οὕτω δὲ τῇ πυκνοσφυξίᾳ τὸν Ἡρόφιλον θαρρεῖν λόγος ὡς βεβαίῳ σημείῳ χρώμενον, ὥστε κλεψύδραν κατασκευάσαι χωρητικὴν ἀριθμοῦ ῥητοῦ τῶν κατὰ φύσιν σφυγμῶν ἑκάστης ἡλικίας εἰσιόντα τε πρὸς τὸν ἄρρωστον καὶ τιθέντα τὴν κλεψύδραν ἅπτεσθαι τοῦ πυρέσσοντος· ὅσῳ δ' ἂν πλείονες παρέλθοιεν κινήσεις τῶν σφυγμῶν παρὰ τὸ κατὰ φύσιν εἰς τὴν ἐκπλήρωσιν τῆς κλεψύδρας, τοσούτῳ καὶ τὸν σφυγμὸν πυκνότερον ἀποφαίνειν, τουτέστι πυρέσσειν ἢ μᾶλλον ἢ ἧττον.

32 "Frequency" as a *differentia* occurs only in Marcellinus' report of Herophilus' water-clock and discussion of Herophilus' prognosis of patients with fever, also from Marcellinus (Herophilus fr. 182.1–8 vS = Marcellinus *De Pulsibus* 263 Schöne). Von Staden 1989: 284 discusses how later Greek physicians employed "speed" and "frequency" as distinct *differentiae*. He concludes that, whatever the relationship between "speed" and "frequency" in Herophilus' thought, Herophilus "did not merely develop 'speed' as a hypothetical differentia which in practice was abandoned in favour of 'frequency,'" as Herophilus' specific classifications of 'speed' show.

beginning of Herophilus' treatise *On Pulses*. On my interpretation the *differentia* "amount" refers to amounts of water at the measurement lines appropriate for the different age-groups in Marcellinus' report of Herophilus' water-clock. The timing device of a water-clock is not the frequency of the individual drops but rather their collective accumulation in vessel underneath. When the vessel underneath is filled, a set period of time has elapsed.[33] The doctor will subsequently compare the number of beats he has counted to the number he would expect in the patient's age-group, the frequency of pulse beat.

The water-clock and sundial were the only devices available to measure time in Greco-Roman antiquity; the water-clock was by far the more precise.[34] All water-clocks operate by allowing gravity to force falling water from one vessel to another. They differ in the amount of water they hold, the rate of the falling water, and whether the falling water is measured in the upper or lower vessel. There are three kinds of water-clocks but only two are relevant for Marcellinus' report about Herophilus: these types are out-flow clocks and in-flow clocks.[35]

In an out-flow clock gravity forces water to flow from a vessel with a small exit hole. The clock's timing is measured by the amount of water that flows out from the vessel. An out-flow clock needs only one vessel, since the water could flow out onto the ground, although to reuse water a lower catching vessel was often employed.[36] The out-flow clock was known in both ancient Greece and dynastic Egypt. The only extant mobile pre-Hellenistic Greek out-flow water-clock was recovered from the Athenian agora.[37] Conical in shape with a flat bottom, the front is marked with the name of an Athenian tribe, two letters to indicate its capacity, and a small hole at its base to release water. It holds 6.4 litres and empties in six minutes; it has no markings on the interior of the vessel.

The oldest extant dynastic Egyptian out-flow water-clock dates from c.1380 BCE and holds 39 liters of water, roughly six times larger than the extant

---

33 The modern analogue is the kitchen timer, whose buzz at the elapsed time records a length of time rather than individual moments, as a stop-watch does.
34 For scholarship on water-clocks in antiquity see Borchardt 1920, Pogo 1936, Cotterell et al. 1986 for dynastic Egyptian clock capabilities and material remains, Young 1939 for Athenian material remains, Drachmann 1948 for Greek theoretical treatises and typologies, Landels 1979 for theoretical and practical issues involved in the production of water-clocks, and Hannah 2009: 98–115, 170–173 for a survey of primary evidence.
35 Drachmann 1948 offers a typology of out-flow, in-flow, and constant-flow. Drachmann's 1948 third kind is not portable and Marcellinus says that Herophilus carried his with him: "entering in to the patient and setting down the water-clock."
36 From the fact that only one vessel was necessary it seems that the Greek term κλεψύδρα, translated as "water-clock", refers only to the vessel by which time was measured: in the case of the out-flow clock κλεψύδρα refers to the upper vessel (if there was a second, lower catching vessel). LSJ *s.v.* κλεψύδρα II.
37 Young 1939 includes photographs.

Athenian clock.[38] As an astronomical device, the outside of the Egyptian clock is covered in hieroglyphs for the twelve months of the calendar. The inside of the clock is covered by twelve series of vertical markings. Set at varying heights, they mark the differing length of hours that have passed of the night for any given month. As the water-flows out, an observer can mark the passing of an hour at different times of the year by inspecting the interior of the out-flowing vessel.

We can distinguish between how the clock functions and the use to which it is put. The Greek water-clock was used for what I have called normative time, or measuring an event by the duration of the clock's emptying. Most Greek water-clocks, like the clock from the agora, were used to limit the length of the speaker's time in court cases such that different types of legal cases had different time lengths established for the prosecution and defense.[39] The duration of the clock establishes the norm against which the event is measured. Without markings on the inside of the clock it is impossible to measure fractions or parts of the time that elapses as the water in the clock flows out.

The Egyptian clock, by contrast, is used for what I have called standardized time, or measuring segments of passing time. The Egyptian clock measures the hours of the passing night for each month, as a way for the astronomer to regulate his own position in the night. The water-clock regulating standardized time measures itself against the visual position of the sun, moon, and stars around the earth. The internal markings of the bowl allow the clock's user to measure standardized fractions of passing time. Effectively then the Egyptian water-clock's standardized time performs the same functions of recording passing time during the night as a sundial would during the day.[40]

The second type of water-clock is an inflow clock, known only from dynastic Egypt.[41] In the in-flow clock gravity forces water flowing from an upper container

---

[38] For an assessment of the Egyptian clock's capabilities, see Borchardt 1920: 14–19. For the best illustration of the clock's exterior see Neugebauer and Parker 1969: plate 2; for the illustration of the clock's interior see Borchardt 1920: plate 3.

[39] Most but not all Greek water-clocks were used for normative time. There are archaeological remains of two large immobile out-flow water-clocks in Oropus and in Athens; see Hannah 2009: 108–109. With a capacity of approximately 1000 liters, these clocks were capable of measuring passing time.

[40] Hannah 2009: 112–113 sees an increasing standardization of time in antiquity, although specifying the precise chronological dates or adoption remains difficult. Remijsen 2007 points to the papyrological evidence of hours marking the arrival of military couriers in a Ptolemaic postal system. Since all times given seem to refer to daylight, these hours were probably indicated by sundial.

[41] Pogo 1936, following a line of older scholarship, argues that the small Egyptian votives represent in-flow water-clocks. Cotterell et al. 1986 dispute this claim. They recognize the stylistic similarity to the large surviving in-flow clock, yet believe that the in-flow clock dates only to the second century BCE after Ctesibius. I follow Pogo's argument. The in-flow clock was not known in Classical Greece. The earliest archaeological evidence for an in-flow clock in Greece dates from the second century BCE; see Hannah 2009: 112 with reference to the Athenian Tower of the Winds.

into the receiving chamber; the timing device is the amount of water filling-in to the receiving chamber. Many small Egyptian in-flow clocks survive; some only nine centimeters high, they are votive offerings intended to represent larger in-flow clocks but are not marked inside their receiving bowls. Painted scenes show an Egyptian pharaoh offering to an Egyptian goddess a votive in-flow water-clock in his outstretched hand.[42] The one large surviving in-flow water-clock from dynastic Egypt, an astronomical device, is marked inside its receiving bowl.[43] Due to the interior markings of the receiving chamber, the Egyptian in-flow clocks, like Egyptian outflow clocks, measure standardized time.

Now Herophilus' water-clock is a normative use of time, as in the style of most Greek clocks. Marcellinus recounts that Herophilus constructed a water-clock to quantify the extent of a patient's fever for each age-group. The measurement bowl of Herophilus' water-clock provides a segment of standardized time during which a normative number of pulse beats is expected. Herophilus' use of the water-clock to measure the pulse against the standardized period of time of the clock, is a normative use of time.

Philological analysis shows that Herophilus' water-clock is also an in-flow water-clock, as in the Egyptian style of clock. Marcellinus' noun "filling up" (ἐκπλήρωσιν) confirms that Herophilus has used an in-flow clock.[44] The action of the water filling up the measuring vessel only takes places in the in-flow clock.[45] Second, Marcellinus' phrase "holding an expressed measurement" (χωρητικὴν ἀριθμοῦ ῥητοῦ) implies that Herophilus has made a receiving bowl capable of measuring the amount of water that should flow in for a specified number of beats for each of the different ages. The adjective χωρητικὴν literally means "capable of having space", as in a capacity for holding or containing with the bowl of the clock.[46] A Roman-period papyrus from Egypt describing how to build a water-clock employs terminology similar to Marcellinus' description of Herophilus' water-clock. "They transmit the measurement of the build of the water-clocks as follows, making the upper conic section twenty-four dactyls, the base twelve dactyls, and the depth fifteen dactyls."[47] The papyrus clock appears to be an out-flow clock on the conic

---

42 Pogo 1936 presents an image of Ptolemaic date; Cotterell et al. 1986: 41 present an image of dynastic Egyptian date.
43 Borchardt 1920: plate 6 illustrates both the exterior and interior of the surviving dynastic Egyptian large in-flow water-clock.
44 LSJ s. v. ἐκπλήρωσις I.
45 Von Staden's 1989: 354 apparatus records Schmidt's conjecture of ἐκκένωσιν for ἐκπλήρωσιν. Schmidt thus understood Herophilus' water-clock to be an out-flow clock. Von Staden's text should be retained.
46 LSJ s. v. χωρητικός 1.
47 P.Oxy. 470.31-38: τὸν δὲ τῶ[ν ὡ]ρολογίων ἀριθμὸν τῆς [κα|τα]σκευῆς οὕτως ἀ[πο|δι]δόασιν, τὸ μὲν ἄνω [| ὀ]λμίσκου δακτύλων [κδ | ποιοῦντες, τὸν δὲ πυθμένα[| ιβ δακτυλων, τὸ βάθος δ[α|κ]τύλων ιη. Borchardt 1920: plates 7–8 offers an edition of the text with emendations to bring the papyrus clock's dimensions into harmony with the known material remains.

frustrum model of the dynastic Egyptian clock. The papyrus seems to use "number" (ἀριθμός) to indicate the dimensions and capacities of the measuring bowl. Applying this sense of "number" (ἀριθμός) to the passage in Marcellinus, "expressed measurement" (ἀριθμοῦ ῥητοῦ) will refer to the dimension of the clock. Altogether "holding an expressed measurement" (χωρητικὴν ἀριθμοῦ ῥητοῦ) will therefore refer to the measurement bowl of the clock. Finally, Marcellinus' verb "overshoot" (παρέλθοιεν) indicates that the measurement bowl will have been marked inside with a level appropriate for each age-group. Suppose then that a line inside the measure bowl appropriate for the age-group of adults corresponds to 50 beats: if the doctor feels 60 beats in the time period of the bowl's filling to the measurement line, the patient's pulse overshoots the norm for the age-group.

We imagine Herophilus entering the house of the patient, setting down an apparatus which allows water to flow into an in-flow water-clock at a predictable rate, holding his patient's forearm to feel the pulse, and counting pulse beats as the water fills the bowl to the measuring lines appropriate for the patient's age-group. Herophilus' attempt to quantify the pulse by water-clock is a new concept: the application of an standardized measurement to normal pulse frequency. The water-clock was an old and traditional tool in both Greek and Egyptian culture, of which the in-flow clock with interior marking seems to have been favored in Egypt, the out-flow clock in Greece. Egyptian clocks were used for standardized time; most Greek clocks were used for normative time. I suggest that we see Herophilus the Greek colonist to Alexandria using an Egyptian tool for a Greek conceptual end.

The comparison between Herophilus' multiple attempts to measure the pulse with some temporal concept shows that both use a normative concept of time. The normative time of Herophilus' water-clock parallels the normative time of Herophilus' appropriation of Aristoxenus' theory of musical rhythm. Aristoxenus' theory of musical rhythm adapts its temporal measurement to the smallest divisible element of the *mensurandum*. Herophilus identified the infant's dilation as the smallest divisible element of arterial pulsation; he constructed the artery's temporal rhythm by using multiples of the temporal unit by which he measured the infant's dilation. Since Herophilus' measurement of the pulse by the primary time-unit is the measurement of a temporal event against an elapsed time, the notion of the primary time-unit is also a normative measurement of time. In both his adaption of Aristoxenus' primary time-unit and utilization of the in-flow water-clock Herophilus employed Greek conceptual understandings of time.

Emergence

I return to the emergence of Herophilus' pulse. Herophilus' new concept of the pulse results from multiple elements, some material, some conceptual. The new concept arose from empirical phenomena such as Herophilus' tactile identification of arterial dilation, the length of time of arterial expansion, and the time of filling-in of Egyptian-style in-flow water-clocks. The new object also arose from conceptual categories: Aristoxenus' primary time-unit, the assumption of equivalence between arterial dilation and contraction in the infant's pulse, the Greek mathematics of proportionality and ratio, the Greek notion of normative time, Herophilus' own mental ability to identify arterial expansion, and so on. The pulse therefore has both an empirical component, the length of time of arterial expansion, and semantic ones, derived variously from language, tradition, practical skill, and the cultural identification of time. The discovery of a new scientific concept does not come without this compound of empirical and mental categories, without phenomena that are ahistorical and phenomena unique to a given time and place. Herophilus did claim that only arteries pulsate; that is scientifically true. Herophilus described those pulsations in a way that depended on contemporary musical terminology; that is a historical science.

The philosophical theme of Herophilus' pulse theory is the emergence of new concepts and ideas in science. We might variously call Herophilus' pulse theory a discovery or invention and thereby signal our philosophical commitment to realism or relativism, respectively. If we say that Herophilus discovered the pulse, we mean that the pulse is a biological reality, a permanent and enduring concept whose existence does not depend on human perception and is subject to the cultural qualifications of human perception. Conversely if we say that Herophilus invented the pulse, we mean strongly that Herophilus' concept – whatever it was – was not only not our concept of pulse but indeed so solely dependent on the cultural conditions of human perception that it exists outside of the natural realm and only inside the cultural conditions of Herophilus' Greek culture in early Alexandria. But I resist framing the debate about Herophilus' pulse theory in the abused dichotomy of realism and relativism.

It would be a mistake to assert baldly that Herophilus' pulse rhythm lacks any biological content at all; instead Herophilus' pulse has complex roots in both empirical and cultural traditions. Herophilus refers his empirical claims to the duration of arterial dilation. To speak of emergence in terms of properties unique to the holistic account, the concept of the pulse emerges from the pulsating motions in space and time of the arterial wall. Descriptive properties supervene on that biological content whose holism is conceptually greater than their individual properties. It is worth underscoring how 'real' a concept Herophilus has assembled. His theory of pulse was supported by physical and temporal phenomena. The culturally accepted existence of the pulse was well established by reference to established ideas, such as movement in space and the particulars of timing devices. Most importantly, the holism of the new object can be identified as a unit. Herophilus sometimes named individual pulses 'gazelle-like' or 'quivering', names that were taken over into the later tradition as the so-called *pulsus caprizans* and *pulsus tremulus*.[48]

---

[48] Herophilus frr. 169–171 vS.

The holistic pulse unit name is a short-hand for a certain syndrome of pathological symptoms. Furthermore, the normative pulse is a criterion for health.[49] It offers the physician an opportunity to peer into the physiological working of the body without opening the body. Pulse took its place alongside the premodern physician's other diagnostic tools for determining health, such as urine and stool. Therefore Herophilus' emergent concept is intellectually salient, allows for further reification, and, for future practitioners, is capable of sustaining a productive intellectual program of further descriptive qualities in its individual components. Premodern scientific concepts can be 'real' and philosophically interesting.

We know indirectly that Herophilus interacted with the Alexandrian court of Ptolemy I Soter and Ptolemy II Philadelphus, since the kings gave him criminals for vivisection.[50] So where might be the court context of Herophilus' pulse? Since it is difficult to see court aesthetics in Herophilus' *differentiae* of size, speed, and vehemence, I suggest that the *differentiae* dealing with pulse timing offer more room for court aesthetics. Aesthetic features of performance and hybridity are characteristic of Alexandrian court aesthetics. After all, Herophilus' appropriation of Aristoxenus' primary time-unit for pulse rhythms is a crossing of different scientific disciplines, the aesthetic discourse of hybridization. If hybridization demand wide-learning by the practicing physician to understand both musical theory and medicine, it also poses a problem for a lay audience: perhaps the lay audience saw hybridization as the entertaining spectacle of science. Herophilus' association with musicians or philosophers at court likely provided the opportunity for a cross-disciplinary approach.

If Herophilus' *differentia* of rhythm intellectually incorporates court aesthetics of hybridization, his *differentia* of amount embodies court aesthetics even more clearly. Herophilus' water-clock arouses wonder in court spectators by its visual and aural spectacle. The water-vessel itself is decorated with all sorts of Egyptian symbols; the water falls and plunks loudly into the receiving bowl. Both the physician and patient must stand close enough to the falling water for the physician to see into the receiving bowl and watch the water level rise to the appropriate measurement mark; the patient too could look into the bowl and wait. The doctor holds the patient's forearm and counts pulse beats in Greek numbers. The doctor pronounces the patient sick with fever or well: Greek medical knowledge comes from an Egyptian cup, on a Ptolemaic patient, with Greek numbers, by the Egyptian waters which surrounded Alexandria. What could be a better medical embodiment for the Ptolemaic cultural project of imposing select Greek practices on

---

49 Herophilus frr. 175, 178 vS.
50 The main evidence remains Celsus *De Medicina* 1.pr.23. Discussions in von Staden 1989: 139–153, Flemming 2003, Lang 2012: 254–258.

Egypt? Herophilus' water-clock ought to takes it historical position alongside other technological objects from antiquity that facilitated exchange between cultures.[51]

I have argued that the development of scientific knowledge was a type of courtly discourse in the Greco-Egyptian court society of the Ptolemaic pharaohs. The study of the social interaction between monarchy and science enlarges the notions of performance and entertainment beyond art and poetry in studies of courtly discourse and integrates the history of science into its social and archaeological context. The timing of Herophilus' pulse also fits these dynamics. The pulse offers Herophilus an opportunity to display his *paideia* in a performance context in front of the assembled court actors, mixes Greek and Egyptian understandings of natural phenomena, and integrates different scientific traditions to create a salient and robust novel scientific object. My reading has emphasized the cultural hybridity in Herophilus' pulse theory. But other readings of Herophilus' pulse theory are also possible that emphasize the Hellenistic fascination for miniaturism or the typically Hellenistic recondite use of a banal pedestrian object.[52] Insofar as our evidence also admits these readings, the sum total strongly suggests that the timing of Herophilus' pulse theory emerged from the social dynamics of the Alexandrian court. Court science apart from the court society under Euergetes and Philopator 246–205 BCE can be located even with a collection of fragmentary evidence.

Archimedes' Mechanized Mathematics

The evidence about Herophilus' pulse suggests that Hellenistic court science can be found beyond the court of Ptolemy 246–205 BCE. But what are the limits of the reading of court science? After all, not all novel Hellenistic science developed from a court context. This section explores the limits of court science. I choose for a case study an ancient work dedicated to a novel scientific method by an author known to have worked and associated with court society: Archimedes' *Method*, a mathematical treatise sent to **no. 136** Eratosthenes. In the *Method* Archimedes slices mathematical objects, balances them on a scale-beam, and by weighing them measures their area or volume.

---

51 See Roby 2016: 66.n9 on cross-cultural technologies and borrowing in antiquity.
52 One may consider the "entertainment" possibilities of court science and its potential for interacting with literature. The best poetic comparison for Herophilus' water-clock miniaturism, the hybridity of Greek and Egyptian cultural elements, and praise of *technē* is Hedylus' praise of Ctesibius' *rhyton*; see Hedylus 4 *HE*. A poetic reception of Herophilus' water-clock is possible and plausible, since Callimachus *In Artemim* 52–54 is best understood as a reception of Herophilus' discovery of the four coats of the eye. Secondly, since the ancient medical technique of prognosis originates from ancient divination practices, it is also possible for ancient lay viewers to understand the doctor's observing the water in the bowl and pronouncement of the patient's health as a divining scene using water; see *RE* 9.1: 79–86 *s. v.* 'hydromanteia' and van der Eijk 2004.

We have already seen some of Archimedes' associations with court society, from his construction of machines resisting the Roman siege of Syracuse, to his design of an enormous ship for King Hiero, to his court science treatise to King Gelon. Plutarch even names Archimedes "a kinsman and friend to Hiero the king".[53] I prefer to understand these terms in biological and social terms, rather than as court titles.[54] At the least Archimedes' intimate association with the royal house of Syracuse is not in doubt.

The *Method* is among the more philologically and mathematically challenging texts in the Archimedean corpus.[55] The text is preserved in a unique palimpsested manuscript, where it lies hidden underneath a surface text of Byzantine prayers.[56] Archimedes' treatise sets out a novel approach to finding volumes of solids of revolution, whereby geometric shapes and the lines that compose them are weighed on a scale-beam. (A solid of revolution is a two-dimensional figure rotated around an axis to create a three-dimensional volume, just as a circle rotated at distance around an axis makes a torus or doughnut.) Mathematical readers have been drawn to the treatise as a premodern anticipation of calculus for its look inside Archimedes' workshop, as it were; historical readers have been drawn to the treatise for its apparently ambivalent attitude towards the mathematical conclusive-

---

[53] Plutarch *Marcellus* 14.12: Ἰέρωνι τῷ βασιλεῖ συγγενὴς ὢν καὶ φίλος.

[54] The evidence gives us pause to read these adjectives as court titles. While the titles συγγενής and φίλος were part of the ranks of courtiers at the Ptolemaic courts, we have no evidence that the Syracusan court imitated the Ptolemies in this respect. Secondly, epigraphic evidence shows that court society expressed the usual title with the genitive φίλος τοῦ βασιλέως (evidence in **nos. 11–18**) but Plutarch *Marcellus* 14 writes the dative.

[55] For Archimedes' surviving and non-extant works and their authenticity see Netz 2004: 10–13. Dijksterhuis 1987 is the best mathematical introduction to Archimedes. The title "Method" for this treatise is in dispute. The manuscript 69.col2.1–3 Netz et al. = 426.1–12 Heiberg transmits ἔφοδος. I stay with the traditional English translation of "Method", although "Journey" is perhaps a better translation. Netz 2009a: 157 argues that *ephodikon* is the genuine Archimedean title by comparing the titles *anaphorikon* of Hypsicles and *okutokion* of Apollonius. Since the ancient mathematician Theodosius of Byzantium composed a commentary on the text according to *Suda* θ142, Theodosius also probably wrote the alternate title transmitted by the manuscript, περὶ τῶν μηχανικῶν θεωρημάτων πρὸς Ἐρατοσθένην "On Mechanical Theorems to Eratosthenes".

[56] A new critical edition of the text remains a desideratum. The text survives in the so-called Archimedes Palimpsest. At the beginning of the twentieth century the manuscript was available to Heiberg who competently read most of it. After a disappearance of nearly seventy years the manuscript reappeared at auction and was made available to scholars once more. The manuscript suffered during its disappearance at the hands of forgers, who painted over extant text with several modern illuminations forged in a Byzantine style. Computer imaging has revealed some details of the text previously obscured but reading the text remains difficult. Heiberg 1913 produced a critical edition; Netz et al. 2011 produced a diplomatic transcription of the manuscript. I have consulted both Heiberg's edition and the transcription. While we await a new critical edition, I produce a cross-reference for cited passages of the *Methodus* to the transcription by Netz et al. 2011 and Heiberg's 1913 edition.

ness of its own demonstrations.[57] I do not intend to "lift the veil" on the *Method* or unravel the text's struggle for certainty.[58] For my purposes the *Method* seems to offer clear evidence for a Hellenistic cultural history of novel scientific ideas that does not emerge from the social conditions of court society. The *Method*'s novel approach for finding areas or volumes is the emergence of a new procedure in a science. We are presented with an integration in Archimedes' *Method* similar to Herophilus' integration of medicine and music. Archimedes' approach integrates the sciences of mathematics and mechanics. My analysis focuses on Archimedes' account of hybridization.

The text marks the different sciences while uniting them. In the opening letter of the *Method* Archimedes immediately introduces a contrast between "mathematics" and "mechanics".

> Seeing that you [sc. Eratosthenes], just as I said, are learned, remarkably preeminent in philosophy, and honoring any science that passes under your eyes, I thought it good to write to you by equipping in the same book a characteristic of a certain manner: it will be possible for one traversing along it to supply starting points for the ability to contemplate some of the things in mathematics through mechanics.[59]

---

[57] Dijksterhuis 1987: 313–322 represents a mathematical reader; Lloyd 1990: 89–95 represents a historical reader.

[58] Netz's 2009a: 79 caution is justified: "Heath's notion – that Archimedes' *Method* was, in his words, 'a sort of lifting of the veil' – is of course nonsense. If the veil was lifted, how come we have spent the last century arguing over what's underneath? No, the *Method* is another, more subtle veil." Much of the argument about certainty has centered on the difficult passage at the end of the earlier propositions. Consider for example the passage at the end of first proposition, Archimedes *Methodus* 77.col1.19–28 Netz et al. = 438.16–21 Heiberg: "So this was not proven by what was now said but still creates a certain impression that the conclusion is true. Therefore seeing that ⟨the conclusion⟩ is not proven but nonetheless suspecting that the conclusion is true, I will test [it] [⟨ἐ⟩τάξομεν scripsi: τάξομεν Netz et al., Heiberg] in respect of the geometrized proof, which was published earlier after I myself discovered it." I make the diagnostic conjecture ⟨ἐ⟩τάξομεν as a Doric future from ἐτάζω "to test" rather than understanding the Koine τάσσω "arrange, put in place". I understand Archimedes to offer a preliminary result through his mechanized approach that he next proves in a mathematically conclusive *reductio ad absurdum*. My mathematical reading of the *Methodus* aligns with the traditional interpretation that Archimedes treats his mechanized approach as a heuristic: the first 11 propositions have been proved elsewhere. Netz 2009a: 131 cautions against any straightforward resolution of the problem, arguing that *Methodus* presents an open-ended puzzle to the reader: "Once again, the end result of having a multipartite text, is to present the reader with an enigma. The goal of the *Method* is not to answer our question, what Archimedes' methodological views were; it is to open this question wide."

[59] Archimedes *Methodus* 71.col1.33–71.col2.8 Netz et al. = 428.18–24 Heiberg: ὁρῶν δέ σε, καθάπερ λέγω, σπουδαῖον καὶ φιλοσοφίας προεστῶτα ἀξιολόγως καὶ τὴν ἐν τοῖς μαθήμασιν κατὰ τὸ ὑποπίπτον θεωρίαν τετιμηκότα ἐδοκίμασα γράψαι σοι καὶ εἰς τὸ αὐτὸ βιβλίον ἐφοδιάσας [scripsi: ἐξορίσας Netz et al.: ἐξορίσαι Heiberg] τρόπου τινὸς ἰδιότητα, καθ᾽ ὅν ἐπιπορευομένῳ [scripsi: ἐπιπορευομένον Netz et al.: σοι παρεχόμενον Heiberg] ἔσται λαμβάνειν ἀφορμὰς εἰς τὸ δύνασθαί τινα τῶν ἐν τοῖς μαθήμασιν θεωρεῖν διὰ τῶν μηχανικῶν. Note the discourse of friendship and honor associated with court *paideia* in Archimedes' praise of Eratosthenes.

Archimedes does not call his approach a 'method' but ambiguously refers to as "a characteristic of a certain manner." His approach is not identified solely with mathematics or mechanics, but rather is a metaphoric road (καθ' ὃν ἐπιπορευομένῳ) and its departure points (ἀφορμὰς) that connect those separate groups. Archimedes does not describe the approach directly in his introductory letter but merely characterizes its heuristic benefits and alludes to his earlier results with it.

The first proposition of the *Method* measuring the area of a parabola is Archimedes' third attempt to measure the parabola by a mechanical procedure. While *Method* introduces a new approach to measuring solids of revolution using mechanical procedures, Archimedes had already attempted to measure the parabola using a kind of mechanical procedure in his earlier treatises *Quadrature of the Parabola* and *Planes in Equilibrium* 2. We have explicit evidence about the chronological sequence of Archimedes' treatises: Archimedes claims to have proven the result of the area of a parabola before the *Method*.[60] In order to understand the *Method*'s differentiation between mathematics and mechanics I suggest treating Archimedes' other mechanical treatises as possible intertexts. I treat intertextuality as a guide to authorial intent, since it signposts the scientific author's sense of the continuity and contrast between his multiple attempts to solve the same problem. Archimedes envisions different audiences for his mechanized mathematics depending on treatise. But he denies the force of reading the mechanized physical world into the *Method*. In effect, Archimedes denies the reader the spectacle of science and so does not produce court science in the *Method*.

Since *Method* has the narrative shape of *Quadrature of the Parabola*, not *Planes in Equilibrium* or *Floating Bodies*, the other Archimedean mechanical treatises, my discussion will focus on *Quadrature of the Parabola*. The mathematical aim of *Quadrature of the Parabola* is exclusively to measure the parabola: the treatise has an introductory letter, several initial propositions (1–5) whose results will be used later, a mechanical argument to measure the parabola (propositions 6–17), and a separate geometrical argument to measure the parabola employing a *reductio ad absurdum* (propositions 18–24). Since the aim of *Method* proposition 1 is to measure the area of a parabola, we can effectively compare *Method* proposition 1 to *Quadrature of the Parabola*'s mixture of the mechanical and mathematical. In *Method* Ar-

---

60 Archimedes *Methodus* 71.col2.32–73.col1.16 Netz et al. = 430.9–22 Heiberg "And for me it happens that the discovery of the theorem now being published happens to have been like the earlier ones. I did wish to publish it, describing the manner, both because of having spoken about it before, in order that I not seem to anyone to have spread an empty rumor, and at the same time being convinced that it would contribute no small usefulness to mathematics. For I suppose that some either of those living or of our descendants will discover other not yet suspected theorems through the proven manner. Therefore I first write what first became clear through mechanics, that any segment of a section of an orthogonal cone is four-thirds of a triangle which has the same base and an equal height."

chimedes invokes the mechanical procedure of *Quadrature of the Parabola* as a discourse about mechanized mathematics.

Netz has argued that a central narrative principle of Archimedes' treatises is literary variety: he suggests that Archimedes' "hybrid treatise", the mixing of mathematical genres, is a generic kind of literary work.[61] *Quadrature of the Parabola* offers a solution to the same problem twice; the reader expects a geometric proof up until proposition 6 and then remembers the mechanical proof when reading propositions 19–24. Netz suggests that the effect is one of denying the sense of finality and completion to either procedure's solution.[62] Archimedes directs the expectations of the reader away from the immediate proposition: the idealized proof – a final, unassailable sequence of argument about the area of the parabola – acts an intertext, even though such a text is not present at all.[63] We readers need not choose which of the two methods of proof are ultimately intended in *Quadrature of the Parabola*.[64]

Reading *Method* with prior knowledge from *Quadrature of the Parabola* suggests that an Archimedean intertext may not even need to exist as a written document for it to be alluded to. Here "text" should be understood in an expansive sense: generic expectations, including the regimented form of Greek mathematical proof, can be understood as a kind of textual discourse, with their own codes and references. An author may mobilize a set of generic allusions that evoke a cultural discourse without reference to a particular written text. Consider the mechanical procedure of *Quadrature of the Parabola* for measuring the parabola. In proposition 6 Archimedes enjoins readers to "imagine a plane perpendicular to the horizontal"

---

[61] Netz 2009a: 115–173.
[62] Netz 2009a: 130 "Here is a hybrid treatise, obtaining the same result twice via different routes. Each of the routes is based on hybrids, on cross-fertilization: abstract conic theory and a concrete theory of the balance, in the first route; abstract conic theory and an even more abstract summation of geometrical progressions, in the second route. But in this sense the hybrids do cross-fertilize and in this way form some kind of organic unity. The relationship within each of the segments – propositions 1–18 and propositions 19–24 – is that of a single mathematical thought. Putting the two segments side by side is a much more radical departure, creating a textured treatise whose two parts are to be read alongside, or against, each other. Is the presence of a more "classical" geometrical proof designed to undercut the first, "mechanical" proof? Or are the two meant to cast light on one another (e.g. in that the mechanical line of thought might explain, in some sense, how one obtains the geometrical one? This certainly is the experience of reading the second following the first). The treatise as a whole throws this kind of meta-mathematical puzzle at the reader and, in a sense, ironically undercuts the very notion of a definitive proof: it highlights, after all, the multiplicity of mathematical routes."
[63] Archimedes' polemics in *Quadratura Parabolae* 262.19 Heiberg "not easily admitted lemmas" can also be read as Archimedes' games with his readers about the finality of proof.
[64] Netz 2009a: 128–129 sees this as a deliberate authoritative strategy by Archimedes. Archimedes shows mastery of both kinds of techniques without explanation in the mystique of the superior intellectual competitor.

and begins to describe a scale-beam constructed in that vertical plane whose fulcrum is the midpoint.[65] At first glance, a scale-beam (ζυγός) is a crude machine that produces the (in)equalities 'equal to', 'less than', and 'greater than'. But the scale-beam is actually a proportional machine that, if always forced to yield the equality 'equal to' such that the machine is balanced, gives an inverse proportion of weights and their distances from the fulcrum.[66] *Quadrature of the Parabola* proposition 6 shows that a weight hung at one end of the scale-beam balances a triangle hung on the scale-beam. Since weights are inversely proportion to their distances from the fulcrum, the ratio between the two distances is equivalent to the ratio between a triangle and a rectangle.[67] Archimedes' mechanical strategy is to balance the machine; his mathematical strategy is to manipulate the various proportions produced by the weights and their distance from the fulcrum.

To set two objects on the scale-beam simultaneously would not be fruitful in measuring their ratio. But if the balanced scale-beam is used as a proportion machine, it can determine the ratio between two objects that do not hang together. In *Quadrature of the Parabola* propositions 6–13 Archimedes proceeds to place triangles and trapezoids upon a balanced scale-beam in order to specify geometric ratios between them. Finally in proposition 14 Archimedes introduces the parabola, diagrammatically specified by triangles and trapezoids. He defines a complex set of equalities between a geometric series of trapezoids to weights unique to larger trapezoids, then defines further ratios between smaller trapezia and their encompassing larger trapezia. Since all figures are now defined in relation to each other, Archimedes is able to prove that the area encompassed by the trapezia circumscribing the parabola is greater than one third the whole triangle and again the area encompassed by the smaller trapezia is less than one third the whole triangle. This form of inequality we recognize as the Archimedean limit: Archimedes proves the result in a double *reductio ad absurdum* in proposition 16 without the scale-beam. In proposition 17 Archimedes restates the theorem in terms of the inscribed triangle, the familiar form of the thesis: the parabola is four-thirds of its inscribed triangle.

What then is the "intuitively gained insight" from the mechanical exposition of the *Quadrature of the Parabola*?[68] Past scholarship regards it as the result of the theorem itself. But the essential point is the geometric progression of the measuring figures. The measurement comes from the division of the triangle by the trape-

---

[65] Archimedes *Quadratura Parabolae* 272.11–12 Heiberg.
[66] That is a proportion of the sort A : B :: C : D where a weight A and D its distance from the fulcrum are on one of side of the balance and the weight B and C its distance from the fulcrum are on the other side.
[67] Archimedes *Quadratura Parabolae* 274.18–21 Heiberg describes the act of weighing and the resulting proportion that the accompanying diagram illustrates.
[68] Disjkterhuis 1987: 336.

zia, not from the scale-beam itself. The scale-beam merely preserves the correct ratio of areas in geometric progression. The scale-beam is a means to an end.

In the *Method* Archimedes alludes to his earlier work on mechanical procedures *Quadrature of the Parabola* and *Planes in Equilibrium*.[69] While both treatises treat the parabola in a mechanical manner, *Quadrature of the Parabola* mixes the mechanical and mathematical in a similar structure to the *Method*. The *Method* repeats as axioms before the first proposition the proofs about centers of gravity from *Planes in Equilibrium*. Archimedes alludes to his earlier treatises for different reasons: he refers readers to *Planes in Equilibrium* for its proofs about centers of gravity and to *Quadrature of the Parabola* for its mechanical procedure.[70] The mechanical procedure of *Quadrature of the Parabola* – the physical act of weighing a mathematical object in a way that is very similar to the weighing of objects in a scale pan against sums of standardized weights – is the mechanical procedure Archimedes expects his readers to know in the reading of the *Method*. In the *Method* Archimedes marks mechanical discourse by directing readers to the textual codes of mechanized mathematics known from *Quadrature of the Parabola*.

Appropriating Mechanical Discourse

What does Archimedes mean by "mechanics" in the *Method*? Mechanics has the Archimedean sense of mechanized mathematics, distinct from the modern sense of applied mathematics, or mathematicized mechanics. In the treatises *Quadrature of the Parabola* and *Planes in Equilibrium* Archimedes finds centers of weight of mathematical forms and positions those mathematical shapes on an idealized balance to find their volumes. The distinctly Archimedean treatment of mechanics processes mathematical figures according to physical principles, such as weight

---

69 Explicit recognition of *Quadratura Parabolae* at *Methodus* 73.col1. 10–16 Netz et al. = 430.19–22 Heiberg; explicit recognition of *De Planorum Aequilibriis* at *Methodus* 75.col2.37–77.col1.1 Netz et al. = 438.2 Heiberg.
70 At *Methodus* 75.col1.27–31 Netz et al. = 436.2–5 Heiberg Archimedes neglects to call readers' attention to the mathematical results of *Quadratura Parabolae*. Archimedes had already proved in a geometric fashion in *Quadratura Parabolae* 5 the inference needed here in *Methodus*. Although a result from *Quadratura Parabolae* stands as possible intertext at this passage in *Methodus*, Archimedes does not in the slightest way allude to it. But in the span of three lines Archimedes draws two mathematical conclusions: the one from a previous mathematician's *Elementa Conica*, the other from his own work *Quadratura Parabolae*. Why then does he refer to the *Elementa Conica* directly and *Quadratura Parabolae* not at all? The theorem from *Quadratura Parabolae* is not a part of the cognitive reference toolkit mathematicians expect of their readers; Netz 1999: 216–239. Archimedes downplays specifically the mathematical content of *Quadratura Parabolae*. *Quadratura Parabolae* stands before all as an intertext in *Methodus* for its mechanical procedure, not its results.

and solidity. There are three elements in Archimedes' discourse of mechanics: the physicalization of mathematical objects, the scale-beam, and centers of weight.

Physicalizing geometric objects endows them with characteristics they do not normally possess. When balanced on the scale-beam, mathematical objects must have mass and thus weight when balanced in the physical world. For example, the text of *Method* proposition 2 imagines that a certain sphere, cone, and cylinder have mass. Their cross-sections are distributed on the scale-beam. The sphere, cone, and cylinder are composed of circles; the figures are solid.[71] The text states that the cylinder balances around the fulcrum point both the sphere and the cone. But it is specifically the cone embedded in the sphere that balances, not the two separate objects, for they must have the same center of weight. That is, the sphere and the cone inhabit the same shared space but, by balancing, possess their individual magnitudes summed on the scale-beam. Only two objects balance in space, not three, since the sphere and the cone form a single strange piece of art on their end of the scale-beam. The fact is not discussed and the text continues to operate mathematically as if two separate shapes occupy the same physical space, capable of individual manipulation and unique equalities on the scale-beam. The *Method* operates as if the sphere and cone are shells, removable from one another, possessing only a shared center of weight. We might be led to think that we have imagined wrongly that the objects are solid; but no, the objects are to be imagined simply as solids in the sense of surfaces in mathematical space. These distinctions do not occur easily to the Greek reader, since Greek mathematics speaks *simpliciter* of geometrical objects as magnitudes, without reference to area, volume, or even physical properties such as mass.[72] In summary, multiple mathematical objects occupy the same space as one object; they have magnitude on the scale-beam and thus mass; they are solid; yet they are separable and treated individually. The discourse of physicalization of mathematical objects does not go very far in the *Method*, acting more as a veneer. Archimedes' geometric objects possess the physical properties of mass and centers of gravity, yet they are not limited by these qualities in the way real physical objects are.

The scale-beam is the second element of mechanical discourse in the *Method*. As we have seen, Archimedes had already used a mechanical method in *Quadra-*

---

[71] Archimedes *Methodus* 79.col2.9–12 Netz et al. = 442.23–25 Heiberg: "Therefore the cylinder, sphere, and cone are filled up [συμπληρωθέντος] by the circles taken". Here "fill" συμπληροῦσθαι is a technical term of Greek mathematics indicating the full construction of a figure for a proof; cf. Euclid *Elementa* 1.8.

[72] An example comes from *Methodus* proposition 4 89.col1.3–13 = 458.1–8 Heiberg: "Therefore when the cylinder and segment of the rectangular conoid have been completed, the cylinder, if it remains there, will balance at point A the segment of the rectangular conoid when it is moved and placed on the scale-beam at point Θ in such a way that the center of weight itself is Θ. Since the aforesaid magnitudes balance at point A, etc." The Greek reader moves from volumes, to mass, to magnitude without qualification.

ture of the Parabola. The style of argumentation about the scale-beam is much different in *Method* from that of the *Quadrature of the Parabola*. While both treatises measure the area of a parabola upon the scale-beam, in *Method* proposition 1 Archimedes guarantees that the center of weight of the triangle lies on the horizontal of the balance, as the law of the lever demands. Treatment of the parabola in *Method* proposition 1 differs from that of *Quadrature of the Parabola* proposition 14. Here Archimedes hangs the parabola by a terminal chord and locates the center of weight underneath the horizontal of the balance: a certain triangle balances on the scale-beam against combined weights. In *Quadrature of the Parabola* Archimedes does not ensure that objects are placed upon the scale-beam by their center of weight as the fundamental theorem of the lever demands. In *Method* Archimedes is more concerned with adherence to the theoretical principle at stake.

The presentation of the scale-beam is also quite different in each treatise. In *Quadrature of the Parabola* we look at the scale-beam as a plane perpendicular to the horizon. Here we readers see each side of the scale-beam and can check that the beam of the scale-beam is balanced along the horizontal. The presentation encourages us to view the mechanical procedure primarily as a weighing procedure. Thus the diagram complements the text, where Archimedes places a sum of weights upon the scale-beam and weighs sum by sum the parabola hung inside a certain triangle.

By contrast in the *Method* we view the scale-beam from above the plane of the horizon, as if we were looking down upon a flat ground. The reader looks down at a horizontal plane from above: all the points in the plane appear at the same depth. From this perspective the reader is physically unable to view the fact that the scale-beam balances. The reader is forced to assume the balancing of objects on the scale-beam. Moreover, in *Method* propositions 6 and 9 no balancing object appears at end of the scale-beam in manuscript diagrams.[73] The act of placing a balancing figure on the scale-beam that the text demands has no diagrammatic referent in the manuscript. In the absence of visual evidence the weighing procedure of the text exists only in the mind of the reader. In addition to differences in the diagrammatic presentation of the scale-beam, the *Method* does not weigh objects against solid weights in the manner of the *Quadrature of the Parabola*. The first manuscript diagram shows a certain triangle balancing a certain line, as if a single line could balance a triangle.[74] The scale-beam in *Method* acts as the mathematical principle of equality rather than physical appearance.

---

[73] Heiberg 1913: 465, 479 drew balancing objects on the scale-beam. Netz et al. 2011: 93, 109 illustrate the manuscript diagrams. The balancing objects M and N in *Methodus* proposition 9 occur alongside the scale-beam, not on it. The diagram denies the reader a picture of a physical weighing procedure.

[74] Netz et al. 2011: 77 illustrate the manuscript diagram.

What does Archimedes gain by drawing the *Method* diagram so that the horizon has vanished? If the horizon is parallel to the ruling of the page as in *Quadrature of the Parabola*, the figures appear to hang suspended or affixed from the balance: as if a physical balancing was occurring. Observing figures suspended from the scale-beam might be a view of the process drawn from daily life, such as the measuring of purchases of grain by merchants in the marketplace. Inspecting the weighing procedure from the side of the balance is an effective attestation of the transparency and justice of the procedure for both the merchant and customer.[75] In contrast to *Quadrature of the Parabola*, by viewing the weighing procedure from above at depth to a plane parallel to the ground, the *Method* denies the observer the visual perspective to check the accuracy of the result. Objects placed at opposite ends of the balance will appear stationary when viewed from above, regardless of the balance or imbalance of those objects.

Archimedes' balancing the parabola on the scale-beam is a potentially performative scientific act, the spectacle of science. *Quadrature of the Parabola* invites the lay reader to picture the performance of weighing the parabola: the act of weighing is pictured from the side of its vertical plane so that the reader can check the accuracy of the scale-beam. The *Method*, conversely, purposely denies the interpretation of the performance of weighing. Since the scale-beam is pictured from above, the reader must assume that the balance of mathematical objects is correct. Furthermore, since the diagrams of the palimpsest show objects only on one side of the scale-beam, the diagrams show no act of weighing at all. The *Method* precludes the reader from interpreting the measurement of geometric objects on the scale-beam as a performative act.

The third element of mechanical discourse is the center of weight. While the concept is nowhere defined in Archimedes' extant works, its meaning is relatively clear: when an object is hung on a balance, the center of weight is the point from which the object remains stationary.[76] The *Method* speaks in two ways of the center of weight: first as a point, a location on the diagram, and second as a magnitude by the fundamental theorem of the scale-beam. The center of weight has a complex relation to the perimeter of a solid object. It is only a point but has magnitude by the fundamental theorem of the scale-beam: centers of weight are inversely proportional to their distance from the fulcrum. Talk about mechanics via a center of weight is transformed into the discourse of geometry. Mechanization is secondary to mathematics in Archimedes' *Method*. The physical world is absent.

---

[75] Cuomo 2001: 4–5 argues that public presentation of numbers and numerical representations, regardless of the numeracy of the populace, is an attempt to present financial or political circumstances openly and transparently.

[76] Dijksterhuis 1987: 299 cites the definition from Pappus *Collectio* 8.5: "I say that the center of weight of each body is some point lying inside [the body], from which the weight, once imagined hung, does not move if carried about and preserves its initial position."

A Close Reading

Since *Method* proposition 1 is the exemplary proposition in the treatise, our best understanding of Archimedes' new approach will come from considering proposition 1 in a detailed close reading.[77] My reading focuses on Archimedes' practice of integrating the mechanical into the mathematical in such a way that Archimedes' authorial control is clear. Archimedes writes in such a way as to prevent a reader from reading the text in a mechanical way except at a precise point.

Archimedes begins the *ekthesis* by drawing the parabola ABΓ and its diameter[78] ΔB, then extends the line segment ΔB to an under-specified point E.[79] Archimedes directs that ΔB be drawn parallel to the diameter of the parabola but, since Δ is the bisection of AΓ, the line segment through Δ and B *is* the diameter. The text orders a construction more general than the text itself gives. The seeming confusion relates to the generality of naming conventions: Archimedes feigns ignorance of the specific diagram in order to enunciate a generalized proposition.[80]

The phrasing of the mathematics of generality in *Method* affects the reader's understanding. The text's move toward generality obscures the immediate placement of the parabola on the scale-beam, which has yet to be introduced. Since the text presented mechanical axioms before proposition 1, readers have been told to expect a mechanical demonstration as if the *Method* is a mechanical treatise. Readers expect, from their knowledge of the application of mechanics to the parabola in *Planes in Equilibrium* and *Quadrature of the Parabola*, that Archimedes will place the parabola on the scale-beam so that its center of weight coincides with the line of the scale-beam. (The center of weight of the parabola lies on the diameter.) Readers might expect that the parabola ABΓ is to lie on a scale-beam, part of which is ΔB, with the point E serving as some as-yet-unspecified point on the horizontal of the scale-beam ΔBE. But we are not given the diameter; we are instead given ΔB. Archimedes diverts our attention from the place of the center of weight of the parabola, as if to downplay the possibility of a mechanical demonstration. The text seemingly precludes the possibility of mechanical performance; only mathematical readings are welcome.

In the *kataskeuē* Archimedes constructs the remaining elements of the proposition.[81] He constructs the scale-beam ("let be imagined ⟨as⟩ a balance the ΓΘ") by extending ΓB to Θ so that the fulcrum point K is a point on the boundary of a particular triangle. But the story is not so simple as a construction. Although

---

[77] Netz 2009a: 75–80 gives a literary reading of this proposition (80) "marked by subtlety, surprise, and authorial playfulness."
[78] This would be called the axis of the parabola in modern terminology.
[79] The *ekthesis* covers *Methodus* 75.col1.8–15 Netz et al. = Heiberg 434.14–17.
[80] Netz 2004: 62.
[81] The *kataskeuē* covers *Methodus* 75.col1.16–24 Netz et al. = 434.20–24 Heiberg.

"imagine" is a regular verb in Greek mathematical writing, Netz has pointed out that a change in gender of the articles of Greek mathematical objects regularly indicates a shift from their reality in the diagram to their imagined existence.[82] Here the line ΓΘ becomes the scale-beam ΓΘ, a shift in gender from ἡ, the feminine article, to ὁ, the masculine article. Readers are invited to see ἡ ΓΘ, the ⟨straight line⟩ ΓΘ, as the scale-beam ὁ ΓΘ. ΓΘ is not even introduced as a line, ἡ ΓΘ; rather the fact of its existence as a line is so taken for granted that Archimedes asks us only to "imagine" it as ὁ ΓΘ, the scale-beam. Archimedes has not so much constructed a scale-beam then but rather constructed purely mathematical objects. In the *Method* the relations between objects deal with the existence of geometric realia in their metrical and topographic properties. The line ΓΘ has the metrical property that it is divided exactly in half. The scale-beam ΓΘ has the same metrical property and it has the mechanically-derived topographic property that it remains stationary. And since the scale-beam ΓΘ remains stationary at K, it cannot yet be seen to have anything balanced set upon it. Archimedes' desire to avoid the mechanical reading is the primary reason, I suggest, why the diagram of the proposition presents no mechanical features. The diagram presents the proposition as it is at the end of the *kataskeuē*. If the *kataskeuē* had positioned the scale-beam such that it had something on one arm and was therefore out of balance, we might have expected a different diagram.

Archimedes next selects a line ΜΞ in the triangle, parallel to the diameter of the parabola. This too is a move toward mathematical generality: any line in the triangle with the property that it is parallel to the diameter of the parabola would suffice for Archimedes' subsequent demonstration. That is to say, the line ΜΞ is an arbitrary (τυχοῦσα) line such that the subsequent procedure is repeatable with a different parallel line. Specifying the line ΜΞ is the final part of the *kataskeuē*. Archimedes has completed the figure drawn in the diagram of triangle ΑΒΓ, inscribed in the parabola ΑΒΓ, in turn inscribed in the larger triangle, whose side bisector has been doubled in length. Archimedes offers a purely geometrical construction. There is the potential of a mechanical demonstration when the line ΓΘ is seen as a balance.

After several mathematical conclusions are drawn in the *apodeixis*,[83] Archimedes has established a complex relationship between an arbitrary line segment of the large bounding triangle, the line ΜΞ, and a particular line segment of the parabola ΑΒΓ, the ΞΟ. He has shown that the ratio of these two line segments is equal to the ratio of two particular line segments, one of which we recognize as one arm of the potential scale-beam. Archimedes now begins to speak explicitly of centers of weight in the mechanical fashion. The pace of argument slows down, math-

---

[82] Netz 2009b.
[83] The *apodeixis* covers *Methodus* 75.col1.24–77.col1.17 Netz et al. = 436.1–438.13 Heiberg.

ematical motivation becomes more explicit, and the proof moves inexorably to its conclusion. Archimedes' new approach becomes apparent.

Archimedes notes that a certain point is the center of weight of ΜΞ, the line segment of the large bounding triangle.[84] The discourse of the mechanical is suddenly foregrounded. Archimedes could have argued geometrically that, since the two line segments were equivalent, their center point is the center of weight. But we have moved from seeing lines and parabolic segments and triangles to seeing them as weights and centers of weight. Archimedes continues in a mechanical manner by placing a copy of a particular line segment at the far arm of the scale-beam so that it balances on its center of weight. This copy and ΜΞ balance by the fundamental theorem of the lever, since the line segment joining their centers is cut inversely proportional by the weights at either end. And if they are balanced weights, their center of weight is the fulcrum of the scale-beam. The balancing is a statement of ratios.

The motivation of arguing after the mechanical fashion now becomes apparent. The ratio of the weights is the same as the ratio earlier proved between an arbitrary line segment of the large bounding triangle, the ΜΞ, and a line segment of the parabola ΑΒΓ. Whereas we had earlier not seen the purpose of the equivalence between complex ratios, we now see it as an individual claim about an arbitrary slice of the large bounding triangle balancing its constituent slice of the parabolic segment ΑΒΓ. The text does not explain how these line segments relate to their respective geometric figures.

In fact no possible intertext of Archimedes' mechanical treatises could prepare us for this result or explain it. *Quadrature of the Parabola* balances whole geometric objects against each other. Intuition about the physical world informs the understanding of balance in *Quadrature of the Parabola*: the experience of scale-beams in the market-place offers intuitive justification that the geometric representations of bar weights and sacks of grain, rectangles and trapezoids, balance each other. Yet what physical reality corresponds to the *Method*'s balancing of line segments? No mechanical intuition drawn from physical reality explains the balancing of an arbitrary slice of the bounding triangle against its constituent slice of the parabolic segment ΑΒΓ. The balancing of line segments in the *Method* is abstraction beyond the physical world.

Archimedes now moves to perhaps the most innovative part of his new approach. However many lines in the bounding triangle are drawn parallel to the arbitrary slice, they will be balanced by their constituent lines of the parabolic segment ΑΒΓ at the far end of the scale-beam. The center of all weights will be the fulcrum point of the scale-beam.[85] Are we to understand the balancing of individ-

---

84 Archimedes *Methodus* 75.col2.1–11 Netz et al. = 436.10–17 Heiberg.
85 Archimedes *Methodus* 75.col2.13–20 Netz et al. = 436.18–23 Heiberg.

ual line segments as a sequential procedure or a grouping of sets?[86] The next sentence is key for solving this question. In the formulaic language of Greek mathematical language, Archimedes makes an inference to reach his conclusion: since (ἐπεί) the bounding triangle is composed of its line segments and the parabolic segment ΑΒΓ is composed of its line segments, therefore (ἄρα) the triangle balances the parabolic segment ΑΒΓ at the far end of the scale-beam.[87] It seems as if Archimedes has made a logical jump somehow from his earlier individual lines segments, the arbitrary slice of both the triangle and parabolic segment, to the whole geometric object. As the subordinate clause of the conclusion states, both geometrical objects are composed of their individual line segments. Archimedes makes a claim about how individual line segments stand, in a deep sense, for their geometric objects in the weighing procedure. His claim is the use of what are called indivisibles in modern scholarship, after its procedural use by Cavalieri in the seventeenth century.

When we read Archimedes' representation of individual line segments for their geometric objects as a grouping of sets, the text presents us with only one inductive claim. Readers are required to assume – it is nowhere stated – that a geometric figure is composed of its set of individual line segments. First Archimedes proves that one member of the set of elements of individual line segments of the triangle and parabola balance. Next Archimedes extends this result and argues that, since one element of the set balances, all elements of the set must balance. This is not an inductive proof but a property of their belonging to the set of elements of composing the triangle and parabola respectively. Finally Archimedes concludes that, since the set of elements of the triangle and parabola balance, the geometric figures themselves balance. Reading Archimedes' claim as a grouping of sets has the advantage that it builds a single logical sequence of thought: from one element of a set, to all elements of the set, to the substitution of the object which stands for all elements of the set. A mathematical question still remains how the triangle and the parabola are composed of the elements of line segments.

I argue that Archimedes takes the view that the line segments which compose the triangle and the parabola are elements of sets after axiom 11 in *Method*.[88] In

---

[86] Dijksterhuis 1987: 321–322 reads the text as a sequential ordering.
[87] Archimedes *Methodus* 75.col2.23–33 = 436.23–30 Heiberg: "And since first the triangle ΖΑΓ has been composed of the ⟨straight lines⟩ in the triangle ΖΑΓ, and second the segment ΑΒΓ has been composed of the ⟨straight lines⟩ taken in the segment similar to the ⟨straight line⟩ ΟΞ, therefore the triangle ΖΑΓ will balance and remain stationary at the point Κ by the segment of the section set around the center of the weight, the ⟨point⟩ Θ, so that the ⟨point⟩ Κ is the center of the weight of the ⟨weight⟩ of both." See Netz 1999: 138, 254–255 on the inference particle ἄρα and its place in the logical form of a Greek mathematical proof.
[88] Archimedes *Methodus* 73.col2.31–75.col1.8 Netz et al. = 434.3–12 Heiberg: "I employ also this theorem: if however many magnitudes equal in multitude to other magnitudes when ordered like-

axiom 11 Archimedes appears to distinguish classes of enunciated magnitudes by their multitude. The argument about moving between elements of sets is an argument about increasing order of multitude, from points, to lines, to areas. Axiom 11 itself is a theorem about the preservation of proportion between elements of sets when the order of multitude of the set is increased, the very theorem we need to justify the above reading.

Netz, Saito, and Tschernetska have argued that *Method* axiom 11 is "based not on a simple assumption of indivisibles but on a very sophisticated argument in proportion theory."[89] Archimedes' argument risks a logical fallacy concerning the axiom's extension to infinite sets. The bounding triangle and parabola are composed of the set of their respective line segments, but there are infinitely many members of this set. Archimedes proved *Method* axiom 11 in his *Conoids and Spheroids* but only under the condition of finite members of sets. How then can Archimedes intend his argument in axiom 11 to be valid for infinite sets in *Method*? The answer is not clear.[90] At any rate, Archimedes has not argued in any way for the reduction of geometric figures into sets of lines. Rather the analysis of the triangle and parabola into their constituent lines happen *simpliciter*, as if there was nothing unusual about it for Archimedes or his readers.

From this point in the *apodeixis* Archimedes' mathematical motivation is clear.[91] Archimedes constructs the center of weight of bounding triangle. By a result proved in his *Planes in Equilibrium*, this point divides a side-bisector of the triangle into a 3:1 ratio. And if the bounding triangle balances the parabolic segment, then the distances of their respective centers of weight must be in inverse proportion. A series of quick substitutions in proportion yields the expected result that the parabolic segment is four-thirds of its interior triangle.

A New Approach

Archimedes' integration of mechanical and mathematical procedures and tools creates the emergence of a new mathematical approach that finds the areas of planar figures and the volumes of solid figures. We might call Archimedes' approach a 'scientific way of seeing'. Just as I argued in the case of Herophilus' pulse theory, the analysis of the emergence of new objects and approaches in science ought not

---

wise two by two have the same ratio and the first magnitudes, either all or some of them, are to other magnitudes in however many ratios and the latter magnitudes are to the corresponding (magnitudes) in the same ratios, all the first magnitudes will have the same ratio to all the (magnitudes) being enunciated that all the latter magnitudes have to all the (magnitudes) being enunciated."
89 Netz, Saito, and Tschernetska 2001: 21.
90 Netz, Saito, and Tschernetska 2001: 17–19 lay out the main positions and offer a critique of each.
91 Archimedes *Methodus* 75.col2.33–77.77.col1.17 Netz et al. = 436.30–438.13 Heiberg.

to be located along the ideological axes of scientific realism and relativism. Rather, scientific emergence depends on the networked structure of pre-existing science. The existence of Archimedes' approach is strengthened by the networked existence of the objects in the different discourses of mechanics and mathematics and by ways of seeing in those separate discourses.

The objects that *Method* proposition 1 considers are mechanical only in one respect. The property of centers of weight is the only mechanical property mathematical objects possess in Archimedes' new way of seeing. Mathematical objects can occupy the same position in mathematical space but physical objects cannot. A line segment and a line segment weighed against each other on the scale-beam balance as mathematical objects, although no physical intuition makes sense of this procedure. Archimedes' approach considers only the physical property of centers of weight.

The text strongly invokes the mathematical potential of mechanical objects. The proof of *Method* proposition 1 depends on the scale-beam and the indivisibles as tools that preserve mathematical proportions. The indivisibles preserve proportion when sets increase in multitude; the scale-beam preserves geometrical proportions. Balancing is a mathematical act, not a physical one. There is no physical or mechanical spectacle in the *Method*'s new approach.

Archimedes directs the reader's attention to the mathematical aspects at all but one point in *Method* proposition 1. The *kataskeuē* is the only formal section of the argument where Archimedes does not limit the reader's interpretative possibilities of a mechanical demonstration. The *ekthesis* limited the possibility of a mechanical demonstration when Archimedes directed that ΔB be drawn parallel to the diameter of the parabola. The reader sees the mathematical as mathematical; the reader also sees the mechanical foremost as mathematical. Only the *kataskeuē*'s imperative to see the line ΓΘ as a scale-beam invokes the mechanical property of the centers of weight.

Archimedes' *Method* pushes the mathematical reader toward a new approach, the integration of the mathematical 'way of seeing' mechanical objects into a mathematical analysis of linear sets that comprise geometric figures. Archimedes exerts authorial control throughout the argumentation. The text alludes to readers' prior expectations of mechanized mathematics. The new approach develops from the selective integration of mechanical properties into mathematical imagination. Archimedes' new approach in the *Method* emerges from a networked structure of the mathematical reader's knowledge of the formal structure of proof in Greek mathematics, the discourse of physical mechanization, the law of the lever and the mathematical consequences of the balanced scale-beam, the mathematical assumption that individual line segments compose their geometric objects as members of sets, and all the numerous unmarked mathematical tools of diagrams, manipulation of ratios and proportions, technical formulae, and so on.

The Limits of Court Science

Archimedes' *Method* offered a new scientific idea that does not arise from the social conditions of court science. The new idea of the *Method* deliberately rejects lay understandings of mathematical objects, balance, and space that depend on the physical world. Perhaps it is unsurprising that new ideas in ancient geometry in particular did not depend on lay justification. After all, Netz reminds us that ancient geometry probably had no public performance: the textual conditions of the embedded diagram and technical formulae conditioned mathematical communication into a written, documented form, transmitted peer-to-peer.[92] Mechanized mathematics that depends on technology, however, is less constrained by its textual form. While the extant textual forms of both *Method* and *Quadrature of the Parabola* adhere to Greek mathematical conventions using mathematical formulae and embedded diagrams, the *Quadrature of the Parabola* presents a mathematical idea drawn from lay insight into the physical world. Archimedes' *Quadrature of the Parabola* invited a reader to check that the weighing of mathematical objects balanced on the scale-beam, using the reader's intuition of the physical world derived from interactions with merchants in the marketplace.

Consequently, this chapter has shown both extensions and limits to the reading of court science outside the court under Euergetes and Philopator 246–205 BCE. In particular I have argued that technology became a locus for interactions between lay audiences and technical audiences in Hellenistic science. Since technology offered possibilities for hybridity between scientific disciplines or cultures, as well as possibilities of performance with motion and sound, technology appealed to Hellenistic court society that valued the aesthetics of hybridity and performance. Not all Hellenistic science arose from a court context, of course. But this caution should give us impetus for further investigation: novel Hellenistic science that depended on technology as a vehicle for communication to a lay audience deserves study as a possible case of court science.

---

92 Netz 2013.

Epilogue

Towards a History of Scientific Interdisciplinarity

This book has been a microhistory. It has unfolded at length the particular cross-disciplinary work of individual scientists. It has considered in detail the rhetorical strategies, aesthetic features, social practices, and people at the court of Ptolemy. As a study in the development of the social belief of scientific knowledge, its thick description has maintained a necessary tension between empiricism and an idealized representation. The asymmetry of social relations led to the ambiguity of friends and flatterers in court friendship. The asymmetry of financial support led to the ambiguity of reciprocity in gift-exchange. The asymmetry of expertise led to the entertaining genre of the court science treatise. Court science was not a space of professionalization, since there were almost no disciplining methods of the court scientist in an autonomous professional sense. Rather it was the original interdisciplinary space, where scientists of different intellectual traditions met and interacted. Court science was not a space of education, in which students imitated teachers or the professions replicated themselves. Instead it was a space of entertainment, where knowledge became a form of spectacle. Court science was a space of social legitimation to a certain kind of knowledge about the natural world, to a rhetoric of play and spectacle and social exclusivity, to a hierarchical social order built on personal trust, loyalty, reciprocity. In a world without the modern social relations of equality and symmetry undergirding the institutions of scientific knowledge – universities, peer-review, or democratic funding – court science was a premodern technique of making knowledge speak the local language of power.

A microhistory paints practices and people frozen in a small moment in time but microhistory does not stand alone. There are larger stories to tell with this material. In the Introduction I suggested that the emergence of scientific knowledge (rather than its closure) can be a significant contribution from the study of premodern science to historical themes at large. Beyond the emergence of scientific knowledge, on which I have dwelt at length, the most obvious story is the larger history of court science. Chapter 5 considered both extensions and limitations to that reading. While this is the earliest chronological material from Greek antiquity I find cast in the language of court science, it is hard to believe that the social arrangement of court science began with the Ptolemaic pharaohs. The social practices of power and the investigation of the natural world have a deeply intertwined past that began before our extant written evidence. The larger history of court science – including the early modern scientists Brahe, Galileo, and even Gauss – is deserving of a historical account richer than a suggestive listing of scientific names.[1] Since the story of court science is not new and is intertwined with the

---

1 I recall again Biagioli's 1993 study of Galileo as a model.

early modern development of scientific institutional bodies, I suggest a different story to which premodern evidence contributes. A different story, so far neglected, is more immediately relevant for contemporary science and intellectual practice. This larger story is the *longue durée* of scientific interdisciplinarity.

Academics have undertaken self-reflexive turns about their own practice as the discourse of "crisis" within the modern academy has spread. Some historians have traced the crises past as the humanities differentiated into the various modern disciplines. Historicization reassures academic readers that *this*, the most recent crisis, has the potential to lead to more productive scholarship.[2] Other historians with more pessimism conclude that the continued rhetoric of anti-proceduralism, anti-labor solidarity, and its concomitant denigration of professional autonomy are leading us to an anti-disciplinary age.[3] A rhetorical call for interdisciplinarity is a common solution to the cultural forces pummeling the current arrangements of disciplinary divisions.[4] Whatever the practical merits to an interdisciplinary arrangement of intellectual specializations, my interest in this epilogue is to historicize the premise of interdisciplinarity.

What is Interdisciplinarity?

Standard handbooks offer two forms of interdisciplinarity, each of which presents a separate history.[5] In the first interdisciplinarity is a unitary vision of knowledge. Unitary knowledge in Western cultures has ancient roots from Plato's and Aristotle's range across disciplines and their varied stances toward the organization and hierarchy of the disciplines, to the ancient Roman encyclopediast Pliny the Elder's vision of *enkyklios paideia* in the preface to his *Natural History*, to the medieval European curriculum of the *quadrivium* and *trivium*, to early modern cultural figures such as the Renaissance man da Vinci, to Leibniz's vision of a *mathesis universalis*, to the Enlightenment *philosophes*' aspiration to embrace all human knowledge in D'Alembert's "Preliminary Discourse" in the *Encyclopédie*. Current exponents of this discursive formation include American liberal arts colleges and the genre of the encyclopedia. Knowledge is simultaneously unified and general, synthetic and integrated. The knower embodies the various disciplines in his or her person and aspires to aristocratic notions of gentility, humaneness, and universalism, reformed in service of the ideals of democratic citizenship. The ideal of gener-

---

2 Turner 2014. As other commentators have noted, Turner's story neglects philosophy. Philosophy, the parent discipline, stands uneasily alongside its daughter disciplines in the liberal arts and sciences and is grouped somewhat unfairly only with humanities.
3 Forman 2012.
4 Jacobs 2014: 1–53 traces this rhetoric.
5 Weingart 2010; Klein 1990: 11–39.

alism is a response to forces external to the content of knowledge itself, whether civic, social, or philosophic.

The second definition of interdisciplinarity is much narrower in intellectual and temporal scope: the twentieth and twenty-first century arrangement of associated area studies and research groups. The arrangement is the consequence of the growth of research universities over the long nineteenth century, the Industrial Revolution and its associated development of technological artifacts, and the increasing specialization of knowledge and its *wissenschaftlich* character. Interdisciplinarity shifted the formation of academic departments over the course of the twentieth century.[6] It engaged humanities and social science departments first in the 1920s to the 1960s and beyond. It exploded in scientific fields in the later half of the twentieth century, a process in the physical sciences accelerated by military needs in the Cold War and in the biological sciences by the coding metaphor of DNA and the possibilities of genomic medicine. Interdisciplinarity became formally institutionalized within the American university system in the 1970s. The discursive formation of interdisciplinary institutes, think-tanks, and associated research groups continues today in research universities. In contrast to the unified vision of knowledge, the second version of interdisciplinarity combines in a single social structure individual knowers of specialized knowledge. The impetus for creating knowledge across disciplinary boundaries is an internal consequence of the specialization of knowledge itself, often with a particular intellectual problem taken as the object of a coordinated interdisciplinary gaze.

Greco-Roman antiquity contains both versions of interdisciplinarity. But it is the second definition of interdisciplinarity that this epilogue engages by generalizing from the case of the ancient sciences. I argue against the supposition that scientific interdisciplinarity is a modern notion. Instead I suggest that scientific interdisciplinarity and disciplinarity are ancient polarities, dependent on each other and generating intellectually productive problems. I recognize scientific interdisciplinarity as a repurposed belated aesthetic, in contrast to its widespread image as an aesthetic of novelty.

Ancient Scientific Interdisciplinarity

Many ancient Greco-Roman scientific treatises open with a review of the origins their discipline. They are heirs to an even older philosophical literature that positioned its specialized authors against predecessors in undifferentiated wisdom, from Heraclitus' personal condemnation of the diverse learning (πολυμαθίη) of Greek sages, to Aristotle's naturalized anthropology of dynastic Egyptian social

---

6 Klein 1990: 32–35.

leisure and natural wonder, to the Hellenistic Peripatetic Aristo of Ceos' character sketch parodying the contemporary "know-it-all" (παντειδήμων), a figure who was ignorant of the then-developed difference between knowledge and craft and even the different disciplines of the sciences themselves.[7] Some of the ancient scientific treatises stress the 'naturalized' origins of their study, just as the Hippocratic *On Ancient Medicine* tells how medicine arose from increasingly elaborate techniques of cooking.[8] Still other treatises present the more familiar picture of a science branching itself from the trunk of philosophy. Hippocrates' contribution, according to the Roman medical encyclopediast Celsus, was that he first separated medicine from philosophy.[9] Celsus' term for Hippocrates' medicine, *disciplina*, eschews the "craft" terminology widespread in the ancient sciences (*technē*, *ars*) and dignifies Hippocrates' practice with the more formal status of "knowledge". Here ancient *disciplina* rises to the level of a discipline.[10] To be sure, scholars have written about the origin of the disciplines in Classical Greece and their ancient historiography.[11] While in antiquity the boundaries between specializations were more fluid and the conceptual contrast between pure and applied science was less apt, it is not wrong to call ancient science "disciplinary" in a meaningful way. As we have seen throughout the present book, individual scientists may be carry the title 'philosophers' at the same time as 'doctor', 'geographer', and so on. Disciplinarity existed but had not yet grown too far from the parent tree of philosophy. As was said in the Introduction, Hellenistic society did not wholly disassociate science from philosophy.

Scholars have told the story of how disciplinary divisions in the ancient sciences took shape over roughly 500–300 BCE, the Classical Period in ancient Greek history.[12] Scientific practices became socially distinct from other craft practices, with recognizably separate professionals. At the same time intellectual developments distinguished bodies of knowledge proper to those professionals, together with a unified set of intellectual problems and methods. Certainly the evidence of Aristotle shows that Greek intellectuals recognized a diverse constellation of disciplinary forms: Aristotle argued that the certainty of intellectual demonstration varied according to field.[13] But with a shift in political fortunes after the death of Alexander the Great (d. 323 BCE), the Hellenistic kingdoms recentered intellectual life away from philosophical academies and craft practices. Social and financial

---

7 Heraclitus fr. B40 DK; Aristotle *Metaphysica* 980a–981b; Aristo frr. 21i, 21k SFOD.
8 Hippocrates *De Medicina Vetera* 3 = 1.574–78L.
9 Celsus *De Medicina* 1.pr.8; von Staden 1999b: 264–267.
10 Important for Celsus' conception of medicine as a *disciplina* within an encyclopedia of *artes* may be Varro's now-lost *Disciplinae*; König and Woolf 2013: 38–40.
11 Lloyd 1979; Zhmud 2006.
12 Lloyd 1979, Lloyd 1991, Meißner 1999.
13 Lloyd 1990: 73–97.

rewards followed the social configuration of elite court actors. Court science arose in response to the political shift from city-states to colonizing kingdoms.

This book has told the story of how interdisciplinary scientific knowledge took shape roughly 300–200 BCE at those Hellenistic courts, for the extant origins of scientific interdisciplinary also lie in antiquity. Court science was the original interdisciplinary space with fruitful exchanges for new innovative science. As we have seen, the court was a space that allowed researchers of different disciplinary backgrounds to meet and interact. Individual knowers were socially grouped together, representing their individual disciplines. Unlike in contemporary scientific interdisciplinary, it was not so much a set problem that demanded integrative attention from multiple knowers as it was an aesthetic discourse that encouraged a particular vision of knowledge. Court science at the court of Ptolemy encouraged cross-disciplinary approaches to scientific knowledge, since hybridity was a privileged discourse among the elite court actors. Social conventions at the court of Ptolemy supported the efflorescence of a type of investigation about the natural world that drew on the already-developed traditions of Greek intellectual life.

Periodizations are schematic as well as suggestive of deep historical changes over a long temporal space. To contrast the disciplinary science of the Classical Period with the interdisciplinary science of the Hellenistic period is simply too reductive. The science of Eratosthenes, Andreas, Herophilus, and Archimedes was interdisciplinary, but disciplinary science continued alongside these authors. Interdisciplinarity did not replace disciplinarity; there was no disruption in scientific practices. Instead court science gave rise to new sets of intellectual pursuits that continued alongside the old. Problems and practices proliferated. Since court science was the original interdisciplinary space, there were no disciplining methods at court. Court science allowed controversies to emerge, not close. We have studied the external social factors that allowed scientific ideas to add to debate.

Fascinatingly the products of interdisciplinary science were reabsorbed into their respective disciplines. Eratosthenes' instrument prompted the revisionary mechanics of the mathematician Nicomedes that his instrument was a-mechanical and bereft of geometric character.[14] Andreas' machine, as we have seen, sits alongside other Hellenistic medical inventions in the doxographical report of the later physician Oribasius. Archimedes' mathematical successors in the next two generations examined the further mechanical properties of parabolas, utilizing the results of Archimedes' mechanically-gained knowledge.[15] The ancient Empiricist physicians, who defined themselves against Herophilus' scientific investigations, alternately rejected and then at last embraced Herophilus' pulse: "Medicine needs this investigation most of all".[16] The tracing of these disciplinary connections supports

---

14 Eutocius *In Archimedis de Sphaera et Cylindro Libros II* 98.5–7 Heiberg.
15 Knorr 1986: 209–292.
16 Galen *De Differentiis Pulsibus* 8.726–728K; Galen *De Experientia Medica* 109 Walzer.

the well-known view of Thomas Kuhn that disciplinary scientific knowledge is a kind of puzzle tradition with exemplary solutions for different groups of practitioners. When the social system of monarchy and patronage that supported Hellenistic court science collapsed due to failing finances and Rome's military conquest, court science seemingly disappeared and with it scientific interdisciplinarity.

Scientific interdisciplinarity did not readily reappear in antiquity, although the new Roman world was ripe for the patronage of Greek intellectuals. Several later works present Greek intellectuals in sympotic settings familiar to their Hellenistic predecessors, discussing intellectual questions of high culture with their Roman patrons. Plutarch's *On the Face in the Moon* discusses learned astronomical knowledge and Athenaeus' *Deipnosophistae* queries in minute literary detail the social practices of the symposium itself, haphazardly embracing medical advice about diet along the way. Lucian of Samosata wrote satirically of Greek intellectuals housed by Roman elites, kept for intellectual entertainment, and brought forth at symposia to answer recherché questions and to display the philosopher's beard and cloak.[17] Knowledge remained a form of spectacle and entertainment in such works but it was not scientific interdisciplinary knowledge. Rather, Roman-period Greek science became more overtly philosophical. For instance, the Greek doctor Galen, born in the Greek east and raised with an aristocratic education who climbed the social ladder in Rome to become court physician to the emperor Marcus Aurelius, owed a significant part of his social success to his preeminence in philosophy and medicine. In these Roman examples Greek science returned its philosophical roots with strengthened practices of argumentation and more formalized criteria for evidence and testimony. In Late Antiquity scientific knowledge became patronized by institutions rather than individuals: Galen's works were codified and taught in medical schools; the mathematicians too became school texts. Knowledge as a form of spectacle was lost; opportunities for cross-disciplinary interaction disappeared when systemization and argumentative completeness accelerated the policing of disciplinary boundaries.

Was interdisciplinary court science merely a blemish on the forward march of disciplinarity in Greco-Roman antiquity? No. Divisions between scientific disciplines were never quite so distinct in antiquity. Many polymaths moved quickly over divisions between the liberal arts and sciences. The Roman Apuleius of Madaura, for one, wrote treatises on plants, rhetoric, and Platonic philosophy, an astonishingly varied set of interests when viewed through the disciplinary divisions of our contemporary university. National divisions, so important to modernist conceptions of history and science, were indistinct in antiquity. I have spoken of Greco-Roman antiquity, but perhaps it would be better to contextualize ancient science with reference to the various states and ethnic groups bordering on the Mediterra-

---

17 Lucian *De mercede conductis*.

nean. Egyptians, Phoenicians, Carthaginians, and Pontics among others pursued scientific knowledge and their works were translated into ancient Greek, the *lingua franca* of intellectual practice around the Mediterranean basin roughly over the millennium from 300 BCE to 700 CE. Insofar as the unitary vision of knowledge united in one's person – to be "educated" (πεπαιδευμένος) in Greek culture – was coordinated with social distinction, the cultural aspiration of general knowledge restrained the forward march of disciplinarity.

Just-so Stories

A larger history of interdisciplinarity embraces the evidence not only from science but also from literature. Ancient scientific interdisciplinarity viewed through a literary form is not a revisionist analysis of demarcation criteria between science and pseudo-science. Neither "science" nor "pseudo-science" are facile stable categories; for that matter nor is "literature". We seek instead an aesthetic form that thematizes causal investigation and at the same time recognizes the potential inexhaustiveness of that explanation. I call this literary form "just-so stories", after Rudyard Kipling's fantastical stories of the same name. Kipling's explanations of select facts of the natural world invoke folk motifs of cunning, punning on language, and the distribution of proportionate justice. "Just-so stories" are a generic category of explanation: Kipling's tales thematize causal explanation with titles such as "How the Camel Got his Hump" and "The Beginning of the Armadillos". While Kipling's animal fables indulge children's curiosity, they also point to a poetics of just-so stories.

Ancient Greeks called just-so stories αἰτία "causes", transliterated as *aitia* or *aetia*, a literary form that has passed over into English as "aetiology".[18] The aetiological literary form represents a literary analogue to children's ceaseless inquisitiveness. The problem with aetiology is that it seems unending: "*aitia* spill over to infect each other"; "the poetry of 'origins' will always face an embarrassment of material"; "the problem with aetiology is not merely that one custom may have different explanations, but that almost everything could require an *aition*."[19] The aetiological writer is faced with a boundless task. So it is useful to the aetiological author that he is a culturally belated author, for all varied cultural discourses – animalian cunning, magic, language as self-reflexive object, medicine, justice, divinity – come prepackaged, as it were. To allude to a certain cultural practice with the codes of cultural discourse is to call to readers' minds a web of cultural associa-

---

[18] I prefer the Latinate transcription that exposes its etymological roots to the more common "etiology".
[19] Fantuzzi and Hunter 2004: 63–64, 65, 83.

tions about categories, explanations, stories. An animalian aetiology brings to mind folk cunning and the triumph of the natural order; a magical aetiology brings to mind stories of magicians and the act of magical rites; a scientific aetiology brings to mind the categories of natural explanation, the patterns of natural action, and learned disciplinary knowledge; a linguistic aetiology brings to mind language's capacity for invention and etymologic revelation; a divine aetiology brings to mind mythological stories of the gods' interests, oracles, and revealed knowledge. The *aitia* of each level function with a resonance of multiple cultural practices, each with their own context and vocabulary. Allusion to cultural discourses allows the author both to embrace his surplus of material and to suggest *aitia* without cataloging or enumerating them.

As an example of this rich vein of just-so stories, I offer a close reading of a text from the court of Ptolemy that stands at the interface of science and poetry while revealing the historical understanding of the interrelation between intellectual practices. This passage is from the court poet **no. 110** Callimachus' book called *Aetia*, a poetiological account of how rituals, rites, and natural phenomena came to be. The present book began with a brief reading of the role of the court scientist **no. 137** Conon, who featured in Callimachus' *Aetia* for his catasterism of **no. 2** Berenice's disappeared lock. Callimachus' erudition, familiarity with court society, and interest in knowledge offer a revealing lens at how the non-scientist elites of Ptolemaic court culture understood the poetics of explanation.

The love story of Acontius and Cydippe is one of the longest and best preserved sections of the *Aetia*, highlighting the integration of multiple disciplinary perspectives.[20] The story narrates how the young man Acontius fell in love with the girl Cydippe by the temple of Artemis. Cydippe unknowingly swore an oath to marry Acontius but fell sick at her failure to do so; the god Apollo at last revealed the destined marriage by an oracle to Cydippe's father. In the poem the present's indebtedness to the past is constantly stressed: Acontius' descendants, the Acontidae, still prosper on Ceos; the prenuptial customs of the Naxians imitate the lovemaking of Hera and Zeus; and Callimachus openly acknowledges his source, the Cean historian and mythographer, Xenomedes. Some commentators have argued that Callimachus directly references Hippocrates' *On the Sacred Disease* in a central passage, a further recognition of the present's indebtedness to the past.[21]

> In the afternoon an evil pallor seized Cydippe and the disease came: we send that disease off to the wild goats and name it sacred in our lies. That grievous disease wasted the girl to the house of Hades. For a second time the marriage couches were spread, for a second time the girl burned with a quartan fever for seven months. For a third time they thought of marriage,

---

20 Callimachus *Aetia*, frr. 67–75e.
21 Lang 2009; Harder 2012: 2.597–598.

for a third time again a deadly cold settled onto Cydippe. Her father did not wait for a fourth time.[22]

If it is correct to argue that Callimachus alludes to Hippocrates' *On the Sacred Disease*, the citation of a prose text would be parallel to Callimachus' use of the historian Xenomedes. But we should not think about Callimachus' references to science in a narrow allusive way. Rather, Callimachus' allusions to science and other contemporary cultural practices are purposely over-determined.[23]

The story of Acontius and Cydippe invokes a cultural discourse of love magic. The story concerns a young man who falls in love at a festival with a girl who does not reciprocate his love. He puts an injunction upon the object of his love; she suffers physically as a result. At last the spell is broken when the goddess who caused the spell is appeased. The most familiar ancient literary story of love magic is Theocritus *Idyll* 2, where the spurned woman Simathea recites rites to draw her absent love. Theocritus has reversed the usual gender roles known from surviving magical spells and papyri.[24] Usually the man invokes the goddess' help to physically harm and torment his love-object until he has her, just as in the story of Acontius and Cydippe. Further, Callimachus' text presents verbal parallels to discourse about love magic. The metaphor of burning, an important part of the magical rite, is present throughout the story.[25] The magical objects of the gods are mentioned. Since the ancient epistolographer copying Callimachus mentioned Aphrodite's magical girdle, Callimachus probably included the girdle in the passages now lost.[26] Moreover, there are references to magical practices that are not love-magic: the apotropism of disease onto goats, the sorcery and witchcraft of the Telchines.[27]

---

22 Callimachus *Aetia* fr. 75.12–20: δειελινὴν τὴν δ' εἷλε κακὸς χλόος, ἦλθε δὲ νοῦσος, / αἶγας ἐς ἀγριάδας τὴν ἀποπεμπόμεθα, / ψευδόμενοι δ' ἱερὴν φημίζομενον· ἣ τότ' ἀνιγρὴ / τὴν κούρην Ἀίδεω μέχρις ἔτηξε δόμων. δεύτερον ἐστόρνυντο τὰ κλισμία, δεύτερον ἡ πα[ῖ]ς / ἑπτὰ τεταρταίῳ μῆνας ἔκαμνε πυρί. / τὸ τρίτον ἐμνήσαντο γάμου κάτα, τὸ τρίτον αὖ[τις / Κυδίππην ὀλοὸς κρυμὸς ἐσῳκίσατο. / τέτρατον [ο]ὐκέτ' ἔμεινε πατήρ …
23 Cf. Netz 2009a: 179.n4.
24 Faraone 1999.
25 In Callimachus *Aetia* fr. 67.2 Acontius burns for Cydippe, in fr. 75.15 Cydippe melts from her disease and struggles with a quartan fever, in fr. 75.30–31 Cyex is advised to meld his family with Acontius', as if smelting the metals electron and gold.
26 In Aristaenetus 1.10 Cydippe's beauty is praised to the point that "Aphrodite adorned her with all her own honors, sparing her magical girdle alone". Callimachus also refers to the magic girdle at *Aetia* fr. 43.53. Importantly, the magic girdle is the object Aphrodite gives Hera to seduce Zeus in Homer *Ilias* 14.214, the story alluded to in *Aetia* fr. 75.4 before the *Abbruchsformel*. Callimachus' endorsement of the sibling marriage of Zeus and Hera is an implicit criticism of Sotades; Harder 2012: 2.583.
27 Callimachus *Aetia* frr. 75.13, 75.64–65.

The story of Acontius and Cydippe also invokes the discourse of medicine. Both vocabulary and context are important. Callimachus would burp out the secrets of Demeter at Eleusis by a metaphor of indigestion.[28] In the central scene quoted above Cydippe becomes ill. The sickness is called an "evil pallor" that "melts" her; she struggles with a quartan fever for seven months. Cydippe's third disease is a destructive chill that takes up residence in her.

Commentators have taken Cydippe's illness to refer to Hippocrates' *On the Sacred Disease*. Callimachus describes the apotropism of the disease on to wild goats and the false naming of the disease as sacred. It is true that the Hippocratic passage refers to an epileptic goat.[29] But the Hippocratic text also states that magicians and sorcerers are those responsible for naming the disease sacred.[30] That we falsely call it sacred, as Callimachus says, echoes the Hippocratic text. And yet the context of the story proves just the opposite: Cydippe's disease is sacred, caused by the goddess Artemis. It is to the subversion of Hippocratic medicine and to the credit of its healing competitor, magic, that we read the Hippocratic text against this passage. Furthermore, scholars have struggled to identify a passage in the Hippocratic Corpus that is a parallel to Cydippe's second and third diseases. I suggest reading Cydippe's progression of illnesses within the discourse of rational medicine: first, an acute, sudden disease (epilepsy); second, a long chronic fever; third, an acute chill. We see here not an allusion to a text – as if Hippocrates' *On the Sacred Disease* or any other Hippocratic text contained these diseases or this sequence of illnesses – but to a kind of cultural knowledge.

So the passage of Cydippe's illness contains not only as medical discourse but also magical discourse. Which is it then? Immediate context argues in favor of medicine; general context in favor of magic. Vocabulary is interdeterminate, favoring neither. I suggest that the passage is overdetermined with meaning, so as to be read either as medical discourse or magical discourse. I argue not so much for one particular version but namely the interaction of the various themes: magic, medicine, revealed and acquired knowledge and the various resonances they form.

Reading contemporary cultural knowledge as discourse in Cydippe's illness reminds us of the poetics of aetiology. Knowledge is an important theme of Callimachus' text, foregrounded by the *technē* "craft, knowledge" Eros teaches Acontius.[31] After all, the narration brings to mind different types of *technē*: there is the *technē erotikē* "erotics" of Eros, the *technē magikē* "magic" of the Telchines, the *technē iatrikē* "medicine" of Hippocrates and Greek science, the *technē chalchikē* "smithing" of the mixing of metals, and so on. Perhaps Cydippe's illness is due to the

---

28 Callimachus *Aetia* fr. 75.7.
29 Hippocrates *De Morbo Sacro* 11.3–4 = 6.382L.
30 Hippocrates *De Morbo Sacro* 1.1–4 = 6.352–354L.
31 Callimachus *Aetia* fr. 67.3.

love magic of Acontius; perhaps it is has a medical aetiology, a type of disease that we can categorize and name; or perhaps it is from the goddess Artemis herself. Callimachus offers multiple levels of *aitia* with which to ground the explanation. The poet himself suggests that "much knowledge is a difficult evil".[32] Can there be any resolution to this surplus of information? It is not simply that there might be an allusion to a scientific explanation, but rather that this possibility is embedded inside a complex of competing aetiological explanations. The aetiological writer is faced with a boundless task abridged only by belated references to disciplinary discourses. Ancient "just-so" stories have lessons to teach us about the history of scientific interdisciplinarity.

A Future History

The picture I have drawn of ancient scientific interdisciplinarity stands at odds with the standard story. A standard story of scientific interdisciplinarity starts in the nineteenth century, with the rise of the research university, and traces the social housing of the disciplines into the departments and specialized research units of the contemporary academic world. Before the nineteenth century, so it goes, scholars and researchers were generalists. Aristotle's hierarchy of theoretical over craft knowledge, combined with Stoic researches into grammar stabilized the disciplines for hundreds of years in Western thought as the *artes liberales*.[33] Without care for practical application, scholars embodied the various disciplines in their person but did not form into social units to direct multiple disciplinary gazes upon an object. But this modernist story is not the whole truth. It privileges the very recent, contemporary world over the patterns of the *longue durée*; it overemphasizes the normative accounts of philosophers instead of the descriptive accounts of practitioners. For the standard story, disciplinarity in the premodern world was as fixed and unmoving as the sphere of the stars. But the truth is a much richer and varied story.

As I have argued, interdisciplinarity has a long history. The disciplines are ancient and have profitably interacted in the deep past. The historical story of generalist knowledge, unified in the individual person, is familiar. The historical story of the interaction between the sciences has ancient forms but has not previously been told. If scientific interdisciplinarity is not so new, why are we rediscovering it just now in history? And if it is not so new, where is it in the historical record

---

32 Callimachus *Aetia* fr. 75.8: πολυιδρείη κακὸν χάλεπον.
33 So Weingart 2010: 3–4. I do not know why the image of premodern science controlled by a strict Aristotelian hierarchy of theoretical and banausic knowledge is so widespread. Berryman 2009: 1–6 is a good antidote to such historiography.

apart from ancient court science? In response to the first question a cultural critic might argue that a history of interdisciplinarity is a reassuring story about the human desire for a general and universal vision of knowledge; but the second question demands a historical response. To move towards a history of scientific interdisciplinarity means recovering a lost history of knowledge production. To confront the challenge I will speculate based on the evidence I have presented, for there are lessons both in court science and in the reading of Callimachus' *Aetia* for the history of interdisciplinarity. My sketch suggests that scientific interdisciplinary needs practices, social spaces, as well a certain aetiological aesthetic, the "just-so" story.

The history of court science might suggest only that historians of interdisciplinarity should search out those institutions where scientists of different disciplinary practice met: the court, the university, the research institute. But the history of scientific interdisciplinarity is much broader. For one, the history of knowledge is not a straightforward branching from the trunk of philosophy. Older relationships – between science and magic, science and philosophy, magic and philosophy – pervade intellectual practice. If Callimachus the poet when describing a causal chain seemingly has divine options open to him (the brother-sister gods Apollo and Artemis) unavailable to his contemporaries, the rational scientists, this is our anachronistic mistake of privileging the scientific over the literary. Just like Conon claimed to see Berenice's hair rising as a new divine constellation in the night sky, so Callimachus' telling of Cydippe's illness and cure partakes of the magical, religious, and scientific. A history of scientific interdisciplinarity needs to not only consider those cultural places where different kinds of scientific practitioners interacted but also where scientists (writ generically) interacted with other proponents of different modes of intellectual practice and explanation.

Second, Callimachus' aetiological poetics gain their force from problematizing otherwise seemingly distinct modes of intellectual practice within a belated aesthetics that privileges tracing causal chains. The difference between magical burning rites and medical cautery exists more in cultural context than direct practice. The aesthetics of aetiological poetics highlight single problems from multiple, competing, pre-existing disciplines. To speak in modernist terms about interdisciplinarity either as a product of the content of knowledge itself or as the object of a coordinated interdisciplinary gaze elevates a puzzle-tradition model of scientific enterprise. Reading interdisciplinary science through an aesthetics of aetiological poetics emphasizes not the puzzle but the disciplinary pursuit. Historical scientific interdisciplinarity focuses not on the problem, but the multiple paths to reach a solution. Callimachus reminds us not of Cydippe's fever (the problem) but rather of the competing disciplinary explanations: Artemis, magic, medical humoralism. A history of scientific interdisciplinarity therefore needs to consider those cultural places where scientific practitioners interacted with literary observers to foreground multiple accounts and explanations. Libraries, coffeeshops, and minor literary journals are possible spaces of scientific interdisciplinarity, where literature meets science and multiple disciplinary perspectives of explanation are privileged.

The historical lesson is that we do not need to have formal institutional spaces in order to have scientific interdisciplinarity. We need separate disciplines, a belated aesthetic preference for multiple disciplinary explanations, and a cultural practice that brings the embodied disciplines together. The history of scientific interdisciplinarity is not necessarily equivalent to the history of media or the history of the public sphere, although the history of scientific knowledge overlaps with those histories. If eighteenth century coffeehouses and twenty-first century Internet forums are potential spaces for interdisciplinary science, these are spaces accessible to many different individuals: they are not public yet not so private, whose participants share common practices. I leave for others to decide whether participants of these spaces valued a poetics of aetiology.

I therefore suggest that interdisciplinarity has a history in the long space of time between the ancient and modern worlds. The evidence surveyed is the beginning of a larger history of intellectual interdisciplinarity. Modernist preoccupations with nation and language may need to subordinated to the *longue durée* of the interaction between disciplines. Probably the history of interdisciplinarity is not a continuous genealogy but rather one of punctuated equilibrium (to borrow a term). Might our contemporary world too profit from an interdisciplinary poetics of aetiology? At present there is a cultural anxiety about intellectual overspecialization and the distance between disciplines. Certainly a glance around the contemporary world finds much overt social and financial support for scientific interdisciplinarity, as well as a desire to reconsider what "interdisciplinary" means in knowledge formation.[34] Regardless whether such institutional support will continue in traditional ways or be subsumed in a new patronage economy emerging from many contemporary forms, interaction between academic disciplines will continue to mark intellectual practice, even if only intermittently. Perhaps future interaction between disciplines will take stock of its own history of interdisciplinarity, stretching back even to Hellenistic science at court.

---

[34] For a perspective from the social sciences, see "interdisciplinary" futures in Fitzgerald and Callard 2015.

Editions of Primary Sources

Materials referenced here are not repeated in the bibliography of secondary sources, unless they are cited independent of their text.

Literary Texts

Achilles Tatius *In Arati Phaenomena* = Maass, E. 1898. *Commentariorum in Aratum reliquiae*. Berolini.
Aelian *Varia Historia* = Dilts, M. 1974. *Claudi Aeliani Varia Historia*. Lipsiae.
Aetius = Diels, H. 1879. *Doxographi Graeci*. Berolini.
Andreas of Carystus = von Staden, H. 1989. *Herophilus: The Art of Medicine in Early Alexandria*. Cambridge.
Antigonus of Carystus *De Mirabilibus* = Giannini, Alexander. 1965. *Paradoxographorum Graecorum Reliquiae*. Milano.
Apollodorus *Bibliotheca* = Wagner, R. 1894. *Mythographi Graeci: Volumen 1: Apollodori Bibliotheca; Pediasimi Libellus de duodecim Herculis laboribus*. Lipsiae.
Apollonius of Citium = Kollesch, J. and F. Kudlien. 1965. *Apollonii Citiensis In Hippocratis De Articulis Commentarius*. Corpus Medicorum Graecorum XI 1,1. Berolini.
Apollonius of Rhodes *CA* = Powell, J. U. 1925. *Collectanea Alexandrina*. Oxonii.
Archimedes *Arenarius* = Heiberg, J. L. 1913. *Archimedis opera omnia*. Vol. 2. Lipsiae.
Archimedes *Quadratura Parabolae* = Heiberg, J. L. 1913. *Archimedis opera omnia*. Vol. 2. Lipsiae.
Archimedes *Methodus* = Netz, Reviel et al. 2011. *The Archimedes Palimpsest*. 2 vols. Cambridge.
Aristaenetus 1.10 = Harder, A. 2012. *Callimachus Aetia*. 2 vols. Oxford.
Aristo of Ceos = Fortenbaugh, W and S. White. 2006. *Aristo of Ceos: Text, Translation, and Discussion*. Rutgers NJ.
Aristotle *Ethica Nicomachea* = Bywater, I. 1890. *Aristotelis Ethica Nicomachea*. Oxonii.
Aristotle *Metaphysica* = Jaeger, W. 1957. *Aristotelis Metaphysica*. Oxonii.
Aristotle *Meteorologica* = Fobes, F. H. 1919. *Aristotelis Meteorologicorum Libri Quattuor*. Cantabrigiae Massachusettensium.
Aristoxenus of Tarentum *Elementa Rhythmica* = Pearson, L. 1990. *Aristoxenus: Elementa Rhythmica*. Oxford.
Asclepius of Tralles *In Nicomachi Arithmetica* = Tarán, L. 1969. "Asclepius of Tralles: *Commentary to Nicomachus' Introduction to Arithmetic*" Transactions of the American Philosophical Society 59.4: 1–89.
Athenaeus = Olson, S. Douglas. 2007–2012. *The Learned Banqueters*. 8 vols. Loeb series. Cambridge MA.
Biton = Marsden, E. W. 1971. *Greek and Roman Artillery: Technical Treatises*. Oxford.
BNJ 3 = Morison, William. 2011. "Pherekydes of Athens (3)" in *Brill's New Jacoby*. I. Worthington *ed*. Brill Online Reference. Accessed 3. 28. 2016.
BNJ 45 = Costa, Virgilio. 2014. "Hegesianax of Alexandria Troas (45)" *Brill's New Jacoby*. I. Worthington *ed*. Brill Online Reference. Accessed 6. 10. 2016.
BNJ 71 = Williams, Mary F. 2013. "Zoilus of Amphipolis (71)" in in *Brill's New Jacoby*. I. Worthington *ed*. Brill Online Reference. Accessed 1. 15. 2016.
BNJ 74 = Williams, Mary F. 2012. "Euphantus of Olynthus (74)" in *Brill's New Jacoby*. I. Worthington *ed*. Brill Online Reference. Accessed 1. 28. 2016.
BNJ 160 = Gambetti, Sandra. 2011. "Anonymous *Belli tertii annales* (160)" in *Brill's New Jacoby*. I. Worthington *ed*. Brill Online Reference. Accessed 2. 22. 2016.

BNJ 161 = Bromberg, Jacques. 2012. "Ptolemy of Megalopolis (161)" in *Brill's New Jacoby*. I. Worthington *ed*. Brill Online Reference. Accessed 3.13. 2016.
BNJ 162 = Ceccareli, Paola. 2008. "Demetrios of Byzantion (162)" in *Brill's New Jacoby*. I. Worthington *ed*. Brill Online Reference. Accessed 3.17. 2016.
BNJ 234 = Roller, Duane W. 2007. "Ptolemy VIII (234)" in *Brill's New Jacoby*. I. Worthington *ed*. Brill Online Reference. Accessed 2.10. 2016.
BNJ 241 = Pownall, Frances. 2009. "Eratosthenes of Cyrene (241)" in *Brill's New Jacoby*. I. Worthington *ed*. Brill Online Reference. Accessed 6.9. 2016.
BNJ 263 = Węcowski, Marek. 2010. "Pseudo-Democritus or Bolos of Mendes (263)" in *Brill's New Jacoby*. I. Worthington *ed*. Brill Online Reference. Accessed 4.7. 2016.
BNJ 266 = Rzepka, Jacek. 2011. "Apollas of Pontus (266)" in *Brill's New Jacoby*. I. Worthington *ed*. Brill Online Reference. Accessed 3.17. 2016.
BNJ 334 = Jackson, Steve and Monica Berti. 2015. "Istros (334)" in *Brill's New Jacoby*. I. Worthington *ed*. Brill Online Reference. Accessed 5.11. 2016.
BNJ 339 = Sickinger, James. 2008. "Asklepiades of Nikaia (339)" in *Brill's New Jacoby*. I. Worthington *ed*. Brill Online Reference. Accessed 4.27. 2016.
BNJ 472 = Engels, Johannes. 2008. "Agathokles (472)" in *Brill's New Jacoby*. I. Worthington *ed*. Brill Online Reference. Accessed 4.28. 2016.
BNJ 627 = Keyser, Paul T. 2014. "Kallixeinos of Rhodes (627)" in *Brill's New Jacoby*. I. Worthington *ed*. Brill Online Reference. Accessed 3.7. 2016.
BNJ 631 = Gambetti, Sandra. 2012. "Satyros of Alexandria (631)" in *Brill's New Jacoby*. I. Worthington *ed*. Brill Online Reference. Accessed 3.24. 2016.
BNJ 632 = Burstein, Stanley. 2009a. "Jason, Περὶ τῶν Ἀλεξάνδρου ἱερῶν ἐν Ἀλεξανδρείᾳ (632)" in *Brill's New Jacoby*. I. Worthington *ed*. Brill Online Reference. Accessed 3.13. 2016.
BNJ 666 = Burstein, Stanley. 2008a. "Dalion (666)" in *Brill's New Jacoby*. I. Worthington *ed*. Brill Online Reference. Accessed 3.13. 2016.
BNJ 667 = Burstein, Stanley. 2009b. "Aristokreon (667)" in *Brill's New Jacoby*. I. Worthington *ed*. Brill Online Reference. Accessed 3.13. 2016.
BNJ 668 = Burstein, Stanley. 2009c. "Bion Soleus (668)" in *Brill's New Jacoby*. I. Worthington *ed*. Brill Online Reference. Accessed 3.13. 2016.
BNJ 669 = Burstein, Stanley. 2008b. "Simonides, de Aethiopia (669)" in *Brill's New Jacoby*. I. Worthington *ed*. Brill Online Reference. Accessed 3.13. 2016.
BNJ 718 = D'Hautcourt, Alexis. 2008. "Basilis (718)" in *Brill's New Jacoby*. I. Worthington *ed*. Brill Online Reference. Accessed 3.13. 2016.
BNJ 783 = Naiden, Fred S. 2008. "Menander (783)" in *Brill's New Jacoby*. I. Worthington *ed*. Brill Online Reference. Accessed 3.13. 2016.
Caelius Aurelianus *Celerum Passionum* = Bendz, G. 1990. *Caeli Aureliani Celerum Passionum Libri III; Tardum Passionum Libri V.* Corpus Medicorum Latinorum V 1. Berolini.
Caelius Aurelianus *Tardum Passionum* = Bendz, G. 1990. *Caeli Aureliani Celerum Passionum Libri III; Tardum Passionum Libri V.* Corpus Medicorum Latinorum V 1. Berolini.
Callimachus *Aetia* = Harder, Annette. 2012. *Callimachus: Aetia*. 2 vols. Oxford.
Callimachus *HE* = Gow, A. S. F. and D. Page. 1965. *The Greek Anthology: Hellenistic Epigrams*. 2 vols. Cambridge.
Callimachus *In Apollinem* = Pfeiffer, R. 1953. *Callimachus: Volumen II: Hymni et Epigrammata*. Oxford.
Callimachus *In Artemim* = Pfeiffer, R. 1953. *Callimachus: Volumen II: Hymni et Epigrammata*. Oxford.
Callimachus *Fragmenta* = Pfeiffer, R. 1949. *Callimachus: Volumen 1: Fragmenta*. Oxford.
Callimachus *In Iovem* = Pfeiffer, R. 1953. *Callimachus: Volumen II: Hymni et Epigrammata*. Oxford.

Cassius Iatrosophista *Problemata* = Garyza, A. and R. Massulo. 2004. *I problemi di Cassio Iatrosofista*. Napoli.
Celsus *De Medicina* = Marx, F. 1915. *A. Cornelii Celsi quae supersunt*. Corpus Medicorum Latinorum I. Lipsiae.
Censorinus *De Die Natali* = Sallmann, N. 1983. *Cesorini de die natali liber ad Q. Caerellium*. Lipsiae.
Cicero *Tusculanae Disputationes* = Pohlenz, M. 1918. *M. Tulli Ciceronis Scripta quae mansuerunt omnia. Fasc. 44. Tusculanae Disputationes*. Lipsiae.
Clearchus = Wehrli, F. 1969. *Die Schule des Aristoteles: Texte und Kommentar III Klearchos*. 2nd ed. Basel.
Clement of Alexandria *Stromateis* = Stählin, O. 1906. *Clemens Alexandrinus. Zweiter Band. Stromata Buch I– VI*. Leipzig.
Quintus Curtius Rufus = Lucarini, C. M. 2009. *Q. Curtius Rufus: Historiae*. Berolini.
Damagetus = Gow, A. S. F. and D. Page. 1965. *The Greek Anthology: Hellenistic Epigrams*. 2 vols. Cambridge.
Demetrius of Phalerum = Wehrli, F. 1969. *Die Schule des Aristoteles: Texte und Kommentar IV Demetrios von Phaleron*. 2nd ed. Basel.
Demetrius of Scepsis = Gaede, R. 1880. *Demetrii Scepsii quae supersunt*. Gryphiswaldiae.
Descartes, René *La Géométrie* = Descartes, René. 1954. *The Geometry, with a facsimile of the first edition*. D. E. Smith and M. L. Latham *trans*. New York.
Dioscorides of Nicopolis = Gow, A. S. F. and D. Page. 1965. *The Greek Anthology: Hellenistic Epigrams*. 2 vols. Cambridge.
Dioscorides *De Materia Medica* = Wellmann, M. 1907–1914. *Pedanii Dioscoridis Anazarbei De Materia Medica Libri Quinque*. Berolini.
Diphilus = Kassel, R. and C. Austin. 1986. *Poetae Comici Graeci. Vol. V: Damoxenus– Magnes*. Berolini.
Diocles = Toomer, G. J. 1976. *Diocles: On Burning Mirrors*. Berlin.
Diodorus Siculus = Oldfather, C. H. et al. 1933–1967. *Diodorus Siculus: Library of History*. 12 vols. Loeb series. Cambridge MA.
Diogenes Laertius = Dorandi, Tiziano. 2013. *Diogenes Laertius: Lives of the Eminent Philosophers*. Cambridge.
Epicurus = Usener, Hermann. 1887. *Epicurea*. Lipsiae.
Erasistratus = Garofalo, I. 1988. *Erasistrati Fragmenta*. Pisa.
Eratosthenes *Geographica* = Roller, Duane W. 2010. *Eratosthenes' Geography: Fragments collected and translated, with commentary and additional material*. Princeton.
Etymologicum Magnum = Gaisford, T. 1848. *Etymologicum Magnum*. Oxonii.
Euclid *Elementa* = Heiberg, J. L. and E. S. Stamatis. 1969–1977. *Euclidis Elementa*. 5 vol. Leipzig.
Euphronius – Powell, J. U. 1925. *Collectanea Alexandrina*. Oxonii; Strecker, Carolus. 1884. *De Lycophrone Euphronio Eratosthene Comicorum Interpretibus*. Greifswald.
Eusebius *Chronicon* = Schoene, A. 1875. *Eusebi Chronicorum Libri Duo. Vol. 1*. Berolini.
Eusebius *Preparatio Evangelica* = Mras, K. 1983. *Eusebius Werke. Achter Band: Die Preparatio Evangelica. Zweiter Teil*. 2 ed. Berlin.
Eutocius *In Archimedis de Sphaera et Cylindro Libros II* = Heiberg, J. L. 1915. *Archimedis opera omnia. Volumen Tertium*. Lipsiae.
Fragmenta adespota TGrF = Kannicht, R and B. Snell. 1981. *Tragoricum Graecorum Fragmenta*. Vol. 2. Göttingen.
FGrH 63 = Jacoby, Felix. 1928–1958. *Die Fragmente der Griechischen Historiker*. Brill.
FGrH 138 = Jacoby, Felix. 1928–1958. *Die Fragmente der Griechischen Historiker*. Brill.
FGrH 585 = Jacoby, Felix. 1928–1958. *Die Fragmente der Griechischen Historiker*. Brill.

FGrH 1026 = Bollansée, Jan. 1999b. "Hermippos of Smyrna (1026)" in *Die Fragmente der Griechischen Historker IV*. G. Schepens ed. Brill Online Reference. Accessed 4.15.2016.
Galen *De Antidotis* = Kühn, K. G. 1827. *Claudii Galeni opera omnia. Tomus XIV*. Lipsiae.
Galen *De Compositione Medicamentorum secundum Locos* = Kühn, K. G. 1826. *Claudii Galeni opera omnia. Tomus XII*. Lipsiae.
Galen *De Differentiis Pulsibus* = Kühn, K. G. 1824. *Claudii Galeni opera omnia. Tomus VIII*. Lipsiae.
Galen *De Dignoscendis Pulsibus* = Kühn, K. G. 1824. *Claudii Galeni opera omnia. Tomus VIII*. Lipsiae.
Galen *De Experientia Medica* = Walzer, R. 1944. *Galen on Medical Experience*. London.
Galen *Glossarium* = Kühn, K. G. 1830. *Claudii Galeni opera omnia. Tomus XIX*. Lipsiae.
Galen *In Hippocratis Epidemiarum Librum III* = Wenkebach, E. 1936. *Galeni In Hippocratis Epidemiarum Librum III Commentaria*. Corpus Medicorum Graecorum V 10, 2.1. Lipsiae.
[Galen] *In Hippocratis De Humoribus* = Kühn, K. G. 1829. *Claudii Galeni opera omnia. Tomus XVI*. Lipsiae.
Galen *In Hippocratis Epidemiarum Librum VI* = Wenkebach, E. 1956. *Galeni In Hippocratis Epidemiarum Librum VI Commentaria*. Corpus Medicorum Graecorum V 10, 2,2. Berolini.
Galen *In Hippocratis De Articulis* = Kühn, C. G. 1829. *Claudii Galeni opera omnia. Vol. XVIII. Pars I*. Lipsiae.
[Galen] *Historia Philosophia* = Diels, H. 1879. *Doxographi Graeci*. Berolini.
Galen *De Pulsibus ad Tirones* = Kühn, K. G. 1824. *Claudii Galeni opera omnia. Tomus VIII*. Lipsiae.
Galen *De Simplicium Medicamentorum Temperamentis ac Facultatibus* = Kühn, K. G. 1826. *Claudii Galeni opera omnia. Tomus XI*. Lipsiae.
Galen *Subfiguratio Empirica* = Deichgräber, K. 1965. *Die Griechische Empirikerschule: Sammlung der Fragmente und Darstellung der Lehre*. 2nd ed. Berlin.
Galen *Synopsis de Pulsibus* = Kühn, K. G. 1825. *Claudii Galeni opera omnia. Tomus IX*. Lipsiae.
Geminus *Isagoge* = Manitius, C. 1898. *Gemini Elementa Astronomiae*. Lipsiae.
Hedylus = Gow, A. S. F. and D. Page. 1965. *The Greek Anthology: Hellenistic Epigrams*. 2 vols. Cambridge.
Hegesander = Müller, K. 1849. *Fragmenta Historicorum Graecorum. Volumen Tertium*. Paris.
Heraclitus = Diels, H. and W. Kranz. 1951. *Die Fragmente der Vorsokratiker. Erster Band*. Berlin.
Herodas *Mimiambi* = Cunningham, I. C. 1971. *Herodas: Mimiambi*. Oxford.
Herodotus = Hude, C. 1927. *Herodoti Historiae. Editio Tertia*. 2 vols. Oxonii.
Herophilus = von Staden, H. 1989. *Herophilus: The Art of Medicine in Early Alexandria*. Cambridge.
Hesiod *Theogonia* = Solmsen, F., R. Merkelbach and M. L. West. 1990. *Hesiodi Theogonia, Opera et Dies, Scutum*. 3rd ed. Oxonii.
Hesiod *Opera et Dies* = Solmsen, F., R. Merkelbach and M. L. West. 1990. *Hesiodi Theogonia, Opera et Dies, Scutum*. 3rd ed. Oxonii.
Hippocrates *De Articulis* = Withington, E. T. 1928. *Hippocrates. Vol. III*. Loeb series. Cambridge, MA.
Hippocrates *De Fracturis* = Withington, E. T. 1928. *Hippocrates. Vol. III*. Loeb series. Cambridge, MA.
Hippocrates *De Medicina Vetera* = Heiberg, J. L. 1927. *Hippocratis Opera*. Corpus Medicorum Graecorum I.1. Berolini.
Hippocrates *De Morbo Sacro* = Jouanna, J. 2003. *Hippocrate Œuvres complètes: La maladie sacrée. Tom. II, 3e partie*. Les Belles Lettres. Paris.
Homer *Ilias* = Monro, D. and T. W. Allen. *Homeri Opera*. 2 vols. 3rd ed. Oxonii.
Hyginus *De Astronomia* = Viré, G. 1992. *Hygini De Astronomia*. Stutgardiae.
Hyginus *Fabulae* = Marshall, P. K. 1993. *Hyginus: Fabulae*. Stutgardiae.

Hypsicles *Elementa* 14 *praefatio* = Heiberg, I. L. and E. S. Stamatis. 1977. *Euclidis Elementa. Vol. V Pars 1. Prologemena Critica, Libri XIV– XV, Scholia in Libros I– V.* Leipzig.
Isocrates *Ad Nicoclem* = Norlin, G. 1928. *Isocrates Volume 1: To Demonicus. To Nicocles. Nicocles or The Cyprians. Panegyricus. To Philip. Archidamus.* Loeb series. Cambridge MA.
Jerome *In Danielem* = Glorie, F. 1964. *S. Hieronymi Presbyteri Opera. Pars I, Opera Exegetica: Commentariorum in Danielem Libri III.* Turnholti.
Josephus *Antiquitates Iudaicae* = Niese, B. 1892. *Flavii Iosephii Opera: Volumen III Antiquitatum Iudaicarum Libri XI– XV.* Berolini.
Justin = Seel, O. 1972. *M. Iuniani Iustini Epitoma historiarum philippicarum Pompei Trogi.* Stutgartiae.
Lucian *Adversus Indoctum* = Harmon, A. M. 1921. *Lucian. Volume III.* Loeb series. Cambridge MA.
Lucian *De mercede conductis* = Harmon, A. M. 1921. *Lucian. Volume III.* Loeb series. Cambridge MA.
3 Maccabees = Rahlfs, A. 1935. *Septuaginta. Id est Vetus Testamentum graece iuxta LXX interpretes.* 2 vols. Stuttgart.
Machon = Gow, A. S. F. 1965. *Machon: The Fragments.* Cambridge.
Marcellinus *De Pulsibus* = Schöne, H. 1907. "Marcellinus' Pulslehre: Ein griechisches Anekdoton" in *Festschrift zur 49. Versammlung deutscher Philologen und Schulmänner.* Basel. 448–472.
Marcian of Heraclea Pontica = Müller, K. 1882. *Geographi Graeci Minores.* 2 vols. Parisiis.
Martianus Capella = Willis, J. 1983. *Martianus Capella: De nuptiis Philologiae et Mercurii.* Lipsiae.
Mnesitheus of Athens = Bertier, J. 1972. *Mnésithée et Dieuchès.* Leiden.
Neoptolemus = Mette, Hans Joachim. 1980. "Neoptolemos von Parion," *Rheinisches Museum für Philologie* 123.1: 1–24.
Nicomachus *Arithmetica* = Hoche, R. 1866. *Nicomachi Geraseni Pythagorei introductionis arithmeticae libri ii.* Lipsiae.
Nicomachus *Harmonica* = Jan, C. 1895. *Musici Scriptores Graeci.* Lipsiae.
Oribasius *Eclogae Medicae* = Raeder, J. 1933. *Oribasii Collectionum medicarum reliquiae. Vol. IV. Libros XLIX– L. Libros Incertos. Eclogas Medicamentorum.* Corpus Medicorum Graecorum VI 2,2. Berolini.
Oribasius *Collectiones Medicae* = Raeder, J. 1933. *Oribasii Collectionum medicarum reliquiae. Vol. IV. Libros XLIX– L. Libros Incertos. Eclogas Medicamentorum.* Corpus Medicorum Graecorum VI 2,2. Berolini.
Pappus *Collectio* = Hultsch, F. 1876–1878. *Pappi Alexandrini Collectionis quae supersunt.* Berolini.
PGR = Giannini, Alexander. 1965. *Paradoxographorum Graecorum Reliquiae.* Milano.
Philodemus *On Poems* 5 = Mangoni, C. 1993. *Filodemo: Il Quinto Libro Della Poetica. PHerc. 1425 e 1538.* Napoli.
Philodemus *De libertate dicendi* = Konstan, D. et al. 1998. *Philodemus: On Frank Criticism.* Atlanta.
Philon *Belopoeica* = Marsden, E. W. 1971. *Greek and Roman Artillery: Technical Treatises.* Oxford.
Philostephanus = Capel Badino, Roberto. 2010. *Filostefano di Cirene. Testimonianze e frammenti.* Milano.
Pindar *Olympia* = Snell, B. and H. Maehler. 1987. *Pindari Carmina.* Stuttgart.
Plato *Phaedrus* = Burnet, J. 1910. *Platonis Opera. Tomus II.* Oxonii.
Plato *Respublica* = Burnet, J. 1910. *Platonis Opera. Tomus IV.* Oxonii.
Pliny *Historia Naturalis* = Mayhoff, C. 1875–1906. *C. Plini Secundi Naturalis Historiae Libri XXXVII.* Lipsiae.
Plutarch *Agis et Cleomenes* = Ziegler, K. and H. Gärtner. 1996. *Plutarchii Vitae Parallelae. Vol. III. Fasc. 1.* Stutgardiae.
Plutarch *Alexander* = Ziegler, K. and H. Gärtner. 1994. *Plutarchii Vitae Parallelae. Vol. II. Fasc. 2.* Stutgardiae.

Plutarch *Amatorius* = Minar, E.L, F. H. Sandbach, W. C. Helmbold. 1961. *Plutarch's Moralia. Vol. IX.* Loeb series. Cambridge MA.
Plutarch *An Seni Respublica Gerenda Sit* = Fowler, H. N. 1936. *Plutarch's Moralia. Vol. X.* Loeb series. Cambridge MA.
Plutarch *Antonius* = Ziegler, K. and H. Gärtner. 1996. *Plutarchii Vitae Parallelae. Vol. III. Fasc. 1.* Stutgardiae.
Plutarch *Aratus* = Ziegler, K. and H. Gärtner. 1996. *Plutarchii Vitae Parallelae. Vol. III. Fasc. 1.* Stutgardiae.
Plutarch *Demetrius* = Ziegler, K. and H. Gärtner. 1996. *Plutarchii Vitae Parallelae. Vol. III. Fasc. 1.* Stutgardiae.
Plutarch *Marcellus* = Ziegler, K. and H. Gärtner. 1994. *Plutarchii Vitae Parallelae. Vol. II. Fasc. 2.* Stutgardiae.
Plutarch *Non Posse Suaviter Vivi secundum Epicurum* = Einarson, B. and P. H. Lacy, De. 1967. *Plutarch's Moralia. Vol. XIV.* Loeb series. Cambridge MA.
Plutarch *Quomodo Adulator ab Amico Internoscatur* = Babbitt, F. C. 1927. *Plutarch's Moralia. Vol. 1.* Loeb series. Cambridge MA.
Plutarch *Regum et Imperatorum Apophthegmata* = Babbitt, F. C. 1931. *Plutarch's Moralia. Vol. III.* Loeb series. Cambridge MA.
Polybius = Büttner-Wobst, T. 1882–1904. *Polybii Historiae.* 4 vols. Lipsiae.
Porphyrio *Commentum in Horatii Flacci Artem Poeticam* = Holder, A. 1894. *Commentum in Hortium Flaccum.* Ad Aeni Pontem.
Proclus *In Primum Euclidis Elementorum Librum* = Friedlein, G. 1873. *Procli Diadochi In Primum Euclidis Elementorum Librum Commentarii.* Lipsiae.
Ps.-Aristeas = Hadas, M. 1951. *Aristeas to Philocrates.* New York.
Ps.- Dioscorides *De Iis Quae Virus Eiaculantur Animalibus* = Sprengel, C. 1830. *Pedanii Anazarbei Tomus Secundus. Medicorum Graecorum Opera quae exstant. Volumen XXVI.* Lipsiae.
Ps.-Scymnus *Periegesis ad Nicomedem* = Marcotte, D. 2000. *Géographes grecs. Tome 1. Introduction générale. Ps.-Scymnus: Circuit de la Terre.* Les Belles Lettres. Paris.
Ptolemy *Harmonica* = Düring, I. 1930. *Die Harmonielehre des Klaudios Ptolemaios.* Göteborg.
Ptolemy *Inscriptio Canobi* = Heiberg, J. L. 1907. *Claudii Ptolemaei Opera quae exstant omnia. Volumen II: Opera Astronomica Minora.* Lipsiae.
Ptolemy *Phaseis* = Heiberg, J. L. 1907. *Claudii Ptolemaei Opera quae exstant omnia. Volumen II: Opera Astronomica Minora.* Lipsiae.
Ptolemy *Syntaxis Mathematike* = Heiberg, J. L. 1907. *Claudii Ptolemaei Opera quae exstant omnia. Volumen I: Syntaxis Mathematica.* 2 vols. Lipsiae.
Quintilian *Institutio Oratoria* = Harder, Annette. 2012. *Callimachus: Aetia.* Oxford. 1: 228.
Rufus of Ephesus *Synopsis de pulsibus* = Daremberg, C. and É. Ruelle. 1879. *Œuvres de Rufus d'Éphèse.* Paris.
Σ.Aratum = Martin, J. 1974. *Scholia in Aratum Vetera.* Stutgardiae.
Σ.Aristophanem *Thersmophoriazusae* = Regtuit, R. F. 2007. *Scholia in Aristophanem. Pars III. Fasciculus 2/3 Continens Scholia in Aristophanis Thesmophoriazusas et Ecclesiazusas.* Groningen.
Σ.Aristophanem *Aves* = Holwerda, D. 1991. *Scholia in Aristophanem. Pars II. Fasciculus III Continens Scholia Vetera et Recentiora in Aristophanis Aves.* Groningen.
Σ.Dionysium Periegetem = Müller, C. 1861. *Geographi Graeci Minores. Volumen Secundum.* Parisiis.
Σ.Iliadem (*Scholia Didymi*) = van Thiel, H. 2014. *Scholia D in Iliadem. Proecdosis Aucta et Correctior.* Köln. <http://kups.ub.uni-koeln.de/id/eprint/5586>
Σ.Pindarum = Drachman, A. B. 1903. *Scholia Vetera in Pindari Carmina. Vol. I: Scholia in Olympionicas.* Lipsiae.

Σ.Nicandrum *Theriaca* = Crugnola, A. 1971. *Scholia in Nicandri Theriaka. Cum glossis*. Milano.
Σ.Apollonium = Wendel, C. 1935. *Scholia in Apollonium Rhodium Vetera*. Berolini.
SH = Lloyd-Jones, H. and P. Parsons. 1983. *Supplementum Hellenisticum*. Berlin.
Soranus *Gynaecia* = Ilberg, J. 1927. *Sorani Gynaeciorum Libri IV; De Signis Fracturarum; De Fasciis; Vita Hippocratis secundum Soranum*. Corpus Medicorum Graecorum IV. Berolini.
Soranus *Vita Hippocratis* = Ilberg, J. 1927. *Sorani Gynaeciorum Libri IV; De Signis Fracturarum; De Fasciis; Vita Hippocratis secundum Soranum*. Corpus Medicorum Graecorum IV. Berolini.
Sotades *CA* = Powell, J. U. 1925. *Collectanea Alexandrina*. Oxonii.
Stephanus of Byzantium = Billerbeck, M. 2014. *Stephani Byzantii Ethnica. Volumen III: K– O*. Berolini.
Stobaeus = C. Washsmuth, C. and O. Hense. 1884–1912. *Ioannis Stobaei Anthologium*. Berolini.
Strabo = Radt, S. 2002–2011. *Strabons Geographika*. 10 vols. Göttingen.
Strato = Desclos, Marie-Laurence and William Fortenbaugh. 2011. *Strato of Lampsacus. Text, Translation, and Discussion*. Rutgers.
Suda = Adler, A. 1928–1938. *Suidae Lexicon*. 5 vol. Lipsiae.
Syncellus, Georgius *Ecloga Chronographica* = Mosshamer, A. A. 1984. *Georgius Syncellus: Ecloga Chronographica*. Leipzig.
Teles *De Fuga* = Fuentes González, P. P. 1998. *Les diatribes de Télès. Introduction, texte revu, traduction et commentaire des fragments*. Paris.
Theocritus = Gow, A. S. F. 1952. *Bucolici Graeci*. Oxonii.
Theophilus *Ad Autolycum* = Grant, R. M. 1970. *Theophilus of Antioch: Ad Autolycum*. Oxford.
Theophrastus *Characteres* = Diggle, J. 2004. *Theophrastus: Characters*. Cambridge.
Vita Arati 1 = Martin, J. 1974. *Scholia in Aratum Vetera*. Stutgardiae.
Vita Arati 3 = Martin, J. 1974. *Scholia in Aratum Vetera*. Stutgardiae.
Vitruvius = Krohn, F. 1912. *Vitruvii De Architecura Libri Decem*. Lipsiae.
Zenobius = Leutsch, E. L. and F. G. Schneidewin. 1839. *Paroemiographi Graeci*. 2 vols. Gottingae.
Zopyrus = Deichgräber, K. 1965. *Die Griechische Empirikerschule: Sammlung der Fragmente und Darstellung der Lehre*. 2nd ed. Berlin.

Documentary Texts: Inscriptions, Numismatics, Papyri

Reference numbers and abbreviations for papyri follow Oates, J. F. et al. 2017. *Checklist of Greek, Latin, Demotic, and Coptic Papyri, Ostraca, and Tablets*. <http://library.duke.edu/rubenstein/scriptorium/papyrus/texts/clist.html>. Accessed 4.9.2017. Remaining documentary texts are as follows:

ABSA 56: 15 no. 39 = Mitford, T. B. 1961. "The Hellenistic Inscriptions of Old Paphos," *Annual of the British School at Athens* 56: 1–41.
Brett 1952: 8 no. 29 = Brett, A. B. 1952. "The Behna Horde of Ptolemaic Coins," *American Numismatic Society Museum Notes* 5: 1–8.
Hadra vase Metropolitan Museum accession number 90.9.37 = Cook, Brian F. 1966. *Inscribed Hadra Vases in the Metropolitan Museum of Art*. New York.
IG = *Inscriptiones Graecae*. Berlin.
OGIS = Dittenberger, W. 1903–1905. *Orientis Graeci Inscriptiones Selectae*. 2 vols. Lipsiae.
P.Rhind = Peet, T. E. 1923. *The Rhind Mathematical Papyrus: British Museum 10057 and 10058*. London.
Prose = Bernard, André. 1992. *La Prose sur Pierre dans l'Égypte Hellénistique et Romaine. Tome 1: Textes et Traductions*. Paris.
Pyramid Texts = Allen, James P. 2015. *The Ancient Egyptian Pyramid Texts*. 2nd ed. Atlanta.

RC = Welles, C. B. 1934. *Royal Correspondence in the Hellenistic Period: A Study in Greek Epigraphy*. New Haven.

Rekhmire, Tomb of = Breasted, J. H. 1906. *Ancient Records of Egypt: Volume 2: The Eighteenth Dynasty*. New York.

Samama = Samama, Évelyne. 2003. *Les médecins dans le monde grec: sources épigraphiques sur la naissance d'un corps médical*. Genève.

SEG = *Supplementum Epigraphicum Graecum*. Leiden.

Syll.[3] = Dittenberger, W. and F. F. H. von Gaetringen. 1915–1924. *Sylloge Inscriptionum Graecarum*. 4 vols. Leipzig.

Udjahorresne, naophorous statue of = Lichtheim, Miriam. 2006. *Ancient Egyptian Literature: The Late Period. With a New Forward by Joseph G. Manning*. Berkeley.

Bibliography

Acosta-Hughes, Benjamin and Susan Stephens. 2011. *Callimachus in Context: From Plato to the Augustan Poets*. Cambridge.
Alexander, Jennifer Karns. 2012. "Thinking Again about Science in Technology," *Isis* 103.3: 518–526.
Andersen, Casper, Jakob Bek-Thomsen and Peter C. Kjærgaard. 2012. "The Money Trail: A New Historiography for Networks, Patronage, and Scientific Careers," *Isis* 103.2: 310–315.
Asmis, Elizabeth. 1992. "Neoptolemus and the Classification of Poetry," *Classical Philology* 87.3: 206–231.
Bagnall, Roger S. 1976. *Administration of the Ptolemaic Possessions Outside of Egypt*. Leiden.
Bagnall, Roger S. 2002. "Alexandria: Library of Dreams," *Proceedings of the American Philosophical Society* 146.4: 348–362.
Baker, Patricia. 2013. *The Archaeology of Medicine in the Greco-Roman World*. Cambridge.
Barton, Tamsyn. 1994. *Power and Knowledge: Astrology, Physiognomics, and Medicine under the Roman Empire*. Ann Arbor.
Benedetto, Giovanni. 2011. "Callimachus and the Attidographers" in *The Brill Companion to Callimachus*. B. Acosta-Hughes, L. Lehnus, S. Stephens eds. Leiden. 349–367.
Bergmann, Marianne. 2006. "Philosophers and Poets in the Sarapieion at Memphis" in *Early Hellenistic Portraiture: Image, Style, Context*. P. Schutz and R. von den Hoff eds. Cambridge. 246–263.
Bernabo, Massimo ed. 2010. *La collezione di testi chirurgici di Niceta*. Roma.
Berrey, Marquis. 2014a. "Early Empiricism, Therapeutic Motivation, and the Asymmetrical Dispute Between the Hellenistic Medical Sects," *Apeiron* 47.2: 141–171.
Berrey, Marquis. 2014b. "Chrysippus of Cnidus: Medical Doxography and Hellenistic Monarchies," *Greek, Roman, and Byzantine Studies* 54.3: 420–443.
Berrey, Marquis. 2015. "Reading Communities and Hippocratism in Hellenistic Medicine," *Science in Context* 28.3: 465–487.
Berrey, Marquis. 2017. "Technology, Performance, Loss: Reconstructing Andreas of Carystus' Surgical Machine," in *Fragments, Holes, and Wholes: Reconstructing the Ancient World in Theory and Practice*. T. Derda, J. Kilder, J. Kwapisz eds. Journal of Juristic Papyrology Supplement XXX. Warsaw. 273–289.
Berryman, Sylvia. 2009. *The Mechanical Hypothesis in Ancient Greek Natural Philosophy*. Cambridge.
Berti, Monica. 2009a. *Istro il Callimacheo. Vol. 1. Testimonianze e frammenti su Atene e sull'Attica*. Roma.
Berti, Monica. 2009b. "Istro e la tradizione dei rapporti fra la Grecia e l'Egitto. Note a *FGrHist* 334 FF43–47" in *Tradizione e trasmissione degli storici greci frammentari in ricordo di Silvio Accame: Atti del II Workshop Internazionale, Roma, 16–18 Febbraio 2006*. E. Lanzillota, V. Costa, G. Ottone eds. Roma. 483–497.
Biagioli, Mario. 1993. *Galileo, Courtier: The Practice of Science in the Culture of Absolutism*. Chicago.
Bikerman, Elias J. 1938. *Institutions des Séleucides*. Paris.
Bingen, Jean. 2007. *Hellenistic Egypt: Monarchy, Society, Economy, Culture. Edited and With an Introduction by Roger S. Bagnall*. Berkeley.
Bollansée, Jan. 1999a. *Hermippos of Smyrna and his Biographical Writings. A Reappraisal*. Leuven.
Bollansée, Jan. 2005. "Historians of Agathokles of Samos: Polybius on Writers of Historical Monographs" in *The Shadow of Polybius: Intertextuality as a Research Tool in Greek Historiography. Proceedings of the International Colloquium, Leuven 21–22 September 2001*. J. Bollansée, G. Schepens eds. Leuven. 237–253.

Borchardt, Ludwig. 1920. *Die Altägyptische Zeitmessung*. München.
Bos, Henk. 2001. *Redefining Geometrical Exactness: Descartes' Transformation of the Early Modern Concept of Construction*. Heidelberg.
Bravo, Benedetto. 2007. "Antiquarianism and History" in *A Companion to Greek and Roman Historiography*. 2 vols. J. Marincola ed. Malden, MA. 2: 515–527.
Calà, Irene. 2012. "Il medico Andreas nei *Libri Medicinales* di Aezio Amideno," *Galenos* 6: 53–64.
Cameron, Alan. 1995. *Callimachus and his Critics*. Princeton.
Capel Badino, Roberto. 2010. *Filostefano di Cirene. Testimonianze e frammenti*. Milano.
Chemla, Karine. 2012. "History and Historiography of Mathematical Proof: A Research Programme" in *The History of Mathematical Proof in Ancient Traditions*. K. Chemla ed. Cambridge. 1–68.
Clarysse, Willy and G. van Derveken. 1983. *The Eponymous Priests of Ptolemaic Egypt (P. L. Bat. 24): Chronological Lists of the Priests of Alexandria and Ptolemais with a Study of the Demotic Transcriptions of Their Names*. Leiden.
Clayman, Dee. 2014. *Berenice II and the Golden Age of Ptolemaic Egypt*. Oxford.
Cotterell, B. et al. 1986. "Ancient Egyptian Water-clocks: A Reappraisal," *Journal of Archaeological Science* 13: 31–50.
'courtier', n. *Oxford English Dictionary Online*. Oxford. Accessed 4.15.2015.
Crook, Zeba. 2013. "Fictive Giftship and Fictive Friendship in Greco-Roman Society" in *The Gift in Antiquity*. M. L. Satlow ed. Malden MA. 61–76.
Cuomo, Serafina. 2001. *Ancient Mathematics*. New York.
Cuomo, Serafina. 2007. *Technology and Culture in Greek and Roman Antiquity*. Cambridge.
Dalby, Andrew. 2000. "Lynceus and the Anecdotists" in *Athenaeus and his World: Reading Greek Culture in the Roman Empire*. D. Braund, J. Wilkins eds. Exeter. 372–394.
Daston, Lorraine. 1998. "The Nature of Nature in Early Modern Europe," *Configurations* 6.2: 149–172.
Daston, Lorraine. 2009. "Science Studies and the History of Science," *Critical Inquiry* 35: 798–813.
Daston, Lorraine and Peter Galison. 2007. *Objectivity*. New York.
Deichgräber, Karl. 1984. "Galen als Erforscher des menschlichen Puls" in *Augewählte Kleine Schriften*. Hildesheim. 288–326.
Dieleman, Jacco and Ian Moyer. 2010. "Egyptian Literature" in *A Companion to Hellenistic Literature*. J. J. Clauss, M. Cuypers eds. Oxford. 429–447.
Dijksterhuis, E. J. 1987. *Archimedes. With a New Bibliographic Essay by Wilbur R. Knorr*. Princeton.
Drachmann, A. G. 1948. *Ktesibios, Philon and Heron: A Study in Ancient Pneumatics*. Copenhagen.
Duindam, Jeroen. 2011. "Royal Courts in Dynastic Empires and States" in *Royal Courts in Dynastic States and Empires: A Global Perspective*. J. Duindam, T. Artan, M. Kunt eds. Leiden. 1–23.
EANS = Keyser, P. and G. Irby-Massie eds. 2008. *Encyclopedia of Ancient Natural Scientists*. Routledge.
Edgar, Campbell Cowan. 1931. *Zenon Papyri in the University of Michigan Collection*. Ann Arbor.
van der Eijk, Philip. 1999. "Antiquarianism and Criticism: Forms and Functions of Medical Doxography in Methodism (Soranus, Caelius Aurelianus)" in *Ancient Histories of Medicine: Essays in Medical Doxography and Historiography in Classical Antiquity*. Ph. van der Eijk ed. Leiden. 397–451.
van der Eijk, Philip. 2001. *Diocles of Carystus: A Collection of the Fragments with Translation and Commentary: Vol. 2 Commentary*. Leiden.
van der Eijk, Philip. 2004. "Divination, Prognosis and Prophylaxis: the Hippocratic work 'On Dreams' (De Victu 4) and its Near Eastern Background" in *Magic and Rationality in Near Eastern and Graeco-Roman Medicine*. H. F. J. Horstmanshoff and M. Stol eds. Leiden. 187–218.

Engberg-Pedersen, Troels. 1996. "Plutarch to Prince Philopappus on How to Tell a Flatterer from a Friend" in *Friendship, Flattery, and Frankness of Speech: Studies on Friendship in the New Testament World*. J. Fitzgerald *ed*. Leiden. 61–79.
Esposito, Elena. 2010. "Herodas and the Mime" in *A Companion to Hellenistic Literature*. J. J. Clauss, M. Cuypers *eds*. Oxford. 267–281.
Fantuzzi, Mario and Richard Hunter. 2004. *Innovation and Tradition in Hellenistic Poetry*. Cambridge.
Faraone, Christopher. 1999. *Ancient Greek Love Magic*. Cambridge MA.
Fitzgerald, Des and Felicity Callard. 2015. "Social Science and Neuroscience beyond Interdisciplinarity: Experimental Entanglements," *Theory, Culture, & Society* 32.1: 3–32.
Flemming, Rebecca. 2000. *Medicine and the Making of Roman Women: Gender, Nature, and Authority from Celsus to Galen*. Oxford.
Flemming, Rebecca. 2003. "Empires of knowledge: medicine and health in the Hellenistic world" in *The Blackwell Companion to the Hellenistic World*. A. Erskine *ed*. Oxford. 449–463.
Forman, Paul. 2007. "The Primacy of Science in Modernity, of Technology in Postmodernity, and of Ideology in the History of Technology," *Technology and History* 23.1/2: 1–153.
Forman, Paul. 2012. "On Historical Forms of Knowledge Production and Curation: Modernity Entailed Disciplinarity; Postmodernity Entails Antidisciplinarity," *Osiris* 27: 56–97.
Fortenbaugh, William. 2011. *Theophrastus of Eresus. Commentary Vol. 6.1: Sources on Ethics*. Leiden.
Foucault, Michel. 1977. *Discipline and Punish: The Birth of the Prison*. A. Sheridan *trans*. New York.
Frankfort, Henri. 1948. *Kingship and the Gods: A Study of Ancient Near Eastern Religion as the Integration of Society and Nature*. Chicago.
Fraser, P. M. 1972. *Ptolemaic Alexandria*. 3 vols. Oxford.
French, Roger. 1994. "General Series Introduction" in *Ancient Astrology*. T. Barton. New York. x–xxiv.
Geertz, Clifford. 1973. *The Interpretation of Cultures: Selected Essays*. New York.
Geertz, Clifford. 1983. *Local Knowledge: Further Essays in Interpretative Anthropology*. New York.
Gehrke, Hans-Joachim. 2013. "The Victorious King: Reflections on Hellenistic Monarchy" in *The Splendors and Miseries of Ruling Alone: Encounters with Monarchy from Archaic Greece to the Hellenistic Mediterranean*. N. Luraghi *ed*. Stuttgart. 73–99.
Geus, Klaus. 2002. *Eratosthenes von Kyrene: Studien zur Hellenistischen Kultur- und Wissenschaftsgeschichte*. München.
Gibson, Sophie. 2005. *Aristoxenus of Tarentum and the Birth of Musicology*. New York.
Gleason, Maud. 2009. "Shock and Awe: The Performance Dimension of Galen's Anatomy Demonstrations" in *Galen and the World of Knowledge*. C. Gill, T. Whitmarsh, J. Wilkins *eds*. Cambridge. 85–114.
Goddio, Frank. 2011. *The Topography and Excavation of Heracleion-Thonis and East Canopus (1996–2006)*. Oxford.
Goodenough, Erwin. 1928. "The Political Philosophy of Hellenistic Kingship," *Yale Classical Studies* 1: 55–104.
Green, Peter. 1993. *Alexander to Actium: The Historical Evolution of the Hellenistic Age*. 2nd ed. Berkeley.
Gutzwiller, Kathryn. 2010. "Literary Criticism" in *A Companion to Hellenistic Literature*. J. J. Clauss, M. Cuypers *eds*. Oxford. 337–365.
Haake, Matthias. 2003. "Warum und zu welchem Ende schreibt man *peri basileias*? Überlegungen zum historischen Kontext einer literarischen Gattung im Hellenismus" in *Philosophie und Lebenswelt in der Antike*. K. Piepenbrink *ed*. Darmstadt. 83–138.
Haake, Matthias. 2013. "Writing Down the King: The communicative function of treatises *On Kingship* in the Hellenistic World" in *The Splendors and Miseries of Ruling Alone: Encounters*

with Monarchy from Archaic Greek to the Hellenistic Mediterranean. N. Luraghi *ed*. Stuttgart. 165–206.

Habicht, Christian. 2006. "The Ruling Class of the Hellenistic Monarchies" in *The Hellenistic Monarchies: Selected Papers*. Ann Arbor. 26–40.

Hacking, Ian. 1999. *The Social Construction of What?* Cambridge MA.

Hannah, Robert. 2009. *Time in Antiquity*. New York.

Harder, Annette. 2012. *Callimachus: Aetia*. 2 vols. Oxford.

Hazzard, R. A. 2000. *Imagination of a Monarchy: Studies in Ptolemaic Propaganda*. Toronto.

Hellmann, Oliver. 2006. "Peripatetic Biology and the Epitome of Aristophanes of Byzantium" in *Aristo of Ceos: Text, Translation, and Discussion*. W. Fortenbaugh, S. White *eds*. New Brunswick. 329–359.

Herman, Gabriel. 1980/1981. "The 'Friends' of the Early Hellenistic Rulers: Servants or Officials?" *Talanta* 12/13: 103–149.

Herman, Gabriel. 1987. *Ritualised Friendship and the Greek City*. Cambridge.

Herman, Gabriel. 1997. "The Court Society of the Hellenistic Age" in *Hellenistic Constructs: Essays in Culture, History, and Historiography*. P. Cartledge, P. Garnsey, E. Gruen *eds*. Berkeley. 199–224.

Hölbl, Günther. 2001. *A History of the Ptolemaic Empire*. T. Saaverda *trans*. New York.

Hunt, A. S. and G. J. Smyly. 1933. *The Tebtunis Papyri. Volume 3*. London.

Hunter, Richard. 2003. *Theocritus: Encomium of Ptolemy Philadelphus*. Berkeley.

IJSewijn, Jozef. 1961. *De Sacerdotibus Sacerdotiisque Alexandri Magni et Lagidarum Eponymis*. Brussel.

Imhausen, Annette. 2003. "Egyptian Mathematical Texts and Their Contexts," *Science in Context* 16.3: 367–389.

Imhausen, Annette. 2010a. "From the cave into reality: Mathematics and cultures" in *Writings of Early Scholars in the Ancient Near East, Egypt, Rome and Greece. Translating Ancient Scientific Texts*. A. Imhausen, T. Pommerening *eds*. New York. 333–347.

Imhausen, Annette. 2010b. "Die Mathematisierung von Getreide im Alten Ägypten," *Mathematische Semesterberichte* 57: 3–10.

Jacobs, Jerry. 2014. *In Defense of the Disciplines: Interdisciplinarity and Specialization in the Research University*. Chicago.

Jaeger, Mary. 2008. *Archimedes and the Roman Imagination*. Ann Arbor.

Jacques, Jean-Marie. 2002. *Nicandre Œuvres: Les Thériaques; Fragments iologiques antérieurs à Nicandre*. Vol. 2. Paris.

Jouanna, Jacques. 1999. *Hippocrates*. M. B. DeBevoise *trans*. Baltimore.

Jouanna, Jacques. 2012. "Egyptian Medicine and Greek Medicine," in *Greek Medicine from Hippocrates to Galen: Selected Papers*. P. van der Eijk *ed*. Leiden. 3–20.

Kehoe, Dennis. 2010. "The Economy: Graeco-Roman" in *A Companion to Ancient Egypt: Vol. 1*. A. Lloyd *ed*. Oxford. 309–325.

Kerferd, G. B. 1981. *The Sophistic Movement*. Cambridge.

Kidd, Ian. 1989. *Posidonius: II. The Commentary: (i) Testimonia and Fragments 1–149*. Cambridge.

Kim, Jaegwon. 2006. "Emergence: Core Ideas and Issues," *Synthese* 151.3: 547–559.

Kinzel, Katherina. 2012. "Geschichte ohne Kausalität: Abgrenzungsstrategien gegen die Wissenschaftssozologie in zeitgenössischen Ansätzen historischer Epistemologie," *Berichtung zur Wissenschaftsgeschichte* 35: 147–162.

Klein, Julie. 1990. *Interdisciplinarity: History, Theory, Practice*. Detroit.

Knorr, Wilbur Richard. 1978. "Archimedes and the Spirals: The Heuristic Background," *Historia Mathematica* 5: 43–75.

Knorr, Wilbur Richard. 1986. *The Ancient Tradition of Geometric Problems*. Boston.

Knorr, Wilbur Richard. 1989. *Textual Studies in Ancient and Medieval Geometry*. Boston.

Koenen, Ludwig. 1993. "The Ptolemaic King as a Religious Figure" in *Images and Ideologies: Self-definition in the Hellenistic World*. A. Bulloch, E. S. Gruen, A. A. Long, A. Stewart eds. Berkeley. 25–115.

König, Jason and Greg Woolf. 2013. "Encyclopediaism in the Roman Empire" in *Encyclopediaism from Antiquity to the Renaissance*. J. König, G. Woolf eds. Cambridge. 23–63.

Konstan, David. 1996. "Friendship, Frankness and Flattery" in *Friendship, Flattery, and Frankness of Speech: Studies on Friendship in the New Testament World*. J. Fitzgerald ed. Leiden. 5–19.

Konstan, David. 1997. *Friendship in the Classical World*. Cambridge.

Konstan, David et al. 1998. *Philodemus: On Frank Criticism*. Atlanta.

Kroll, Wilhelm. 1924. "Kreuzung der Gattungen" in *Studien zum Verständnis der Römischen Literatur*. Stuttgart. 204–224.

Kuriyama, Shigehisa. 1999. *The Expressiveness of the Body and the Divergence of Greek and Chinese Medicine*. New York.

Kwapisz, Jan. 2014. "Kraters, Myrtle, and Hellenistic Poetry" in *Hellenistic Poetry in Context*. Hellenistica Groningana 20. M. A. Harder, R. F. Regtuit, G. C. Wakker eds. Leuven. 195–215.

Landels, J. G. 1979. "Water-clocks and time measurement in Classical Antiquity," *Endeavour* 3.1: 32–37.

Lang, Philippa. 2009. "Goats and the Sacred Disease in Callimachus' *Acontius and Cydippe*," *Classical Philology* 104.1: 85–90.

Lang, Philippa. 2012. *Medicine and Society in Ptolemaic Egypt*. Leiden.

Langslow, D. R. 2007. "The Epistula in Ancient Scientific and Technical Literature, with Special Reference to Medicine" in *Ancient Letters: Classical and Late Antiquity Epistolography*. R. Morello and A. D. Morrison eds. Oxford. 211–234.

Latour, Bruno. 1987. *Science in Action: How to Follow Scientists and Engineers Through Society*. Cambridge MA.

Latour, Bruno. 2000. "On the Partial Existence of Existing and Non-Existing Objects" in *Biographies of Scientific Objects*. L. Daston ed. Chicago. 247–269.

Latour, Bruno. 2005. *Reassembling the Social*. Oxford.

Latour, Bruno and Steve Woolgar. 1986. *Laboratory Life: The Construction of Scientific Facts*. 2nd ed. Princeton.

Lennox, James. 1985. "Aristotle, Galileo, and the 'Mixed Sciences'" in *Reinterpreting Galileo*. W. Wallace ed. Washington, D. C. 29–51.

Lennox, James. 1998. "Review: Roger French *Ancient Natural History*," *International Journal of the Classical Tradition* 4.3: 470–472.

Leventhal, Max. 2017. "Eratosthenes' Letter to Ptolemy: The Literary Mechanics of Empire," *American Journal of Philology* 138.1: 43–84.

Lewis, Orly. 2015. "Marcellinus' *De pulsibus*: A Neglected Treatise on the Ancient 'Art of the Pulse'," *Scripta Classica Israelica* 34: 195–214.

Lewis, Orly. 2016. "The Practical Application of Ancient Pulse-Lore and its Influence on the Patient-Doctor Interaction" in *Homo Patiens: Approaches to the Patient in the Ancient World*. G. Petridou, C. Thumiger eds. Leiden. 345–364.

Lewis, Orly. 2017. *Praxagoras of Cos on Arteries, Pulse, and Pneuma: Fragments and Interpretation*. Brill.

Lloyd, G. E. R. 1979. *Magic, Reason and Experience: Studies in the Origin and Development of Greek Science*. Cambridge.

Lloyd, G. E. R. 1987. *The Revolutions of Wisdom: Studies in the Claims and Practice of Ancient Greek Science*. Cambridge.

Lloyd, G. E. R. 1990. *Demystifying Mentalities*. Cambridge.

Lloyd, G. E. R. 1991. "The Invention of Nature" in *Methods and Problems in Greek Science*. Cambridge. 417–434.

Lloyd, G. E. R. 2002. *The Ambitions of Curiosity: Understanding the World in Ancient Greece and China*. Cambridge.

LSJ = Liddell, H. G. and R. Scott. 1996. *A Greek-English Lexicon*. 9th ed. Revised by H. S. Jones, supplement edited by P. G. W. Glare. Oxford.

Manning, J. G. 2003. *Land and Power in Ptolemaic Egypt: The Structure of Land Tenure 332–30 BCE*. Cambridge.

Manning, J. G. and Ian Morris. 2005. "The Economic Sociology of the Ancient Mediterranean World" in *The Handbook of Economic Sociology*. 2nd ed. N. Smelser and R. Swedberg. Princeton. 131–159.

Mansfeld, Jaap. 1998. *Prolegomena Mathematica: From Apollonius of Perga to Late Neoplatonism*. Leiden.

Marganne, Maire-Hélène. 2006. "À la recherche de l'oeuvre perdue d'Héliodore" in *Ecdotica e ricezioni dei testi medici greci*. V. Boudon-Millot, A. Garyza, J. Jouanna, A. Roselli eds. Napoli. 67–82.

Marsden, E. W. 1971. *Greek and Roman Artillery: Technical Treatises*. Oxford.

Massar, Natacha. 2005. *Soigner et Servir: Histoire sociale et culturelle de la médecine grecque à l'époque hellénistique*. Paris.

Mauss, Marcel. 1990. *The Gift: The Form and Reason for Exchange in Archaic Societies*. W. D. Halls trans. New York.

Meißner, Burckhardt. 1999. *Die technologische Fachliteratur der Antike. Struktur, Überlieferung und Wirkung technischen Wissens in der Antike (ca. 400 v. Chr. – ca. 500 n. Chr.)*. Berlin.

Mendell, Henry. 2016. "Archimedes: Sand-Reckoner (Translation and Commentary)" <http://web.calstatela.edu/faculty/hmendel/Ancient%20Mathematics/Archimedes/SandReckoner/SandReckoner.html>. Accessed 5.13. 2016.

Mette, Hans Joachim. 1980. "Neoptolemos von Parion," *Rheinisches Museum für Philologie* 123.1: 1–24.

Mol, Annemarie. 2003. *The Body Multiple: Ontology in Medical Practice*. Durham, NC.

Montana, Fausto. 2015. "Hellenistic Scholarship" in *Brill's Companion to Ancient Greek Scholarship*. 2 vols. F. Montanari, S. Matthaios, A. Regnakos eds. Leiden. 60–183.

Montanari, Franco. 1988. *I frammenti dei grammatici Agathokles, Hellanikos, Ptolemaios Epithetes: in appendice i grammatici Theophilos, Anaxagoras, Xenon*. Berlin.

Mooren, Leon. 1975. *The Aulic Titulature in Ptolemaic Egypt: Introduction and Prosopography*. Brussel.

Moyer, Ian. 2011. "Court, *Chora*, and Culture in Late Ptolemaic Egypt," *American Journal of Philology* 132.1: 15–44.

Murray, Jackie. 2012. "Burned After Reading: The So-Called List of Alexandrian Libraries in P.Oxy.X.1241," *Aitia* 2 <http://aitia.revues.org/544>. Accessed 1.25. 2016.

Murray, Jackie. 2014. "Anchored in Time: The Date of Apollonius' *Argonautica*" in *Hellenistic Poetry in Context*. A. Harder ed. Leuven. 247–284.

Murray, Oswyn. 1996. "Hellenistic Royal Symposia" in *Aspects of Hellenistic Kingship*. P. Bilde et al. eds. Aarhus. 15–27.

Murray, Oswyn. 2007. "Philosophy and Monarchy in the Hellenistic World" in *Jewish Perspectives on Hellenistic Rulers*. T. Rajak, S. Pearce, J. K. Aitken, J. Dines eds. Berkeley. 13–28.

Murray, Oswyn. 2008. "Ptolemaic Royal Patronage" in *Ptolemy II Philadelphus and His World*. P. McKenchnie, P. Guillaume. Leiden. 7–24.

Netz, Reviel. 1998. "The First Jewish Scientist?," *Scripta Classica Israelica* 17: 27–33.

Netz, Reviel. 1999. *The Shaping of Deduction in Greek Mathematics: A Study in Cognitive History*. Cambridge.

Netz, Reviel. 2002. "Counter Culture: Towards a History of Greek Numeracy," *History of Science* 40: 321–352.

Netz, Reviel. 2003. "The Goal of Archimedes' *Sand-Reckoner*," *Apeiron* 36.4: 251–290.

Netz, Reviel. 2004. *The Works of Archimedes: Translated into English, together with Eutocius' commentaries, with commentary, and critical edition of the diagrams. Vol. 1: The Two Books On the Sphere and the Cylinder*. Cambridge.

Netz, Reviel. 2009a. *Ludic Proof: Greek Mathematics and the Alexandrian Aesthetic*. Cambridge.

Netz, Reviel. 2009b. "Imagination and Layered Ontology in Greek Mathematics," *Configurations* 17.1–2: 19–50.

Netz, Reviel. 2013. "Authorial Presence in the Ancient Exact Sciences" in *Writing Science: Medical and Mathematical Authorship in Ancient Greece*. M. Asper *ed.* Berlin. 217–253.

Netz, Reviel, Ken Saito, and Natalie Tschernetska. 2001–2002. "A New Reading of *Method* proposition 14: Preliminary Evidence from the Archimedes Palimpsest," *SCIAMVS* 2: 9–29; 3: 109–125.

Neugebauer, Otto and Richard Parker. 1969. *Egyptian Astronomical Texts*. 3 vols. Providence RI.

Nielsen, Inge. 1988. "Royal Banquets: The Development of Royal Banquets and Banqueting Halls from Alexander to the Tetrarchs" in *Meals in a Social Context: Aspects of the Communal Meal in the Hellenistic and Roman World*. I. Nielsen, H. S. Nielsen *eds.* Aarhus. 102–133.

Nussbaum, Martha. 1994. *The Therapy of Desire: Theory and Practice in Hellenistic Ethics*. Princeton.

Nutton, Vivian. 1988. "Archiatri and the Medical Profession in Antiquity" printed as Chapter 5 with original pagination in *From Democedes to Harvey*. London.

Nutton, Vivian. 2012. *Ancient Medicine*. 2nd edition. New York.

O'Conner, Timothy and Hong Y. Wong. 2012. "Emergent Properties," *The Stanford Encyclopedia of Philosophy*. E. N. Zalta *ed.* <http://plato.stanford.edu/archives/spr2012/entries/properties-emergent/>. Accessed 11.7.2013.

Pàmias, Jordi. 2004. "Dionysus and Donkeys on the Streets of Alexandria: Eratosthenes' Criticism of Ptolemaic Ideology," *Harvard Studies in Classical Philology* 102: 191–198.

Pàmias, Jordi and Klaus Geus. 2007. *Eratosthenes Sternsagen (Catasterismi). Griechisch-deutsch; herausgegeben, übersetzt und kommentiert*. Oberhaid.

Pearson, Lionel. 1990. *Aristoxenus Elementa Rhythmica. Aristoxenean Rhythmic Theory*. Oxford.

Peterson, Erik. 1929. "Zum Bedeutungsgeschichte von Παρρησία" in *Reinhold-Seeberg-Festschrift 1: Zur Theorie des Christentums*. W. Koepp *ed.* Leipzig. 283–297.

Pfeiffer, Rudolph. 1968. *The History of Classical Scholarship: From the Beginnings to the End of the Hellenistic Age*. Oxford.

Pfeiffer, Stefan. 2004. *Das Dreket von Kanopus (238 v. Chr.): Kommentar und historische Auswertung eines dreisprachigen Synodaldekretes der ägyptischen Priester zu Ehren Ptolemaios' III und seiner Familie*. München.

Pogo, A. 1936. "Egyptian Water-Clocks," *Isis* 25: 403–425.

Porter, James. 2000. *Nietzsche and the Philology of the Future*. Palo Alto, CA.

PP = Peremans, Willy and E. van't Dack. 1950–1981. *Prosopographia Ptolemaica*. 9 vols. Louvain.

Raaflaub, Kurt. 2004. *The Discovery of Freedom in Ancient Greece*. Chicago.

RE = Pauly, A., G. Wissowa, et al. 1893–1978. *Paulys Realencyclopädie der classischen Altertumswissenschaft: Neue Bearbeitung*. Stuttgart.

Rehm, A. and E. Schramm. 1929. *Bitons Bau von Belagerungsmaschinen und Geschützen*. Abhandlugen der Bayerischen Akademie der Wissenschaften: Philosophisch-historische Abteilung Neue Folge 2. München.

Remijsen, Sofie. 2007. "The Postal Service and the Hour as a Unit of Time in Antiquity," *Historia: Zeitschrift für Alte Geschichte* 56.2: 127–140.

Ribbeck, Otto. 1883. *Kolax: Eine ethologische Studie*. Leipzig.

Ritner, Robert. 1992. "Implicit Models of Cross-Cultural Interaction: A Question of Noses, Soap, and Prejudice" in *Life in a Multi-Cultural Society: Egypt from Cambyses to Constantine and*

Beyond. J. Johnson *ed*. Chicago. 282–290. Restored edition available at <http://oi.uchicago.edu/pdf/Chapter34.pdf>.

Robins, Gay and Charles Schute. 1987. *The Rhind Papyrus: an ancient Egyptian text*. London.

Roby, Courtney. 2016. *Technical Ekphrasis in Greek and Roman Science and Literature: The Written Machine Between Alexandria and Rome*. Cambridge.

Roller, Duane W. 2010. *Eratosthenes' Geography: Fragments collected and translated, with commentary and additional material*. Princeton.

Roselli, Amneris. 1998. "Tra practica medica e filologia ippocratica: il caso della περὶ ἄρθρων πραγματεία di Apollonio di Cizio" in *Sciences exactes et sciences appliquées à Alexandrie*. G. Argoud, J-Y. Guillaumin *eds*. Sainte-Étienne. 217–231.

Roselli, Amneris. 2015. "'According to both Hippocrates and the Truth': Hippocrates as Witness to the Truth, from Apollonius of Citium to Galen" in *Ancient Concepts of the Hippocratic: Papers Presented at the XIIIth International Hippocrates Colloquium. Austin, Texas, August 2008*. L. Dean-Jones, R. Rosen *eds*. Leiden. 331–344.

Saito, Ken and Nathan Sidoli. 2012. "Diagrams and arguments in ancient Greek mathematics: lessons drawn from comparisons of the manuscript diagrams with those in modern critical editions" in *The History of Mathematical Proof in Ancient Tradition*. K. Chemla *ed*. Cambridge. 135–162.

Scarborough, John. 1995. "The Opium Poppy in Hellenistic and Roman Medicine" in *Drugs and Narcotics in History*. R. Porter, M. Teich *eds*. Cambridge. 4–23.

Scarborough, John. 2012. "Pharmacology and Toxicology at the Court of Cleopatra VII: Traces of Three Physicians" in *Herbs and Healers from the Ancient Mediterranean through the Medieval West. Essays in Honor of John M. Riddle*. A. van Arsdall, T. Graham *eds*. Farnham. 7–18.

Schatzberg, Eric. 2012. "From Art to Applied Science," *Isis* 103.2: 555–563.

Scheidel, Walter. 2004. "Creating a metropolis: a comparative demographic perspective" in *Ancient Alexandria between Egypt and Greece*. W. V. Harris, G. Ruffini *eds*. Leiden. 1–31.

Schepens, Guido and Kris Delcroix. 1996. "Ancient paradoxography: origin, evolution, production, reception" in *La Letteratura di Consumo nel Mondo Greco-Latino. Atti del Convengo Internationale: Cassino, 14–17 Settembre 1994*. O. Pecere and A. Stranaglia *eds*. Cassino. 373–460.

Schiefsky, Mark. 2007. "Art and Nature in Ancient Mechanics" in *The Artificial and the Natural: An Evolving Polarity*. B. Bensaude-Vincent, W. R. Newman *eds*. Cambridge MA. 67–108.

Schiefsky, Mark. 2015. "*Technē* and Method in Ancient Artillery Construction: The *Belopoeica* of Philo of Byzantium" in *The Frontiers of Ancient Science. Essays in Honor of Heinrich von Staden*. B. Holmes, K.-D. Fischer *eds*. Berlin. 615–653.

Schlier, H. 1952. "παρρησία, παρρησιάζομαι" in *Theologisches Wörterbuch zum Neuen Testament* 5: 869–884.

Schöne, Hermann. 1896. *Apollonius von Kitium: Illustrierter Kommentar zu der Hippokrateischen ΠΕΡΙ ΑΡΘΡΩΝ*. Leipzig.

'scientist', n. *Oxford English Dictionary Online*. Oxford. Accessed 2. 28. 2016.

Selden, Daniel. 1998. "Alibis," *Classical Antiquity* 17.2: 289–412.

Sidebotham, Steven. 2011. *Berenike and the Ancient Maritime Spice Route*. Berkeley.

Shapin, Steven. 1992. "Discipline and bounding: The history and sociology of science as seen through the externalism-internalism debate," *History of Science* 30: 333–369.

Shapin, Steven. 1994. *A Social History of Truth: Civility and Science in Seventeenth Century England*. Chicago.

Shapin, Steven. 1996. *The Scientific Revolution*. Chicago.

Spawforth, A. J. S. *ed*. 2007. *The Court and Court Society in Ancient Monarchies*. Cambridge.

von Staden, Heinrich. 1982. "Hairesis and Heresy: The case of the *haireseis iatrikai*" in *Jewish and Christian Self-Definition III: Self-Definition in the Greco-Roman world*. B. F. Meyer, E. P. Sanders *eds*. London. 76–100, 199–206.
von Staden, Heinrich. 1989. *Herophilus: The Art of Medicine in Early Alexandria*. Cambridge.
von Staden, Heinrich. 1995. "Anatomy as Rhetoric: Galen on dissection and persuasion," *Journal of the History of Medicine and Allied Sciences* 50: 47–66.
von Staden, Heinrich. 1998. "Andréas de Caryste et Philon de Byzance" in *Sciences exactes et sciences appliquées à Alexandrie*. G. Argoud, J.-Y. Guillaumin *eds*. Saint-Étienne. 147–172.
von Staden, Heinrich. 1999a. "Rupture and Continuity: Hellenistic Reflections on the History of Medicine" in *Ancient Histories of Medicine: Essays in Medical Doxography and Historiography in Classical Antiquity*. P. van der Eijk *ed*. Leiden. 143–188.
von Staden, Heinrich. 1999b. "Celsus as Historian?" in *Ancient Histories of Medicine: Essays in Medical Doxography and Historiography in Classical Antiquity*. P. van der Eijk *ed*. Leiden. 251–294.
Stephens, Susan. 2003. *Seeing Double: Intercultural Poetics in Ptolemaic Alexandria*. Berkeley.
Strecker, Carolus. 1884. *De Lycophrone Euphronio Eratosthene Comicorum Interpretibus*. Greifswald.
Strootman, Rolf. 2011. "Hellenistic Court Society: The Seleukid Imperial Court under Antiochos the Great, 223–187 BCE" in *Royal Courts in Dynastic States and Empires: A Global Perspective*. J. Duindam, T. Artan, M. Kunt *eds*. Leiden. 63–90.
Strootman, Rolf. 2014. *Courts and Elites in Hellenistic Empires: The Ancient Near East After the Achaemenids, c.330–30 BCE*. Edinburgh.
Susemihl, Franz. 1891. *Geschichte der Griechischen Literatur in der Alexandrinerzeit*. Vol. 1. Leipzig.
Taub, Liba. 2008. "'Eratosthenes Sends Greetings to King Ptolemy': Reading the Contents of a 'Mathematical' Letter," *Acta Historica Leopoldina* 54: 285–302.
Taub, Liba. 2013. "On the Variety of 'Genres' of Greek Mathematical Writing: Thinking about Mathematical Texts and Modes of Mathematical Discourse" in *Writing Science: Medical and Mathematical Authorship in Ancient Greece*. M. Asper *ed*. Berlin. 333–366.
'technology', n. *Oxford English Dictionary Online*. Oxford. Accessed 8.14. 2014.
Toomer, G. J. 1976. *Diocles: On Burning Mirrors*. Berlin.
Too, Yun Lee. 2010. *The Idea of the Library in the Ancient World*. Oxford.
Tsouna, Voula. 2007. *The Ethics of Philodemus*. Oxford.
Turner, James. 2014. *Philology: The Forgotten Origins of the Modern Humanities*. Princeton.
Tybjerg, Karin. 2003. "Wonder-making and philosophical wonder in Hero of Alexandria," *Studies in the History and Philosophy of Science, Part A* 34.3: 443–466.
Veyne, Paul. 1990. *Bread and Circuses: Historical Sociology and Political Pluralism*. B. Pierce *trans*. London.
Vitrac, Bernard. 2008. "Ératosthène et la théorie des médiétés" in *Ératosthène: un athlète du savoir*. C. Cusset, H. Frangoulis *eds*. Saint-Étienne. 77–103.
Weber, Gregor. 1993. *Dichtung und höfische Gesellschaft: Die Rezeption von Zeitgeschichte am Hof der ersten drei Ptolemäer*. Stuttgart.
Weingart, Peter. 2010. "A Short History of Knowledge Formations" in *The Oxford Handbook of Interdisciplinarity*. R. Frodeman *ed*. Oxford. 3–14.
Welles, C. Bradford. 1934. *Royal Correspondence in the Hellenistic Period: A Study in Greek Epigraphy*. New Haven.
West, M. L. 1992. *Ancient Greek Music*. Oxford.
Winter, Eric. 1987. "Weitere Beobachtungen zur 'Grammaire du Tempel' in der Griechisch-Römischen Zeit," in *Tempel und Kult*. W. Helck *ed*. Wiesbaden. 61–76.
Woolf, Greg. 2013. "Approaching the Ancient Library" in *Ancient Libraries*. J. König, K. Oikonomopoulu, G. Woolf *eds*. Cambridge. 1–20.

Young, Suzanne. 1939. "An Athenian Clepsydra," *Hesperia* 8: 274–284.

Zhmud, Leonid. 2006. *The Origin of the History of Science in Classical Antiquity*. A. Chernoglazov *trans.* Berlin.

Index Locorum

Literary Texts

Achilles Tatius *In Arati Phaenomena*
19   75.n248, 120.n132
22   60
33   76.n250

Aelian *Varia Historia*
13.22   39.n55

Aetius
2.31.3 (p.362.25–363.4 Diels)   76.n250

Andreas of Carystus
1   80.n274, 187.n84
2a   84, 185.n76, 188.n87
2b   84.n303
5   81.n275
7   81.n277, 82.n287
8   82.n287
9   81.n281, 82, 128.n9, 185.n76
11   81.n282, 183.n72
12   81.n282
13   81.n282, 183.n72
14   81.n282
15   81.n282
16   81.n282
16b   80.n273
17   81.n282, 184
18   81.n281, 81.n282, 83.n293
19   81.n282
21   81.n282, 83.n292
22   81.n275, 81.n282
23   81.n282
24   81.n282, 84.n302
25   83.n295
26   81.n282
27   81.n282
28   81.n282
29   81.n282
30   81.n282
31   81.n282
32   81.n282, 81.n285
33   81.n282
34   81.n282, 81.n284
35   81.n282
36   81.n282
37   81.n282, 81.n283, 185.n76
38   81.n275, 81.n282
39   81.n282
40   81.n282
44   82.n291, 185.n76
45   81.n281, 82.n289, 185.n76
46   82, 185.n76, 188.n88
47   81.n281, 83, 188.n88
47b   80.n273, 84.n302
48   81.n282
49   81.n282

Antigonus of Carystus *De Mirabilibus*
19   186.n79

Apollodorus *Bibliotheca*
3.17–20   135.n29

Apollonius of Citium
10   188.n85
10.3   127
10.3–8   138, 154.n94
10.13–14   138.n40
10.17   155.n97
10.15–12.1   127
12.1–5   150
14.8–11   143–144
16.10–11   94.n20, 155.n96
20.1–5   140.n48
28.1–16   156.n99
38.7–15   141
42.29–44.5   144
74.12–15   142
80.15–16   155.n96
112.7–9   141

Apollonius of Rhodes
pp.4–8   57.n148

Archimedes *Arenarius*
216.1–9   133
216.15–218.7   156
218.5–7   157
218.31–220.7   157.n102
220.20–25   157.n103

222.3–6   148.n72
222.11–14   149
222.16–17   149
226.24   148.n74
226.12–232.26   148.n73
236.3–16   149.n78
236.17–22   130.n14, 153–154
240.19–20   148
256.25–30   147.n70
258.1–5   157.n102
258.5–12   134
258.7–10   157

Archimedes *Quadratura Parabolae*
262.2–13   77.n257
262.19   213.n63
272.11–12   214.n65
274.18–21   214.67

Archimedes *Methodus*
69.col2.1–3 = 426.1–12   210.n55
71.col1.33–71.col2.8 = 428.18–21   74, 153.n90, 211
71.col2.32–73.col1.16 = 430.9–22   212.n60
73.col1.10–16 = 430.19–22   215.n69
73.col2.31–75.col1.8 = 434.3–12   222.n88
75.col2.1–11 = 436.10–17   221.n84
75.col1.8–15 = 434.14–17   219.n79
75.col1.16–24 = 434.20–24   219.n81
75.col1.24–77.col1.17 = 436.1–438.13   220.n83
75.col1.27–31 = 436.2–5   215.n70
75.col2.13–20 = 436.18–23   221.n85
75.col2.33–77.77.col1.17 = 436.30–438.13   223.n91
75.col2.37–77.col1.1 = 438.2   215.n69
77.col1.19–28 = 438.16–21   211.n58
79.col2.9–12 = 442.23–25   216.n71
89.col1.3–13 = 458.1–8   216.n72

Aristaenetus
1.10   235.n26

Aristo of Ceos
21i.11–38   114, 230.n7
21k   230.n7

Aristotle *Ethica Nicomachea*
1107b.21–30   101.n55
1108a26   107.n82

1124b   106.n80
1127b   100.45

Aristotle *Metaphysica*
980a–981b   230.n7
982b   26.n56

Aristotle *Meteorologica*
352b–353a   20.n48

Aristoxenus of Tarentum *Elementa Rhythmica*
2.9–11, 6–8   197

Asclepius of Tralles *In Nicomachi Arithmetica*
1.86   75.n245

Athenaeus
1.4b   115.n113
1.16d–e   113.n109
2.61c   39.n53
4.128c–130d   109.n95
4.142c–f   122.n138
4.144e   46.n89
4.155b   115
5.177b   112.n106
5.189e   29
5.196b–197c   110.n97
5.204c–d   80.n271
5.206e–209b   123.n140
6.246c   53
6.251d   47.n94, 47.n95, 86.n307
6.255d   105
6.255c–257d   106.n81
7.276a–c   40
7.279d   70.n212
7.289c–d   129
7.312e   185.n76
8.364f–365e   115.n113
10.457c–e   111–112, 158.n107
10.488c   112.n105
11.478a   65.n188
12.552c   69.209
13.577a   55.136
13.583a–b   87.n309
14.620d   55.n138
14.621a   116.n118
15.689a   36.n31

Biton
43   160.n117
44.1–3   160

44.3–6   152
44.6   160.n117
48.1–2   145.n64
48   160.n117
48–49.1   153
51.4   145.n64
52.1   160.n117
52–53.2   151
53.4   160.n117
55.1   160.n117
56.6–7   145.n64
57   160.n117
61   160.n117
61.1   145.n64
64.2–3   145.n64
65   160.n117
67.3–4   145.n64
67.5   160.n117
67.5–9   152
68.1   145.n65

BNJ 3 (Pherekydes of Athens)
F59   83.n298

BNJ 45 (Hegesianax of Alexandria Troas)
all   31.n8
T3   32.n11
T4a   115.n115

BNJ 71 (Zoilus of Amphipolis)
all   32.n10

BNJ 74 (Euphantus of Olynthus)
F1A   86.n306

BNJ 160 (Anonymous *Belli Syrii tertii annales*)
all   34.n21

BNJ 161 (Ptolemy of Megalopolis)
all   68.n201
T2   68.n204
T3   68.n205
F2   53

BNJ 162 (Demetrios of Byzantion)
all   68.n206

BNJ 234 (Ptolemy VIII)
F11   39.n53

BNJ 241 (Eratosthenes of Cyrene)
all   75.n242
T1   87.n312, 87.n313, 88.n315
T7   34.n19

BNJ 263 (Bolos of Mendes)
all   88.n317

BNJ 266 (Apollas of Pontus)
F1, 9   67.n199
F6   67.n200

BNJ 334 (Istros)
all   64.n182
F1–10   186.n80
F6   64.n184
F43–47   65.n185

BNJ 339 (Asklepiades of Nikaia)
all   71.n218

BNJ 472 (Agathokles)
all   71.n216

BNJ 627 (Kallixeinos of Rhodes)
F1   67.n196
F2   110.n97

BNJ 631 (Satyros of Alexandria)
F1   63.n176, 74.n232, 88.n316, 95.n24

BNJ 632 (Jason)
F1   55.n138

BNJ 666 (Dalion)
all   87.n311

BNJ 667 (Aristokreon)
all   87.n311

BNJ 668 (Bion Soleus)
all   87.n311

BNJ 669 (Simonides *de Aethiopia*)
all   87.n311

BNJ 718 (Basilis)
all   87.n311

BNJ 783 (Menander)
all   88.n314

Caelius Aurelianus *Celerum Passionum*
2.136 = 236.2   72.n224

Caelius Aurelianus *Tardum Passionum*
2.34 = 564.10   72.n224
3.55 = 710.22   72.n224
5.50 = 884.4–6   128.n9

Callimachus *Aetia*
2   39.n53
43.53   235.n26
54–54a   36.n33
54–60j   57.n149
67–75e   234.n20
67.2   235.n25
67.3   236.n31
75.4   235.n26
75.7   236.n28
75.8   237.n32
75.12–20   234–235
75.13   235.n27
75.15   235.n25
75.30–31   235.n25
75.64–65   235.n27
110   1.n1, 2.n5, 2.n6, 57.n150
178   111.n101

Callimachus *In Apollinem*
61–69   166.n10

Callimachus *In Artemim*
52–54   209.n52

Callimachus *Epigrammata*
3   57.n150
51.3   136.n32
52.3   137.n33

Callimachus *fragmenta* Pfeiffer
203   56.n144
384   45.n84, 46.n89, 57.n150
384a   45.n84, 46.n89
392   116.n118
404, 412–459   57.n151
407–411   57.n152

Callimachus *In Iovem*
55   137.n33
79–90   95.n23

Cassius Iatrosophista *Problemata*
58   81.n275

Celsus *De Medicina*
1.*pr*.8   230.n9
1.*pr*.23   208.n50
5.*pr*.1   81.n277
8.20.4   183.n72

Censorinus *De Die Natali*
18.5   120.n130

Cicero *Tusculanae Disputationes*
5.64–65   167.n12

Clearchus of Soli
17   107.n82
19   105, 106.n81
63   111–112
84–95   112.n105

Clement of Alexandria *Stromateis*
1.14.61.1   66.n194

Quintus Curtius Rufus
3.6   95.n26

Damagetus
1   40.n49

Demetrius of Phalerum
63   116.n117

Demetrius of Scepsis
7   115.n114

Descartes, René *La Géométrie*
2   176.n46

Dioscorides of Nicopolis *Epigrammata*
2   55.n139, 186.n81
7   59.n155
14   58.n153
16   59.n155
17–24   186.n80
24   58
33   58.n153, 59
34   58.n153
35   186.n81
36   186.n81

37 58.n153
38 59.n155
39 58.n153

Dioscorides of Anazarbus *De Materia Medica*
3.126–127 81.n283, 185.n76

Diphilus
97 29

Diocles *On Burning Mirrors*
3–6 78.n259

Diodorus Siculus
1.1.1 98.n39
1.70.2 97.n37
1.70.4 103.n66
1.70.6 98, 100.n46, 159.n112
1.70.9 98
3.18.4 42.n69, 80.n270
17.31.4–7 95.n26

Diogenes Laertius
3.46 70.n212
5.47 96.n30
5.49 96.n30
5.60 66.n193
7.177 70, 187.n83
7.178 69.n210
7.186 29.n4
8.88 84.n300

Epicurus
5 112
20 112
56 112.n106

Erasistratus
35 85.n304
268 128.n9

Eratosthenes *Geographica*
2 61.n166
6 60.n164
8 76.n252
15 19, 77.n255

Etymologicum Magnum
βιβλιαίγισθος 84, 185.n76, 188.n87
γάλλος 63.n178

Euclid *Elementa*
1.8 216.n71
13 131.n16

Euphronius
pp.176–177 *CA* 63

Eusebius *Chronicon* I
p.251.26 35.n26

Eusebius *Preparatio Evangelica*
15.53.3 76.n250

Eutocius *In Archimedis de Sphaera et Cylindro Libros II*
78.13–80.24 175.n39
88.4 153
88.5–13 135, 171
88.11–13 174.n33
90.4–11 159.n114
90.8–11 171, 174.n33
90.11–13 153, 174.n34
90.13–27 166
90.14 174.n34
90.16 175.n39
90.17–26 174.n36
90.18 174.n34
90.21 174.n34
90.27 174.n34
92.27–94.3 176
94.4–7 158.n110, 178
94.8–14 146
96.6–8 176.n46
96.10–13 136.n30, 174.n34, 174.n36
96.10–15 166
96.10–27 172
96.14–15 177.n48
96.16–19 159.n114
96.20 165.n8, 178.n52
96.20–21 177.47
96.21 174.n34
96.22–27 136, 158–159
96.27 147.n69
98.5–7 231.n14

FGrH 63 (Euhemeros von Messene)
all 98.n41

FGrH 138 (Ptolemaios Lagu)
all 34.n21

FGrH 585 (Sphairos der Borsythenite)
T1, T3a   69.n210

FGrH 1026 (Hermippos of Smyrna)
all   65.n187
F1, 9, 10, 13, 17, 21–27   84.n300
F90–91   65.n188

Galen *De Antidotis*
14.150K   128.n9

Galen *De Compositione Medicamentorum secundum Locos*
12.765K   81.n285

Galen *De Differentiis Pulsibus*
8.501K   196.n16
8.726–728K   231.n16

Galen *De Dignoscendis Pulsibus*
8.853K   195.n14
8.959–860K   194
8.869K   195.n14
8.913K   198.n24, 200.n30
8.915K   199.25

Galen *De Experientia Medica*
109   231.n16

Galen *Glossarium*
9.105K   81.n275

Galen *In Hippocratis Epidemiarum Librum III*
17A.606–607K = 79.8–22   35.n24

[Galen] *In Hippocratis De Humoribus*
all   76.n250

Galen *In Hippocratis Epidemiarum Librum VI*
17B.145K = 203.18–26   84.n299

Galen *In Hippocratis De Articulis*
18A.338–339K   183.n72

[Galen] *Historia Philosophia*
72   76.n250

Galen *De Pulsibus ad Tirones*
8.454K   196.n17

Galen *De Simplicium Medicamentorum Temperamentis ac Facultatibus*
11.795–796K   84.n302

Galen *Subfiguratio Empirica*
8 = Empiricist fr. 10b, p.69.10–20   80.n273, 84.n302

Galen *Synopsis de Pulsibus*
9.453K   195.n14
9.463–465K   199

Geminus *Isagoge*
8.24   75.n248, 120

Hedylus *Epigrammata*
4   79.n268, 209.n52

Hegesander
5   129

Heraclitus
B40   230.n7

Herodas *Mimiambi*
1.26–32   18

Herodotus
1.47   134.n20
2.20   64.n184
2.109   173.n32

Herophilus
157   195.n13
158   195.n13
162.77–86   194
163a   195.n11
163b   195.n11
169–171   207.n48
174   198.n24, 199.n26, 200.n30
175   208.n49
177   195.n12, 200.n29, 200.n30
178   198.n23, 208.n49
180   195.n14
181   195.n14
182   202
183   195.n13, 197.n18, 198.n23, 199, 199.n28
184   195.n14
186   195.n15

Hesiod *Theogonia*
96 95

Hesiod *Opera et Dies*
38–39 95
639–640 59

Hippocrates *De Articulis*
19 = 4.132L 145
72 = 4.296–300L 142, 143.n56, 182.n67, 182.n68

Hippocrates *De Fracturis*
13 = 3.466L 182.n67
30 = 3.524L 182.n67

Hippocrates *De Medicina Vetera*
3 = 1.574–78L 230.n8

Hippocrates *De Morbo Sacro*
1.1–4 = 6.352–354L 236.n30
11.3–4 = 6.382L 236.n29

Homer *Ilias*
2.214 95
14.214 235.n26

Hyginus *De Astronomia*
2.14, 578–584 135.n28
2.24, 1003–1007 2

Hyginus *Fabulae*
136 135.n29

Hypsicles *Elementa* 14
praefatio 98.n38

Isocrates *Ad Nicoclem*
1–2 116–117
3 106.n73
11 96.n32
15 101.n51
19 102
20 104
22 99.n43
25 101.n56
27 103
28 105, 158.n109
29 98.n31
31 98.n31

36 101.n54
53 106

Jerome *In Danielem*
11.3–7 = 905.982–984 45.n82
11.13 = 907.1034–1035 55.n135

Josephus *Antiquitates Iudaicae*
12.157–171 43.n74, 44.n76
12.170–71 44
12.185 44.78
12.211–13 54.132

Justin
26.3.3–8 36.n29
27.1.9 34.n17
30.1.9 55.n135
Prologus 27 35.n26

Lucian *Adversus Indoctum*
15 38.n46

Lucian *De mercede conductis*
all 232.n17

3 Maccabees
all 38
1.4 40
2.25–26 53–54
2.29 63.n178

Machon
18 87.n309

Marcellinus *De Pulsibus*
255–267 202

Marcian of Heraclea Pontica
1.565 42.n69

Martianus Capella
6.598 173.n32

Mnesitheus of Athens
20, 45 129.n11

Neoptolemus
2a 60
5 61
6a.39–41 61.n167

Nicomachus *Arithmetica*
1.13.2–19   75.n245

Nicomachus *Harmonica*
11.6   75.n246

Fragmenta adespota
166   135

Oribasius *Eclogae Medicae*
53.5 = 215.8   72.n224
59.3 = 224.27   72.n224

Oribasius *Collectiones Medicae*
49.4 = 6–10   183.n71
49.4.5 = 6.19   185.n75
49.4.8 = 6.27–28   185.n75
49.4.12 = 6.32   185.n75
49.4.19 = 7.17   185.n75
49.4.53–55 = 4.9–10   80.n273, 183.n69
49.5.1–5 = 10.12–27   184

Pappus *Collectio*
3.20–23   75.n244, 164.n4, 165.n8, 176.n46
7.3   75.n247, 176.n46
7.22   75.n247
7.29   75.n247
8.5   218.n76

PGR
24–28   61.n168

Philodemus *On Poems* 5, *PHerc.* 1425
col.XVI, 143–144   61.n167

Philodemus *De libertate dicendi*
col.XVIb   107.n83
col.XXIIb.10–13   107.n84
col.XXIIIa.1–6, col.XXIIIb6–15   107

Philon *Belopoeica*
50.14–29   80, 175.n40
50.26   67.n197, 94.n20, 123.n142
51.10–20   79.n267, 175
56.14   79.n267
67.28–68.3   79.n267
72.26–27   159.n116

Philostephanus
T3   66.n195

F4   66.n190
F11   66.n195
F18   66.n191
F29   66.n189
F35   37.n39, 66.n195, 67
F36   66.n195
F37   66.n195

Pindar *Olympia*
2.1–2   90.n2
2.98   134.n19

Plato *Phaedrus*
265d–e   19.n45

Plato *Respublica*
3.390a–b   113.n109

Pliny *Historia Naturalis*
6.29.183   87.n311
7.57.207–208   67
11.89.219   195.n15
20.76.200   81.n284
22.13.31   83.n294
29.1.4   83.n297

Plutarch *Agis et Cleomenes*
23.2   69.n210
53–54   122.n138
54.1–2   37, 55.n136
54.5   46.n90
56.3   53
57.2–3   48.n101, 109.n94
58.9   48.n101
60.2–3   39.n53

Plutarch *Alexander*
70.3   96.n27

Plutarch *Amatorius*
753D   55

Plutarch *An Seni Respublica Gerenda Sit*
790A   116.n120

Plutarch *Antonius*
28.7–9   110–111

Plutarch *Aratus*
12.6   35.n22

Plutarch *Demetrius*
43.5   67.n196

Plutarch *Marcellus*
14.7–15   75.n244, 123.n139, 157.n104, 165.n8, 210.n53
19.9   1.n3

Plutarch *Non Posse Suaviter Vivi secundum Epicurum*
1095C   112

Plutarch *Quomodo Adulator ab Amico Internoscatur*
51C–D   108.n87
55E   104–105
56C   108.n89
58A   94.n20
59E–F   108.n93
60A   113
62B   108.n89
64F   108.n89
66A   108
66D   108
66E–74E   108.n88

Plutarch *Regum et Imperatorum Apophthegmata*
189D   116

Polybius
2.56   68.n203
4.48.5   25.n53
5.34   37.n38
5.36.1   36.n35
5.56.4, 7–8   118
5.56.10   103.n66
5.62.4   37.n41
5.65.9   37.n41
5.81   1.n2
5.107   37.n42
9.1.4   64
12.25e   98.n39
15.25–34   46.n91, 46.n92
15.25.9   41.n64
15.25.12–27   50.n114
15.25.13   51
15.25.16   51.n120
15.25.21–22   47.n96
15.25.32   47.n95
15.27.7   51.n120
15.27.37   51.n120
15.28.2   29.n4
15.31–34   47.n98
15.31.12   55.n137
15.32.6–8   46.n87
15.33.8–9   55.n137
15.33.11   46.n92
15.34.1   68
15.34.3–4   46.n88
18.55.7   68.n204

Porphyrio *Commentum in Horatii Flacci Artem Poeticam*
1   61.n165

Proclus *In Primum Euclidis Elementorum Librum*
68.15–17   131.n15

Ps.-Aristeas
2   98.n39
12   189.n90
40   189.n90
125   106
183   110.n98
187–300   187.n83
188   100
190   100
206   94.n20, 99
208   100.n49
209   98.n41, 102.n57
210   98.n41
211   98.n41
215   98.n41
224   95.n23
227   99, 189.n89
228   102.n59
229   98.n41
230   102
239   114
242   102
246   103, 115.n113
270   104
286–287   114, 124.n144, 159.n112, 174.n35
290   95.n25, 113.n107
304   103.n66, 118.n123

Ps.-Dioscorides *De Iis Quae Virus Eiaculantur Animalibus*
26.49K   85.n304

Ps.-Scymnus *Periegesis ad Nicomedem*
all   127.n3

Ptolemy *Harmonica*
2.14   75.n246

Ptolemy *Inscriptio Canobi*
149.1   167.n13

Ptolemy *Phaseis*
67.7–8   2.n8

Ptolemy *Syntaxis Mathematike*
7 = 1.2.100–101.9–19   4.n13

Quintilian *Institutio Oratoria*
11.2.14   67.n200

Rufus of Ephesus *Synopsis de pulsibus*
4 = 223–225   200.n29, 200.n30

Σ.Aratum
146   3.n10

Σ.Aristophanem *Thesmophoriazusae*
1059   38.n48

Σ.Aristophanem *Aves*
1403b   63.n173

Σ.Dionysium Periegetem
289   66.n189

Σ.Iliadem
2.145 (*Scholia Didymi*)   66.n191

Σ.Pindarum O.13
27c   66.n194

Σ.Nicandrum *Theriaca*
823   82, 185.n76

Σ.Apollonium
2.1052–1507   87.n310

SH
125–129   61.n168
309   63.n180
397   77.n255, 167.n3
465–470   31.n8

568–596   83.n295
604a   62
712   31
979   39, 136.n32

Soranus *Gynaecia*
3.37 = 117.15–17   72.n224
4.1 = 131.4–7   81.n227, 82, 185.n76

Soranus *Vita Hippocratis*
1   85.n305
4   83, 188.n88

Sotades
1   116

Stephanus of Byzantium
Κύπρος   66.n190

Stobaeus
1.26.5   76.n250

Strabo
1.1.8 = 5C.13 (10.13 ff)   60.n163
1.1.10 = 7C.2 (14.2)   61.n166
1.2.2 = 15C.19–22 (34.19–22)   132.n17
1.2.15 = 24C.10–12 (56.10–12)   60.n164
1.3.4 = 49C.1–50C.20 (122.1–124.20)   19
1.4.9 = 66C.29–31 (166.29–31)   96
2.3.5 = 100C.22–23 (244.22–23)   60.n161
7.3.6 = 299C.21–22 (260.21–22)   76.n252

Strato
84–86   66.n193

Suda
α3910   128.n8
α3933   68.n208
β147   97
ε2898   74.n237, 74.n238, 87.n312, 87.n313, 88.n315
θ142   210.n55
κ227   57.n151
π3040   54.n133

Georgius Syncellus *Ecloga Chronographica*
103.6–10   76.n254

Teles *De Fuga*
23.4–12   51.n116, 106.n79

Theocritus
15.100–143   38.n49
16.1–2   89.n3
17.3–8   89
17.13–50   95.n24
17.77–94   62.n171
17.79–80   59.n157
17.82–84   90.n6
17.95   90.n6
17.108–116   89
17.128–135   116.n118

Theophilus *Ad Autolycum*
2.7   63.n176, 95.n24

Theophrastus *Characteres*
2, 5   107.n82

Vita Arati 1
all   31, 188.n85
Vita Arati 3
16   79

Vitruvius
1.6.9–11   76.n250
9.*pr.*9–10   123.n139
9.*pr.*13–14   75.n244, 165.n8

Zenobius
3.94   36.n35, 38.n47, 39.n53, 102.n60, 104.n69

Zopyrus
267   128.n9

Documentary Texts: Inscriptions, Numismatics, Papyri

ABSA 56: 15 no.39
all   51.n118

BGU 6.1211
all   63.n177

Brett 1952: 8 no.29
all   46.n90

Hadra vase, Metropolitan Museum accession number 90.9.37
all   52

IG 2².838
all   45.n79

IG 7.3166
all   46.n90

IG 12.8.156
all   50.n115

OGIS 9
all   45.n81

OGIS 42
all   72.n223

OGIS 54
all   63.n176

1–6   34.n18, 95.n24, 102.n60
9–13   43.n72

OGIS 56
all   178.n54
1   102.n60
5, 7, 8, 18, 21–23, 25, 31–33, 36, 44, 46–47, 54, 58, 75   101.n52
8   101
15–18   123.n141
16–20   36.32
36–46   119.n27
38   122
41–42   122
44–46   122
46–73   37.n37
50–54   74.n234

OGIS 65
all   53.n128

OGIS 79
all   45.n84
1–6   73.n228

OGIS 80
all   45.n84, 46.n90

OGIS 81
all   48.n105

OGIS 82
all   43.n73

OGIS 86
all   43.n73

OGIS 90
all
3–4   137
4–6   53
5–6   68.n205
12   100.n50

OGIS 104
all   48.n102

P.Cairo.Zen. 3.39355
all   48.n102

P.Cair.Zen. 4.59571
all   73.n226
12–14   73

P.Enteux 60
all   48.n102

P.Grenf. 2.14b
all   49.n108

P.Mich.Zen. 55
all   50.n112, 72.n221

P.Oxy. 470
31–38   205

P.Oxy. 1241
all   34.n19

P.Petrie 3.53l–m
all   49.n107

P.Petrie 3.58c
all   43.n75

P.Rhind
no. 41   167

P.Teub. 3.793
all   173.n32

P.Teub. 3.860
all   46.n90

P.Tebt. 703
recto col.ii, 29–40   169
recto col.iii, 70–87   170
verso col.i, 183–191   170

Prose
no.12–14   40.n58, 98.n41, 102.n60
no. 15   73.n231

Pyramid Text
Utterance 222   137

SEG 27.1114
all   48.n102

SEG 33.671
all   72.n223

SEG 33.672
all   72.n220

Rekhmire, Tomb of
§§698, 707   169
§§718–728, 730–745, 748–749   170

RC
nos. 11–13   42.n66
no. 71   25.n55

Samama
no. 132   72.n223
nos. 133, 233   118.n125
no. 393   73

SB 6.9302
all   124.n145

*Syll.*³ 2.585.
136–137, 190–191   48.n103

Udjahorresne, naophorous statue of
all   179.n55

Index Rerum

aesthetics 6, 8, 27–28, 31–32, 55–69, 82–83, 111–113, 132–139, 164, 186–187, 190, 208–209, 225, 236–238
aetiology (see: genre)
agonism 5.n15, 28, 79, 84–85, 106, 108, 111–112, 159, 172, 188
Alexandria 1–2, 18, 32–33, 37, 42, 44, 48, 50–53, 55, 57–59, 65, 67, 69–72, 74–75, 78–81, 86, 110, 114, 116, 123, 150–151, 164, 167, 169–170, 173, 175, 178, 189.n89, 191, 206–208
Alexander the Great 18, 34.n21, 38, 45, 47, 52–53, 95–97, 103, 105–106, 129, 151, 230
Andreas of Carystus 1, 6–8, 18, 20–22, 26, 28, 38.n44, 74, 80–85, 128, 138, 179–190, 231
Athens 35, 42, 45, 52, 56, 64, 71, 74–75, 106, 204.n39
Antiochus III the Great 32, 37, 50.n114, 51, 115, 118–119
antiquarianism 31–32, 55–56, 58, 63–72, 83–84, 88, 138–139, 186, 188
Apollonius of Citium 127–128, 130, 133, 138–145, 150–151, 153–156, 161, 187, 190
Apollophanes of Seleucia 118–119, 187
Aratus of Soli 31–32, 38, 41, 79, 187
Archimedes of Syracuse 1–2, 4, 6–8, 17–18, 22, 26–28, 74, 77–79, 123, 127, 130, 133–135, 140, 147–150, 153–154, 156–158, 161, 167, 187, 209–225, 231
– Method 18, 209–225
– Quadrature of the Parabola 77–78, 148, 212–219, 221, 225
– Sand-Reckoner 127, 133–135, 140, 147–154, 156–158
Aristotle of Stagira 7.n18, 10, 20.n48, 26.n56, 71, 95, 100–102, 105–106, 107.n82, 156, 229–230, 237
asymmetry 13–14, 26, 90, 131–132, 150, 161, 189.n89, 227
authorship 13, 24–25, 28, 31–32, 39, 56–58, 64, 69, 71, 91, 96, 127–133, 138, 140, 154, 161, 171, 173, 188–189, 212, 229–230, 233–234
authority 24–26, 28, 91, 94, 117, 148–161, 188–189

balancing, mechanical 7, 209, 216–222, 224
belatedness 6, 8, 27–28, 31–32, 56, 59, 62, 132–139, 161, 164–165, 174.n37, 186, 188, 229, 233–234, 237–238
beneficence (see: virtues, beneficence)
Biton 127, 130, 139, 140.n46, 145–146, 151–153, 159–161
book learning 18, 32, 34–35, 56, 58, 64–66, 70–71, 75, 77, 84, 98, 114, 116, 123, 160

Callimachus of Cyrene 2–4, 32, 36, 39, 45, 56–58, 64–67, 70, 76.n252, 137, 166, 178, 186, 209.n52, 234–238
– Aetia 2–3, 36, 39, 57, 234–238
Clearchus of Soli 105–107, 111–113, 158
colonialism 18, 42, 59, 65–66, 96, 187, 206
conceptual holism 22–23, 193, 207–208
Conon of Samos 1–6, 28, 36, 57, 77–79, 118, 119–120, 234, 238
court society (see also: friendship) 15, 18, 23–26, 29–31, 40, 42, 70, 85–86, 90–95, 102–103, 115, 121, 178, 185–188, 190, 209–211, 225, 234
court science treatise (see: genre)
courtiership (see also: friendship; virtues) 8, 90–91, 103–109, 154–161

dependency 8, 29, 59, 64, 84, 86, 104–106
diagram (see also: image in text) 1, 22, 139–148, 167, 217–220, 225
disinterestedness (see: virtues, disinterestedness)
dioiketes 45, 47–50, 77, 167, 169, 173, 177
disciplinarity 6–7, 10.n27, 21, 26, 72, 92, 114, 174–175, 180, 190, 225, 227–234, 237–239

Egypt, dynastic (see also: colonialism; mathematics)
– monuments 19–20, 39, 42, 178–179, 203–205
– practices 3, 37, 43, 56, 66, 76–77, 118–121, 137, 167, 169–171, 203–206
– priests 121, 163
emergence 6–8, 20–23, 189–191, 206–208, 211, 223–224, 227
entertainment (see also: performance; spectacle) 4–5, 8, 23, 26, 28, 58, 61, 83,

117, 131–132, 158, 161, 164.n7, 178–179, 188, 209, 227, 232
Epicureans 107, 113
epigram (see: genre)
epistemic closure, lack of 15–16, 21, 23, 227
equality, social 26, 41–42, 92–94, 106, 109–110, 129, 160–161, 227
Eratosthenes of Cyrene 6–10, 18–21, 26, 28, 36, 40–41, 56, 60–63, 70, 74–79, 84–85, 87–88, 92, 96, 103, 113, 119–121, 128, 130, 132–133, 135–138, 140.n46, 146–147, 153–154, 158–159, 161, 163–179, 188–190, 209, 211, 231
– *Geography* 19, 76–77
– *Hermes* 60, 77, 164
– *Letter to Ptolemy* 9, 128, 130, 133, 135–137, 140.n46, 146–147, 153–154, 158–159, 161, 164–168, 170–179
– *On the Eight-Year Cycle* 75, 120
– *On the Measurement of the Earth* 76, 173
ethical self 8, 11, 26, 91, 119, 129, 131, 133, 154–161
experience 59, 68, 98, 150–154, 160–161, 221
experiment 4–6, 9, 11–12, 16, 20, 25, 80, 94, 117, 148–150, 158, 175–176, 190
– handling 149, 154, 158
– reproducibility 149–150, 158
expertise 8, 26, 28, 117, 127, 130–132, 150–154, 160–161, 227
Eudoxus of Cnidus 27.n60, 78, 84.n300, 157, 172, 175, 187
euergetism (see also: beneficence; patronage) 101, 123–124

friendship 8, 24–26, 30–31, 41–42, 44–45, 48–49, 51, 54, 67, 72, 77.n257, 80, 89, 91–95, 99, 101–104, 106–110, 112–119, 123–125, 127, 129, 131, 136, 138, 141, 145, 154–161, 172, 174, 178, 189–190, 210, 211.n59, 227
– friend-of-the-contest 58
– friend-of-doctors 94, 127, 141, 155
– friend-of-honor 44, 94, 99, 101, 105, 109
– friend-of-humanity 100–101, 115, 123
– friend-of-learning 94, 113–114, 117, 155, 174
– friend-of-lectures 64, 114
– friend-of-the-Muses 112
– friend-of-renown 80, 94, 107, 123
– friend-of-spectacle 112
– friend-of-*technē* 67, 80, 94, 123, 158, 178
– friend-of-truth 94, 99–100

– *philos* 24–25, 38, 41–48, 51, 53–55, 73, 79, 81, 85, 87, 93–94, 100, 104, 108, 115–116, 118, 128, 165.n104, 185, 187, 210

Gelon II of Syracuse 127, 133–135, 148, 153–154, 156–158, 210
genealogy, historical 6–7, 10–16, 21–23, 182, 193, 239
genre (see also: authorship; readership)
– aetiology 57, 186.n80, 233–239
– court science treatise 8, 28, 82, 127–161, 163–164, 210, 227
– epigram 31, 57–61, 136–137, 146–147, 158–159, 165–169, 172–174, 186
– *historia* 133, 137–139, 160, 164.n6
– paradoxography 27, 56–58, 61–62, 64.n184, 65–66, 72, 76, 82–83, 88, 160, 186
– *peri basileias* "On Kingship" 70, 96, 112.n106
– *Symposium* 112.n106
gift-exchange 3, 5–6, 8, 23–26, 31–32, 41–42, 44–45, 77, 82, 86, 89–91, 102, 115–125, 127, 129, 131, 155, 159, 161, 163, 227
great-souled (see: virtues, great-souled)

hedonism (see: luxury)
Herophilus of Chalcedon 6–8, 17–18, 20, 22, 26, 28, 81, 84.n299, 186, 191–209, 231
Hiero II of Syracuse 1, 123, 157, 210
Hippomedon of Sparta 50–51, 106.n79
historia (see: genre)
honesty (see: virtues, honesty)
honor 3, 33, 67, 73, 74, 80, 94, 99, 104–105, 109, 115, 121, 123–125, 211
hybridity 3, 5–6, 27–28, 39, 62, 64–65, 80–82, 147, 163, 167, 174, 186, 188, 208–209, 211, 213, 225, 231

image in text (see also: diagram) 132, 139–150, 161
interdisciplinarity 8, 28, 228–239

kingship (see also: friendship; virtues; war) 25–26, 34–35, 37–39, 42, 69–70, 80, 89–102, 109–117, 112–125, 128–137, 154, 158–159, 161, 173–174, 190, 208
– building 39, 79, 123, 135–137, 166, 174

lay person 5, 23–26, 94, 127, 130–134, 140, 142, 146, 148–149, 155, 161, 171, 208, 209.n52, 218, 225

lever (see: scale-beam)
library (see: book learning)
loyalty (see: virtues, loyalty)
luxury   35, 37, 47–48, 53–54, 63–64, 70, 110, 113, 139

machine   6–7, 21–22, 79–80, 87, 123–124, 138–139, 145–147, 151–153, 160, 166, 175–176, 179–182, 190, 210, 214
– Archimedes' (see: scale-beam)
– Eratosthenes'   6–7, 21, 75, 147, 153, 165–166, 168, 174, 176–178, 231
– for joint reduction: Andreas'   6–7, 21–22, 81, 180, 183–186, 188, 231
– for joint reduction: Hippocrates'   81, 138, 142–143, 182
mathematics   1, 6–7, 15, 27, 75, 77–80, 84, 92, 131, 134–136, 140, 143.n56, 146–149, 153, 159, 165, 167–168, 171–178, 180, 209–225, 231–232
– Egyptian   167–169, 173
– proof-style   168, 173
mechanics   6–7, 10, 18, 75, 78–80, 87, 92, 139, 142, 145–146, 151–153, 160–161, 165.n8, 168, 174–186, 188, 211–225, 231
medicine   6–7, 17, 72–74, 80–85, 88, 92, 108, 110–111, 118, 129, 140–141, 149–151, 154–155, 164.n2, 180–190, 192–208, 229–233, 236
mildness (see: virtues, mildness)
mixing (see: hybridity)

novelty   7, 9, 12–13, 16–18, 23, 56, 67, 124, 186, 188, 229

objectivity (see: virtues, disinterestedness)
oikoumene   42.n69, 60–61, 76, 133, 158

paideia   28, 47, 90, 95, 111–113, 117, 139, 141, 145, 147, 154–158, 161, 209, 211.n59, 228
paradoxography (see: genre)
parrhēsia   28, 91, 93–94, 99, 104–109, 115–117, 119, 129, 154–156, 158–161, 171, 189
patronage   1, 3–6, 8, 15, 24–26, 30, 32, 34, 38–39, 42, 65, 67, 76, 79–80, 90–91, 93–94, 97, 108–109, 117–118, 121, 123–125, 127, 132, 154–155, 160–161, 178, 185, 189.n89, 232, 239
peer-review   4.n14, 24–25, 124, 150, 227

performance   6, 27, 55–56, 59, 110–111, 116.n116, 174, 178, 181–190, 208–209, 218, 225
Peripatetics   9.n23, 20, 65, 67, 70, 94, 96, 105–107, 111, 113–114, 116, 197, 230
persona (see also: courtiership; kingship; virtues)   6, 10.n27, 31–32, 91–95, 109, 117, 127, 129–132, 138, 150–161, 187, 189.n89
Philon of Byzantium   79–80, 123, 152, 159.n116, 160, 166, 175–176, 184
philos (see: friendship)
physicalization   216
physis   13, 19–20, 32, 36, 62, 102, 189.n89, 201–202, 207
piety (see: virtues, piety)
praise   8, 27, 32, 39, 44, 56–57, 62, 67, 89–90, 96, 101, 105, 109, 133–137, 145, 154, 157, 159, 172, 209.n52, 211.n59
priests   8, 42, 44, 52–53, 98, 104, 114, 121, 178, 187
progress (see: novelty)
proportion   6–7, 75, 79, 136, 147, 152, 165–166, 168, 174–178, 199, 207, 214, 223
Ptolemy I Soter   34, 52, 116, 131, 166, 192, 208
Ptolemy II Philadelphus   30, 34, 49, 58, 66, 69, 74.n235, 79, 87, 89, 192, 208
Ptolemy III Euergetes   1–4, 34–35, 42–45, 49, 50–51, 54–58, 61, 68–69, 71–74, 77, 79, 85–86, 98, 101, 117–119, 121–123, 127, 135–137, 147, 153, 158–159, 163, 166–167, 172, 178, 187
Ptolemy IV Philopator   1, 29.n3, 30–32, 37–39, 42, 45–48, 50–52, 54–55, 58, 63, 67–71, 75, 79–80, 85, 87–88, 98, 102, 104, 109, 113, 115, 118, 163, 174.n37, 180, 187–188, 189.n89
Ptolemy V Epiphanes   41, 43, 53, 68–71, 75, 87, 137
pulse   6–7, 17, 22, 191–209, 231

realism, scientific   140–142, 193, 207, 224
readership   27, 56, 58, 61, 68, 91, 112, 127, 130–132, 140, 145–146, 149, 153–154, 160–161, 183–184, 196, 210, 213, 215, 217–220, 224, 233
relativism   14–15, 18, 91–92, 193, 207
renown (see: honor)

scale-beam   7, 209–210, 214–222, 224–225
Sciences Wars, American   14, 193

self-control (see: virtues, self-control)
slavery (see: dependency)
social constructionism   10, 13–18, 21, 23, 224
Sosibius of Alexandria   37, 45–47, 49, 50.n114, 51, 57.n150, 68, 73, 81–82, 96, 104, 128, 185
spectacle   3, 26, 58, 83, 127, 134, 178, 190, 208, 212, 218, 224, 227, 232
symposium   41, 44–45, 53–55, 91, 93, 95, 100, 103, 106–107, 109–117, 122–123, 154, 174, 187, 190, 232
– *eranos*   115–117, 127, 155
– *symbolē*   115

technē   84, 178, 180, 182, 188–189, 209.n52, 230, 236
technology (see also: machine)   6, 9, 11–12, 18, 26, 56, 66, 123, 139, 142, 166, 180–182, 186, 188, 190, 225
thick description   18, 23, 31, 93, 161, 163, 227
trust   6, 12, 15, 24–26, 94, 104, 117, 132.n17, 134–135, 138, 148–149, 150, 227

universities   4.n14, 8, 227, 229, 232, 237–238
utility   114, 117, 123–124, 127, 132, 136, 139, 141, 148, 153, 159–160, 165–166, 174, 178–179, 193, 212.n60

virtues   (see also: friendship; *paideia*; *parrhēsia*)   6, 8, 11–13, 16, 20–21, 94, 96–104, 106, 109–110, 113–115, 117, 123, 129, 154, 156, 161
– beneficence   31–36, 38, 41, 89–90, 94, 96, 98, 100–102, 109, 123–125, 174, 178
– disinterestedness   4, 9, 11–13, 16–17, 20, 100, 114, 123, 196, 202
– epistemic   11–13, 16, 20, 22, 117, 154, 161
– great-souled   98, 101–102, 105–106, 109, 123
– honesty   15, 104, 106, 109, 114–116, 123
– loyalty   25.n55, 42–43, 48, 51, 92–93, 95, 100, 104–105, 109, 115, 117, 119, 161, 187, 189.n89, 227
– mildness   98–99, 129
– piety   89, 98, 129
– self-control   98–99, 129

water-clock   17, 195, 196, 202–209
war   1, 25, 34, 38–39, 42–43, 45, 67–69, 79, 119, 123, 127, 138–139, 145, 151–153, 160, 166, 184, 186, 188
wit   44, 54, 110, 112, 129

Zopyrus of Alexandria   128, 150–151